The MACHINIST'S **THIRD** BEDSIDE READER
More projects, new ideas, and shop know-how

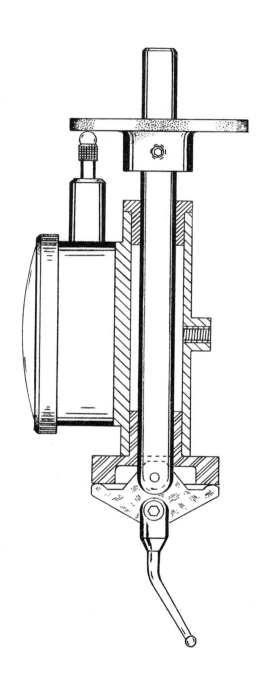

D1227418

The MACHINIST'S **THIRD** BEDSIDE READER

More projects, new ideas, and shop know-how

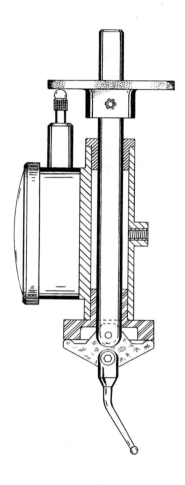

written and published

by

Guy Lautard

2570 Rosebery Avenue
West Vancouver, B.C.
CANADA V7V 2Z9
(604) 922-4909

Website: www.lautard.com

Published by Guy Lautard. **Printed in Canada**

6th Printing (January 2002)

Canadian Cataloging in Publication Data

Lautard, Guy, 1946-
  The machinist's third bedside reader

  ISBN 0-9690980-9-X
  1. Machine-shop Practice, 2. Machinists'
tools. I. Title.
TJ1160.L38 1993 670.42'3 C93-091768-5

DISCLAIMER

The mention by name of a particular maker's product(s) in this book
is not intended as an endorsement, nor should it be construed as such.

WARNING

Metalworking is an inherently dangerous activity. Both hand and power operated tools can inflict serious and/or permanent and/or fatal injury. It is your responsibility to know and USE safe working practices and procedures. If you don't know how to do a job safely, don't try to do it until you find out how to do it safely in your particular shop and shop situation. Machine tools can maim or kill without even noticing that the injured person was in the vicinity. Watch out for your own safety at all times. No one else can do this for you.

The Machinist's THIRD Bedside Reader
More projects, new ideas, and shop know-how

TABLE OF CONTENTS

ACKNOWLEDGEMENTS

Because of the number of people who contributed to the contents of **The Machinist's THIRD Bedside Reader**, it is not practical to mention them all individually on the Acknowledgements page. However a heartfelt "thanks" is extended herewith to all those named throughout this book for their contributions, photos, drawings, and their prompt, complete and patient replies to my letters requesting more info, or that they review the edited version of their material in draft form. Without these peoples' enthusiasm and willingness to share their knowledge and ideas, this book would not be what it is. If there are any shortcomings in this book, they are mine.

Thanks is also extended to

-- F. Brownell & Son, *Publishers*, for permission to utilize information on sandblast equipment which appeared in the book **Gunsmith Kinks, Vol. II**, and

-- the publishers of *American Machinist* Magazine, for permission to use the item on taper turning at pages 16 and 17.

A special thanks is again due my wonderful wife Margaret, without whose never-ending help on many fronts, my efforts would not be possible.

Finally, thanks to all of you, I have what is for me the ideal job: writing books for the best bunch of people in the world - my customers - machinists, friends and fellow enthusiasts.

INTRODUCTION

I will repeat little of what I said in the Introductions to **The Machinist's Bedside Reader (TMBR#1), TMBR#2**, and **"Hey Tim, I gotta tell ya..."**, because just about everything therein still applies. (You *have* read those books, haven't you? If you haven't, what you're missing is what you don't know.)

If you've read my previous books, and have put some time into making some of the items shown therein, you should by now have a reasonably well tooled shop. Of course nobody's shop is ever fully tooled, and there are lots of things a fella might like to have that he either doesn't need, or can't afford, or both, in either order. However, when you get a new shop supply catalog in the mail, and can go through it from end to end and find only a few items you want to buy, you know your shop must be pretty well equipped.

And then of course comes the question, "Now what to do with all this jammy stuff?!" The answers to that question, of course, are pretty diverse, and range from steam locomotive models to cars to guns to home-built aircraft, etc.

Not all of us can afford the time to pursue some of the ideas we might like to, either because of job or family obligations, or the money just isn't there to (a) really tool up the shop the way we'd like it, or (b) thereafter indulge in some particular pursuit that attracts us.

This book may solve one or more of the above problems for you. In the pages that follow, you will find projects to improve your shop's tooling, projects that will make enjoyable use of what time you may have for workshop pursuits, and a variety of ideas on how to make a buck in your shop.

The latter will not interest everyone. But if your shop activities must pay their way, or put bread on the table, you may find some interesting ideas herein. If you don't care to make a buck in your shop, some of the very same projects and ideas may interest you from a completely different perspective, and/or other ideas may (hopefully will) have you fairly exploding with the desire to tear off to your shop to start making something.

If this book helps you have fun and/or make money in your shop, it will have succeeded in its purpose.

In choosing what to include in this book (and its predecessors), I wonder to myself, "Will the guys like this, or will they not?"

Obviously nobody is going to like everything I decide to include. But even if there's a chunk - even a big chunk - on something that's of no interest to you, hopefully you will find a lot of other things about which you will say, "Now *that's* worth the price of the whole book ten times over!"

BREVITY

My writing is not characterized by brevity. One customer wrote from Australia with the comment, "Your detailed instructions must drive machinists nuts, but I love it!" My readership ranges from the beginner in his basement to master machinists. I want to make sure the beginner gets sufficient instructions to enable him to do a job. And while I had no inkling at the outset that professional machinists would find my books of interest, one of my great satisfactions is the calls and letters I have had from men far my senior in age and experience, some of them master tool and die makers, who have told me variously that they now use ideas they found in my books, or that while they didn't find much that was news to them, my books have revived their love of the trade.

In any case, back to the matter of brevity. When I write something, my objective is to provide you with as much useful information as possible, and I want that information to be understandable to any interested reader on first reading. (You may chose to read it 4 or 5 times before you try to carry out the work, but I want my **meaning** to be clear the *first* time through.) If at times you feel that I am belaboring a point, you will likely find that what I'm trying to do is make my meaning crystal clear.

By the same token, you would do well to read what I say very carefully. Even the placement of a comma can significantly alter the meaning of a sentence, and if you find one, it's there for a reason.

TERMINOLOGY
One term you will find in this book frequently is the term "shop-made." I prefer this to "home-made" because the latter may be construed to mean "2nd rate." Something that is shop-made is something we would make in our own shops, and it may be far superior to something you might find in a store.

ABBREVIATIONS
You will find some abbreviations herein. "ϕ" means diameter. If you find it associated with a dimension on a drwg, it means the part is round at that point. ₵ means centerline. CRS = cold rolled steel; HRMS = hot rolled mild steel; OAL = overall length; shcs = socket head cap screw.

"hsm" means a home shop machinist, or what the Brits would call a model engineer. My shop is in my basement, so I tend to think of a hsm as a guy "doing something in his basement." HSM refers to *Home Shop Machinist Magazine*. By the same token, ME means *Model Engineer Magazine*; LS is *Live Steam Magazine*, and *FWW* is *Fine Woodworking Magazine*. Margaret deplores my use of "drwg," and frequently changes it to "drawing" when proofreading. If you find this handy abbreviation, you will know I caught it in time to fix it properly. She also frowns upon me knockin' the "g" off the end of a word. I tell her it's a cultural difference - I read too many Will James books as a kid.

PATIENCE
From some of the letters I get, I believe there is a tendency among many hsm's - or at least some - to be too impatient: we want to see it working RIGHT NOW!

Sometimes this is ok, or even necessary. It would be nice if I could have had the items in this book written, edited, illustrated, and all ready to go to the printer RIGHT NOW - but it's taken me a couple of years to put it all together. Happily, I enjoy the work involved. With patience and work one enjoys, a good end product should result - which, in the case of this book, is important to both you and me.

I will later make mention of an interesting book that speaks, in part, about having (or cultivating) the right attitude of patience and the desire to do a good job. It is a book that may change forever the whole approach you take to doing some of the things you do. It is about doing a job - any job - right, and thoroughly, and knowing how to do it, or studying it until you see how to do it right.

One of my guys - we'll call him George - built a clock movement. Every so often he writes to me with news of progress in his shop, either generally or with respect to the clock. He gets in a rush and either botches something, which he then has to do over, or rushes off and buys something before satisfying himself that it is worth buying, or even satisfactory to his purpose, or otherwise hurries up and gets himself further behind then ever. However, in spite of these self-inflicted wounds, in a recent letter he reported that his clock was now complete and coming to very accurate time, and he says - note this - he is **enjoying** the task of "regulating" the clock (so that it runs at the correct rate), and consequently *is taking his time and doing it right*.

Every tool you make, and every job you do in your shop will increase your ability to do similar and future jobs, so don't begrudge the time it takes to do it right. Some jobs do not merit the expenditure of great care and fussy finish, but even so, they should be done in a workmanlike manner.

George doesn't have a knurling tool. I asked him why he didn't make one like the one in TMBR#1. His reply was that it would take too much time. When I made mine, it did take me a fair amount of time, but it turned out quite nice, and it has served me ever since on many projects, where its "stamp" of fine quality, uniform knurling, readily executed, is a pleasure in every way - it looks good on the job, and I do it with a tool I made, with care, years ago - an old friend who still helps me. I could make one in half the time today, because of greater experience and better tooling. But I got that experience and tooling through the very process of making stuff like that knurling tool.

So if you are inclined to be impatient to see the wheels go around, no matter how crudely you may have made the wheels in your impatience to see them stuffed into place on ill fitting axles, etc., let me encourage you to slow down, enjoy the satisfaction of doing each part of a project right because (and so that) when you finish it, it will be right from end to end, and will thereafter be a source of pleasure and pride rather than the cause of comments like, "Well, I was in kinda a hurry, and it don't work quite the way it should".

BRANCHING OUT
Although my shop is small (12' x 12'), it is reasonably well equipped. And the more I study possible projects, or things I might want to make, the more I see what one can readily do, and also, I see what it makes sense not to do. It makes little sense to make something that you can easily buy ready made, unless the ready made item costs too much to suit your pocket, or is, for one reason or another, unsatisfactory.

On the other hand, in a recent issue of *Fine Woodworking Magazine* I saw an article on making a jig to enable both solid and open spiral woodturning. This I like! No such item is offered commercially, but I could make a beautiful job of making one of these!

"But you don't do woodwork, Lautard. You said so. And you don't have a wood lathe." Well, I bought one a while back, and I'm beginning to see how to achieve better results in woodwork than I have heretofore. As a result, you'll find drawings for a turned wood lamp herein, plus a wad of info on how to make better wooden (tool) storage boxes than those made from Medite which I spoke of in TMBR#1. Probably if I'm progressing thusly in my skills and interests, others are too, or will, as a result of their expanding interests.

Not everything in this book is 100% practical. Some, I guess, are quite impractical - Bill Fenton's little spar tree logging blocks, for example. You can buy blocks capable of handling any hoisting job you are likely to dream up in your shop. But if you happen - as I do - to like the idea of making pulley blocks, and rigging up some elaborate system of block and tackle to do something necessary or entirely un-necessary, well, why not? You will never see a commercial pulley block as cute as the ones Bill made - his are simply too appealing for words.

So much for introductory remarks. Be patient, do the job right, and to a standard that pleases you, and/or of which you feel proud. Enjoy what you do.

SOME THINGS ARE THE SAME....
I said in the beginning of **TMBR#2** that I sort of view us as a bunch of guys sitting around a big stove talking. Well, naturally, I don't want to do all the talking, so...

THIS BEDSIDE READER IS A LITTLE DIFFERENT...
You will find a slightly different format in some places in TMBR#3. Many of you have sent me useful ideas and info for dissemination to the brethren, and some of this material appears herein as letters to me. If nothing else, it makes for a slight change of scene, since - as you may have noticed - the Bedside Readers are not broken down into chapters.

As a kid in high school I discovered a book called "Cache Lake Country," by John J.

Rowlands. The thread of the book was this: The author gets asked to keep an eye on a large tract of forest land. He goes in with a canoeload of supplies, and puts up a log cabin on the shore of a lake which he names "Cache Lake". Pretty soon he's set for the winter, snug as a bug in a rug. An old Indian chief lives a mile or so down the lake one way, and an artist lives about as far away in the other direction. The three of them become good friends. The Chief shows the other two how to make a toboggan, and snow shoes. A crew of prospectors comes through, and an engineer with the group designs a crystal radio they can make - the author has some parts sent in to "the Settlement", 30 or 40 miles away, and hikes out to get them. They also build a heliograph, which the Chief thinks is great.

Two or three years ago I came across a copy of "Cache Lake Country" in a used book store. I sent it down to Tim. He seemed to like it too. I told BB about it, and he said, "I like it already, and I haven't even seen it!"

I'd like to think there's a certain similarity between my books and "Cache Lake Country": a bunch of guys doing stuff of possibly mutual interest, and sharing their ideas.

"Who's BB?" you may wonder.

One day BB was where his most creative thinking is often done, reading a outdated magazine, when he came across an ad for a book he thought might be interesting, so he looked to see who the advertiser was, and his interest turned to astonishment...

"Is this the source of **The Machinist's Bedside Reader**?" Yes. "Do you still have copies available?" Yes. "Well, I'd like to get a copy."

"Where are you calling from?" I asked.

"Straight down the hill from you about 6 blocks!"

And so it was that I met **Bryan Boutry** - architect, artist, and model aircraft builder *par excellence*. We have become good friends.

Some while later BB sold his house, and designed and built a new one about half a block west. I can see his roof from my living room. His new place has plenty of shop space, and he began to fill it up with some very fun stuff.

You'll meet BB more than once later herein. We've had a lot of fun doing things back and forth. When my new milling machine was delivered, BB helped get it into the shop. When he installed a DRO unit on his Colchester lathe, he brought the top of the saddle up and I milled a chunk out of it to accommodate one of the measuring spars. When I had several chunks of aluminum channel to saw up, BB put them through his horizontal bandsaw for me.

BB took a quarter off me in a bet one day - I said that a Starrett vernier protractor is called a vernier protractor. BB said, "No, it's called a universal bevel protractor." He was right, too.

RESEARCH METHODS
....or How to Find Out Almost Anything

Libraries are a wonderful source of almost any kind of info you might want to have, and most librarians will go out of their way to help you track down whatever you want. Here are some possibilities which you may not realize exist for your delectation.

Tom Cat
BB was looking for a source of some rubber grommets he needed for a model aircraft engine mount he designed, and now manufactures. He asked me if I had any ideas on how to track

down a maker of such things? I suggested he look in Thomas' Register, which he would find in the library. That evening he buzzed me. He hadn't found what he was looking for, but Thomas' Register had more or less unhinged his mind. "I never even knew such a thing existed. You ought to put that in your book - the guys should know about that!"

Thomas' Register - often referred to simply as "Tom Cat" - is a listing, by category, and by company name, of many large and small US manufacturers, plus a compendium of some of these companies' catalogs, or partial catalogs. Look up "punches," for example, and you will find pages and pages of listings of makers of different types of punches. Springs, cutters, lathes, machine tools, bending equipment, hose, etc. etc. The whole set runs to about 28 volumes, each about 2-1/2 - 3" thick. Guys like us will just about go bananas looking in there, and finding such a ready means of accessing info on who makes what!

Virtually any small town library would have Tom Cat, if not this year's, then probably last year's. Go take a look.

Inter-Library Loan
At various places throughout this book, I cite magazine articles, books, etc. Yes, it would be nice to be able to put everything between these covers, but it's neither practical nor possible to do so.

There is a service known as Inter-Library Loan. If a book you want is not held by your local library, your librarian should be able to borrow it for you from another library, near or far, that does have it (see below). This service is well to know about. If a library is not willing to release a rare or valuable book/magazine from its collection, even to send to another library, it is often possible to get photocopies of specific pages or articles you want.

You may encounter a cost of a few dollars to get a particular book, depending on various factors, but $5 or $10 to have the use of a book which you cannot see otherwise, and which contains info you want or need, is money well spent.

Reader's Guide to Periodical Literature
To locate published info on a topic, you can look in the Reader's Guide to Periodical Literature. These are hefty volumes which list, by subject, and by author, articles published in a broad cross-section of magazines in a given block of time (e.g. Jan/June 1982; the next volume will deal with the next block of time in the same way). Your librarian can point out where they keep these books in your library, and show you how to use them.

Union List of Serials
The Union List of Serials is a set of books, available in most libraries, which lists those libraries which have holdings of various periodicals. Want to know what libraries have collections or partial collections of *American Machinist, Home Shop Machinist, Live Steam,* or *Model Engineer Magazine*? This is how to find out.

A Computerized Index to *Fine Woodworking* Magazine
If you ever want to look up something in *FWW*, or in any of their related books, you will want to know about "PC Index to FWW", which is a computerized Index to *FWW* put out by Meredith Associates, P.O. Box 792, Westford, MA 01886. If your library has a collection of FWW back issues, tell them about it - they will probably want to get a copy.

Added to the 5th printing:
Added to the 5th printing:

(Now I know this last item will come as a surpirise, particularly for all you guys who thought I was still using a manual typewriter...)

And then of course there is the Internet.. If you are not on the Internet, you are missing the most amazing informational resource imaginable.

Even more amazing is that you can now find *me* on the Internet, at **www.lautard.com**

MORE ON COLOR CASE HARDENING
from **Captain Kurt A. Bjorn, USAF, Alamogordo, NM**

In August of '91 I got a nice letter from Kurt Bjorn, one of my guys down in New Mexico. Kurt has a degree in Chemistry, and was at the time a fighter pilot with the USAF. Kurt sent along a couple of nice color photos of two very handsome blackpowder rifles he'd made - sort of antidote to the high-tech world of jet fighter aircraft.

After I got through looking at the photos, I dug into the letter, and my face fell for a minute until I realized he wasn't really mad at me..... As I got further into the letter, I also realized that here was a gold mine of info that just **had** to go into **TMBR#3**... So here's what Captain Kurt had to say, and believe me, it's well worth studying:

Dear Guy:

I'd like to alternately congratulate and curse you...

First the curses: Your story, "The Bullseye Mixture," in **TMBR#2**, definitely piqued my interest in heat treatment and color casehardening. Being a hsm/gunsmith type, the mental gears begin to rotate, forming up happy images of (my) beautifully machined and polished parts being retrieved from the quench tank aflame with brilliant blues and reds. And so began a quest from which I have not yet returned... (drum roll, please!) If you wanted a title for what follows, you might call it ('nuther drum roll, guys).....

Safe Color Casehardening with Predictable Results

Lately, I've been doing a good bit of machine work, as well as gunsmithing. My equipment (before reading "The Bullseye Mixture") consisted of a 8" x 18" lathe, c/w a milling head attachment, and a drill press.

Most of my muzzleloading rifles are built over a period of several months, from sand castings and/or finished parts. I taught myself basic machine work, dividing, etc., so I could make as many of the parts as possible, and I've had pretty good success so far. When I read "The Bullseye Mixture", I decided I must also acquire the equipment to do my own color case hardening.

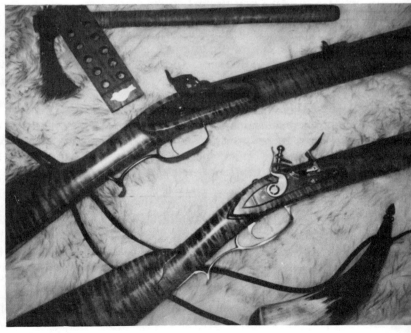

Out came Brownell's catalog, for a look at their "electric oven" offerings. Yessir..... that computer controlled multi-cycle Neycraft furnace would be just the ticket. Plus, it'd be great for precise heat treatment of high carbon steel, and my wife could fire her ceramics in it as well.... ("Whatever you want, Dear!") With consent and reason in hand (or was reason out the door?!), the furnace was quickly ordered.

ext, I needed containers. I made one thick-walled little crucible from clay, but it cracked and
oke after three firings. I now use small flower pots - cheap and disposable - which I seal
ith clay. More on these below.

ext, charcoal. For that, I saw no reason not to follow your recipe for same in the "The Bulls-
e Mixture", so I told my wife, "Sweetheart, save every nut shell and fruit pit that comes into
e house, and put 'em in this bucket." I added pounds from eating thousands of nuts. One
y, as I was buying more nuts, I noticed the dried bean bin. Hmmm... them little devils look
st like fruit pits... so I bought ten pounds! All went into a coffee can, which was sealed and
ed over a Coleman stove outside. I wrecked the stove, but made a great batch of charcoal.

hat wrecked the stove was the jets of greasy flaming smoke issuing from the vent holes; these
ry nearly melted the lid of the stove, as well as depositing carbon everywhere.

an charcoal works very well indeed, and will give you a first class case that is very hard and
ep. However, the downside is that wherever individual charcoalized beans are in contact
ith the work, you may get an unattractive coloration - sort of salt-and-pepper dotted sprin-
ing. (This *might* be eliminated by crushing the bean charcoal finer before using it. GBL)
owever, bone meal (see below) is so superior that I have stuck with it from the first day I
ed it.

ext, the "secret powders": Sodium carbonate? No sweat - I got some at the supermarket, just
e that guy Lautard said.

rium carbonate? No thanks! As a chemist, I bristled when I read of its use. Elemental
etals, with few exceptions, are horrifically and insidiously toxic. Iron, gold, silver, copper
d aluminum are tolerated reasonably well. Most others will accumulate in and rot your
ain/liver/kidneys/add-your-favorite-organ-here, and never leave your body. I'd rather kiss a
tlesnake than mess with barium. Call me paranoid, but I want to be makin' chips when I'm
.

spent months modifying basic parameters, quench additives, surface preparation, etc. The
st I ever got was a very dull coloration, predominantly dark. But I will say this, Guy: the
el was incredibly hard - you couldn't ask for more on that score!

e dark coloration of the steel meant oxidation, so I figured too much oxygen was getting to
y workpieces prior to quenching. Unnecessarily high temperatures will produce a similar
fect in the presence of even small amounts of oxygen. Hmm... Off to the library to do some
search.

ost references speak of the use of cyanide, as we both know. My aversion to barium would
ly be exceeded by my disinclination to mess with cyanide. One writer mentioned bone meal
s you also did in "The Bullseye Mixture"). One thing I noticed was that recommended
mperatures varied. Your story and "Tubal Cain"* recommend 1650-1700°F. An OLD
insmithing text recommended 1400-1450°F. I decided to give this lower temperature a try...
d... it worked! A miniature fly-cutting tool holder I machined up from CRS was not only
oroughly case hardened, but also given a brilliant coat of blues and reds. Makes me proud
st to look at it.

* The pen-name of an English writer whose work often appears in
Model Engineer Magazine. GBL

ere's a summary of my procedures:

a garden shop, buy several small flower pots, and a 10 lb. bag of bone meal. Also, pick up
me washing soda on the way home. (The washing soda is optional, but you will sacrifice
rdness if you leave it out, especially at the lower temperatures I would recommend.)

When you get home, wash the bone meal thoroughly. You must NOT ignore this step! Dump the whole bag of bone meal in a bucket, take it outside, and run gallons of water through it, stirring it slowly all the while with your hand or a stick. The overflow of water from the bucket will carry off unwanted impurities from the bone meal.

Next, char the bone meal as described in "The Bullseye Mixture".

> **Caution**: this will stink like a charnel house, so do it away from other people. When I do it, I put a batch in a large coffee can, hammer another coffee can on top to seal, and punch half a dozen holes in each end with a nail for gas venting, and - having learned that trying to charcoalize bone meal on my Coleman stove has about the same effect on it as charring beans - I bury this critter in a nice pile of firewood in the fireplace. After making sure that I have a large quantity of firewood on hand, I light the fire. About 3 or 4 hours of vigorous firing seems to work fine. When charred thus in an indoor fireplace, the stench (and believe me, it is truly vile!) - so long as your chimney draws properly - will be carried up and out of the house, and will not hang about. Otherwise, get thee into the desert, a dump, or somewhere where a campfire is legal.

Wash the bone charcoal again in a bucket, flooding it with clean water and agitating it with your hand. The desirable fraction of the charcoal will sink, and very little will be carried off by the wash water, which is carefully poured off from the rim of the bucket.

After several washings, reduce the level of the water so that the bone charcoal (picture a quantity that would fill a 2-1/2 lb. coffee can) is of the consistency of a slurry, and throw in somewhere between a couple tablespoons and a handful of washing soda. (I know, that's deplorably imprecise, coming from a chemist, but that's about as fussy as you need to be on this aspect of the job!) Stir it about, or knead it together with gloved hands, to get the washing soda evenly dispersed, then spread it out on a large garbage bag in the sun. It will dry thoroughly in about two days.

> **NOTE**: When Kurt first wrote to me, he didn't know of a source for ready-made bone charcoal. He said the bone meal he got was "about like oatmeal," and pointed out that bone meal from another supplier might not be the same. For that reason, he was desirous to learn of a source of ready-made bone charcoal. I was more than pleased to be able to help him out on that score, in view of all the good info he'd sent me. Kurt said in a follow-up letter that for anyone but the absolute diehard do-it-yourselfer, ready-made bone charcoal would definitely be preferable. For a source of same, see below. GBL

You will also need some fireclay, or, alternatively, partially hardened ceramic slip (= clay + water + nuthin' else). This stuff is easy to find: just go into a ceramic supply shop (look in the Yellow Pages) and ask for some "clay". Any clay that will air harden will work fine.

Next, the flower pots: the ones I use do not have a drain hole - they're probably intended for use as a tray or dish to hold a full sized flower pot. You will use two of these pots at a time. If either pot has a hole in it, seal it with clay.

Fill one pot/dish with some of the bone charcoal. Set the thoroughly polished and degreased workpiece in the charcoal, and cover it completely with more charcoal. Pack it down TIGHT, and then invert a second flower pot over the first. If there are no holes in your top dish, you probably won't be able to fill your two-piece container completely full, but that doesn't seem to matter: just put a few strips of brown paper on top of the charcoal. The paper will burn up during firing, and consume any excess oxygen in the pack. I use plain brown paper (from a grocery bag) to avoid any bleaches, dyes, inks, etc. which might stain or darken my workpiece. This may not matter, but it can't hurt, either. Make sure you use enough paper: if you use too little paper, you may not purge all the oxygen; if you use more than enough, it will do no harm, as any excess will simply turn to charcoal once the excess oxygen is consumed.

Seal the joint between the pots with clay. Don't worry about leaving a vent hole - even if you tried, you couldn't seal it well enough to have it burst.

Now, heat the whole shebang at about 250°F for 3 hours. This will drive off the moisture from the clay seals. Then take 'er up to 1425°F and let 'er roast for 5 to 6 hours.

Prepare the quench water: Put some cold water in a bucket, and throw in about 3/4 of a pound of potassium nitrate* per gallon of water. This will give you a saturated solution - i.e. one holding all the potassium nitrate it can.

    * Cheap source of potassium nitrate:
        Hagenow Labs, Inc. 1302 Washington St., Manitowoc, WI 54220.

Note: the "industrial grade" stuff is fine for our purposes; there's no need to spring for the much more expensive "Reagent-" or "Laboratory-grade" stuff. (And by the by: molten Potassium Nitrate (without the water) is a super tempering agent, and will impart a beautiful blue to polished steel. Ever heard of "Nitre Blueing"? This is it.)

As you explained in "The Bullseye Mixture", Guy, you need bubbles. I thumped one end of some 1/4" copper tubing with a hammer to seal it shut, bent it into a coil, drilled about twenty 1/8" holes in it, and set it in the bottom of the bucket. Yours truly's lungs provided the air supply. A helper on a bicycle pump, or a supply of low pressure compressed air, would definitely be a convenience. If you use lung power for your air supply, make sure you have enough pipe to keep your face a reasonable distance away from the quench. (also: Wear GOGGLES and a FACE GUARD, and HEAVY GLOVES or MITTS, and DON'T suck!) Try oil on the surface if you want - I think it darkens the colors excessively.

Now the fun begins! When you're ready, use tongs to remove your flower pot "pack box" and poise it right over the quench pot. Get the bubbles going, hard. Break the top of the pot with a hammer, and pour EVERYTHING inside into the quench. Even a few seconds of exposure to the air is too much.

Fish around among the charcoal and pieces of flower pot for your workpiece, and remove it. I have used the same quench five or six times in a row, with no cleaning of the quench or removal of the cruddy, used bone charcoal.

NOTE
DON'T try to salvage and re-use the charcoal from the quench tank! Unless you washed it VERY thoroughly and painstakingly, *you would be setting yourself up for an EXPLOSION!* Reason: Charcoal and potassium nitrate (from the quench tank) is almost blackpowder. I can guarantee that the mixture would burn beautifully at hardening temperatures!

After months of frustrating fun, these techniques - combined with the information so palatably and enjoyably conveyed through your story, "The Bullseye Mixture," in TMBR#2 - have proven to be consistent and reliable at producing colors. If you need a very hard, deep case, take the temperature to 1675°F or so, but don't expect as much color.

I've truly enjoyed both "Bedside Readers". My father read my copy of **TMBR#1**, and was especially intrigued with the story therein of the lathe the Allied POW's conjured up from almost nothing (while in the hands of the Japanese), as he was among those soldiers slated for the invasion of the Japanese home islands. He ended up in Japan with the Occupation forces, and while there learned the basics of machine work, which he has passed on to me.

                                    Kurt

The undernoted Company can supply ready made bone charcoal:

Ebonex Corporation, P.O. Box 3247,
2380 Wabash Street, Melvindale, MI 48122
Phone: (313) 388-0060.

I have a letter from Ebonex dated October 11, 1977, in which their General Manager warns me most strongly that **there is a very great danger of an explosion when charcoalizing bones one might get from the butcher** and boil in an attempt to remove at least some of the "natural juices." Kurt concurs, and says the smart thing to do would be to order a bag of "10 mesh x 28 mesh Bone Black" from Ebonex; at time of writing it would cost about $16 for a 20 lb. bag, plus shipping cost to your location.

A final word of warning: The information contained herein should **NOT** be regarded as imparting sufficient know-how to make it wise - or safe - to undertake the case hardening and/or heat treatment of critical parts like receivers, breech blocks, rifle bolts, etc. without first acquiring **a lot of experience.** Also, shrinkage and/or warpage can occur when heat treating even simple parts, and the consequences of such shrinkage or warpage can be disappointing or worse. Stress relieving before and after machining, before heat treatment, may help prevent such problems.

Above all, THINK, and WORK, *safely*. Nothing could be more tragic than burns or other injury arising from the pursuit of what was supposed to be fun.

Approximations of π

Most who read this book will have some sort of pocket calculator, many of which have π built in at the push of a button. Those who do not may find the following of more value than those who do, but hopefully you will find it interesting, whether you need it or not.

1) The Bible gives, as an approximation for π, the value of 3 (1st Kings, chapter 7, verse 23) and for some classes of work - including some machine shop calculations - this is entirely close enough.

2) An encyclopedia gives π as 3.141,592,654.

3) A more common approximation of π is the value 22/7.
    However, 22/7 = 3.142,857.143 =/= (does not equal) π.
    The error in this approximation amounts to about 1 in 2485.

4) Another approximation of π is 355/113.
    Now 355/113 = 3.141,592,920 =/= π either, but it's a lot closer.
    In fact, the error in this approximation is only about **1 in 12 million!**

So if you ever want a handy, close approximation of π, **355/113** is the one to use.

A COLLET CHUCK SYSTEM FOR ANY LATHE

Accurate collets in the lathe spindle nose provide a level of centering accuracy that is generally not had from even the better class of lathe chucks. Of course there are extremely accurate 3-, 4- and 6-jaw chucks, but these are too expensive to get through the door of the average basement machinist's shop.

There have been several articles in ME over the years on how to make draw-in collets for a lathe. Two or three of these were truly excellent, and went into every possible detail of the matter, and were therefore rather long - I doubt that even I could have lengthened them at all. Perhaps we can go into this topic in that kind of depth in a future Bedside Reader, but for now, here are a couple of briefer ideas.

Last Fall, Bill Fenton gave me a Rubberflex tapping collet holder and a single Rubberflex collet that will handle work from 3/16 - 3/8"φ. I'd like to see Myford bring out a new item for their lathes using Rubberflex collets, a Myford backplate, and a nose cap to close the collets.

For some reason, Myford #2MT collets (which go up to 1/2"φ capacity) seem (today) to be priced at an unusually high level compared to most other collets*, and I suppose they were back when Bill made a collet chuck outfit for his Myford ML7 many years ago.

> * (about $40 each, compared to 5C and R8 collets from about $7 to $25 each, depending on whether you want the "fine quality import" or the "top quality US made".

Bill made a fitting to screw onto the lathe's spindle nose in place of the regular chuck and bored it to take collets with a self-releasing taper (see **TMBR#2**, page 92). He then made a closing piece to screw onto this spindle nose piece, and a set of collets to take work up to 3/4"φ.

If you flip to page 39 in **TMBR#2**, you will find a drwg there that pretty well details what Bill made, although the dimensions and proportions would be different. Bill's tapers were probably steeper - in fact the front face of the collets could be flat, or very nearly so, if a fella wanted it that way. To interpret the drwg in **TMBR#2** in terms of Bill's collet outfit, look at it this way....

At the extreme left, where you see the centerline symbol and '5/16"φ', would be (in Bill's case) the piece that screws onto the lathe's spindle nose - let's call that the Spindle Nose Piece. The sectioned portion with the male thread to the right of that would be integral with the Spindle Nose Piece. The collets would be dimensioned to fit inside this, as they do on the drwg in **TMBR#2**. The closing piece would be analogous to the 1"φ x 1.117" long piece in the drwg, but might have a much steeper taper on the outer conical section on the extreme right. As I recall, this piece, in Bill's outfit, was machined from about 2-1/2" a/f CRS hex stock, but it was not proportionately longer. Either a hex or round knurled closing piece would work well for tightening up on a collet by hand or with a special wrench.

Bill's collet outfit cost him nothing but his time, and - as mentioned above - it handles work up to 3/4"φ.

There are a couple of other angles on this collet business.

One thing you might not have noticed is that some of the mail order machine shop supply houses sell #2 (and #3) MT collets that close via a draw-bolt inserted from the outboard end of the lathe spindle. These collets run about $23 each at date of writing (May '93). The draw-bolt would be easy to make, and - for Myford owners - a lot easier to swallow than the price of Myford collets.

Another idea worth considering, for small work, is to buy a taper shank (i.e. a drill chuck arbor) to fit your lathe spindle nose socket, and to then machine the big end to take either shop-made or commercial collets (e.g. watchmaker's lathe collets), and a shop-made nose piece to push the collets into the tip of the taper shank item - again, refer to page 39 in **TMBR#2** for a general idea of the collet/nose piece portion.

Finally, here's an obscure point you may have overlooked in drooling through the Starrett catalog. Tim Smith pointed it out to me, and I may have mentioned it somewhere else in my writings, but if I did, I cannot now find where. Unlike their #162 pin vises, Starrett's #240 series pin vises have a turned (unknurled) portion on the handle, just back of the knurling behind the chuck. You can grip a #240 pin vise in a lathe chuck or collet, and use the pin vise as a collet for small diameter work. (Tim says they're dead accurate, too.) The set of 4 pin vises in the #240 series will handle work of any diameter from 0 to 0.200"ϕ, and they'll do it for a lot less money than a series of conventional collets that would only cover specific sizes in that same range. Plus you can use them as conventional pin vises too, so they'll do double duty.

One last point: In a letter in the March/April '84 issue of HSM, at page 7, Bruce Jones, of Sacramento, CA pointed out that the Jacobs people offer 4 sizes of drill chuck with the back end threaded 1"-8, 1"-10, or 1-1/2"-8. These are intended to be screwed onto a lathe spindle having the same thread, giving the user accurate chucking capabilities up to 3/4"ϕ, depending on which chuck it is. (**Added to the 6th printing**: So far as I know, these chucks are no longer offered.GBL)

Compressed Air on Tap Anywhere in the Shop

I commented in TMBR#2 about compressed air being a no-no in any shop my friend Bob Haralson ran. **Howard Bradley**, of Boise, Idaho, wrote to say "... tell that friend of yours who is so dead set against compressed air around machine tools that it just depends on how good a fella is as to whether it's ok or not. (That oughta light his fire!)"

Compressed air should not be used to clean machine tools, because it can drive fine swarf into places where you don't want it. However, there is nothing to prevent a guy from using a little discretion. My friend Terry has compressed air on tap all over his shop, via a "linear reservoir" soldered up from half inch copper water pipe. Adaptors to go from copper pipe to compressed air fittings are readily available, and you can have outlets anywhere just by putting - or pulling - a Tee in the pipe, and putting in a branch line, and/or a coil-up plastic air hose.

I recall a reader's letter in *FWW* pointing out the danger in using PVC pipe for the same purpose: if pressurized PVC pipe is accidentally whacked with something, the pipe may shatter, sending sharp pieces of rigid plastic pipe flying through your shop in a most unhealthy manner. Copper pipe with soldered joints is much the better choice.

Added to the 5th printing:
A file for producing a nice finish in the lathe...

One of the tips in **"Hey Tim..."** concerned how to modify a file for use in producing a nice finish in the lathe. This is done by going over the face of the file with a small abrasive stone, to even out any tooth-to-tooth irregularities, whereafter it will give a smoother cut. The idea came from a 94 year old machinist, and was passed on to me by my friend **Walt Warren**. I had a call one night from **Fred Emhoff**. Fred told me in stern and serious tones that he'd ruined a good file on the strength of my reputation, but then laughed and said he sure liked the results it produced thereafter. (The modified file will work fine on round stock, but is more or less useless on flat surfaces.) Although **"Hey Tim, I Gotta Tell Ya"** contained a lot of good information, **it is now out of print**. However, we plan to bring out a much-revised version of that book, with a whole bunch of new content, and possibly even including our out-of-print fiction story, **"Strike While the Iron is Hot."** Watch our website, or otherwise keep in touch.

Here's another idea: if you have a 2-flute end mill that's had one tooth chipped off, you can make a small flycutter from it. It's supposed to work well.

SOME INTERESTING NOTES ON TAPER TURNING

The first portion of these notes is based on a 1985 letter from Fred Jackson, of London, England, to Tubal Cain (the well known pen name of a long time contributor to ME Magazine). Mr. Jackson sent me a copy of his letter, as part of a very pleasant correspondence we had some years ago. GBL

.... During all the years that I have been mulling over ME and dreaming about the workshop I hoped one day to possess, I have realized that I would want numerous #2 Morse Taper shanks for all sorts of purposes. I have quailed at the thought of setting over the top-slide and using hand feed, and even more so at the idea of setting over the tailstock in order to use power feed, with the consequent wear on misaligned centers, and the difficulties of resetting the tailstock back to zero. And so, about a year ago, at work, I started to make an adjustable tailstock center similar to the one described in **TMBR#2**, page 89. One thing and another prevented completion of the job for a while, and in the meantime, I noticed in ME, in connection with this very matter, the mention of ball-ended centers. Like a bolt from the blue, light dawned! Suddenly, all was clear!

"Ball-ended centers," I thought. "What a wonderful idea." Incredibly, I had never seen them mentioned before - or at least thought I hadn't: browsing through Guy Lautard's magnificent Index to M.E. (blush!GBL) and re-reading many previous articles on taper turning, I found that they *have* been mentioned several times --- but it just hadn't penetrated!

Applying myself to the problem of making ball-ended centers, I suddenly remembered an old, dismantled Ford auto engine that had been lying around at work for some months. I rescued two pushrods from the debris, and there were two accurate-looking semi-circular seatings which were surely concentric with the .221"ϕ parallel shank. I cut off a short piece from each pushrod. Then I plugged a soft center into the lathe mandrel, machined off the cone, drilled 5.5mm (0.2165") x 1" deep, and reamed 7/32" with a brand-new machine reamer which I hoped would cut a couple of "tenths" big. I was now able to press fit the .221" shank into this hole, stick a 5/16" ball bearing into the hemispherical seating with epoxy, and there was my headstock ball-ended center.

I made the adjustable tailstock center pretty much as described in ME (and later in **TMBR#2** GBL) by "Mac" Mackintosh, including pressing the taper shank and working center into the two parts of the tool: I pressed home the second piece of Ford pushrod, and stuck a ball-bearing on as above.

I was now in a position to start attempting to produce #2 Morse tapers. Having "cheated" already, by using two existing soft centers, I "cheated" again, by buying a taper shank of the type used for a tailstock Die Holder. These are about 5-1/2" long, with a 5/8"ϕ parallel nose 2-1/2" long. I plugged this into the lathe mandrel, put an indicator on it, and to my delight found it running true within 0.0002". (Perhaps I was lucky!)

I then checked the center hole, using a "wobbler" (**TMBR#1**, page 22), and found this also very accurate. It therefore seemed safe to open up the center hole to roughly 1/32" less than the diameter of the 5/16" ball, having first taken a very light facing cut across the end of the shank, for absolute accuracy. (Prior to this operation, I had made a small brass stop collar, a tight fit on the body of the Slocombe center-drill, and held by a grub-screw for extra security. This enabled me to regulate the depth of the center hole.)

I then installed my 4-jaw chuck on the lathe, reversed the bought shank, and, gripping same by the parallel portion, I indicated the latter portion true, and then checked the concentricity of taper and center hole. Finding them fine, I faced the end, and opened up the center hole using my brass stop.

My master setting piece was now complete, and I was ready for production!

I cut off ten pieces of 3/4" mild steel bar a little over 5-1/2" long. Using the usual methods of a stop in the mandrel taper, and a saddle dead-stop, I faced each of these pieces to exactly the same length as my master, and center-drilled to exactly the same depth at both ends. (I found this operation irritating and time-consuming; due to the brass stop collar, the swarf packed up into the flutes of the center-drill, and had to be continually poked out and brushed clean. Had I to do much of this type of work I think I would make or buy the Myford lever-operated tailstock attachment. This has a simple depth-drilling adjustment which would obviate the need for the brass stop.)

> (Or use another center drill, without a stop collar, to rough drill the holes nearly to depth, and/or recess the stop collar to give some clearance at the working face of the drill, and/or make the collar differently, perhaps with an offset stop pin set into the collar - this would allow one to set the collar further back on the center drill body, leaving space for the escape of chips. GBL)

I then rough turned my ten pieces to a little over 5/8"ϕ for 2-1/2" long, between centers, reversed the work, and finish-turned the remaining 2-1/2" length to 0.720"ϕ.

Now I was ready to use my ball-ended centers. I plugged one into the headstock, the adjustable one into the tailstock, and indicated the slide of the adjustable center fixture so that it was exactly horizontal, and inserted my "master" between them, with the tapered end towards the tailstock.

Some while before, I had made myself a holder which allowed me to readily mount an indicator dead on lathe center height. Using this with an indicator graduated in "tenths", plus the saddle traverse, I adjusted the t/s center until I got a constant reading over the whole length of the taper. Anyone doing this might as well know from the start that a "tenths" indicator is extremely sensitive. I did find the operation difficult, but ultimately I arrived at a setting which appeared satisfactory.

Now I removed the "master", inserted one of my ten pieces, and replaced the top slide (compound rest) with my "Gibraltar" toolpost*, and fitted a 5/8" sq. knife tool --- and immediately hit snags. Due to the projection of the adjustable tailstock center slide, not even the "Gibraltar" would allow the toolbit to get close enough to the right-hand end of the work!

> * (Mr. Jackson, as you will soon see, thinks very highly of his "Gibraltar" tool post. The "Gibraltar" is essentially a massive block of cast iron, of truncated conical form, with one flattened side. This attaches to the cross slide in place of the compound rest. It was designed by Tubal Cain, to whom, as noted at the outset, the author of this letter was writing. GBL)

I solved this problem by revolving the toolpost through 180°, so that the Gibraltar's tool-slot was on the right-hand side, and thus the thing nearest to the adjustable t/s center. However, I now found that, towards the large end of the taper, the leadscrew swarf-guard was going to run into the lathe's Q/C gearbox. Loathe though I was to remove this guard, I did so, taking some pains to avoid disturbing the adjustment of the screw for the clasp-nut gib.

Everything now appeared to be in order, and I could probably have just about managed without further ado, but as an extra precaution I cut about 3/16" off the tail of the standard Myford 3/4" driving dog, and replaced the square-headed screw therein with a shorter one. There was now no danger of the carrier fouling the toolpost, and the ultimate test was at hand.

It went like a dream! I roughed out the first blank, tested it with a #2MT socket I had in hand, and could feel no shake. I re-honed the tool, and took a 3 thou finishing cut.

Lacking experience, I had no faith in my ability to check the taper with three chalk-lines, or toolmakers' blue, but 45 years ago I operated a universal grinder (with some tool-and-cutter grinding thrown in) for two years, and we used to check tapers with a sine bar. So I stopped here, and took this one completed taper to work with me the next day.

Unless my arithmetic is faulty, a #2MT has an included angle of 2° 51' 40". Machinery's Handbook gives the sine bar constant 0.2486 for 2° 51', and 0.25005 for 2° 52'. So I settled for 0.22498" as my constant, and at this setting my taper came out 0.000,1" fast. The British Standard tolerance is 0.000,2", so this was good enough for me.

I returned home, roughed out my remaining nine blanks, carefully re-honed the tool, and again took a 3 thou finishing cut, removed the work, returned the saddle to its starting position by reversing the lathe rather than releasing the half-nuts, left the cross-slide index untouched, and completed the remainder of the blanks.

A couple of days later I took them all to work, subjected them to the sine bar test, and the results were really quite remarkable. So far as I can see, I now have ten absolutely identical and perfect #2MT shanks. To maul the English language slightly, I'm "over the moon".

No doubt some praise is due to my brand-new Super 7, but I am convinced that the accuracy achieved is very largely due to the "Gibraltar" Toolpost; it is a very fine idea and design.

So far, I have only used two of my shop-made shanks, finish-turning two of the parallel shanks down to 0.625"ϕ.

I expect to have many uses for much shorter ones, but I shall simply cut off the excess, and not consider the off-cut a waste of material - it will be used for something. Before I alter the setting of the adjustable center I shall take Mr. Mackintosh's advice, and dowel the slides in the #2MT position. It will be interesting to see whether I can repeat the performance.

Fred

Setting a Taper Attachment to ±0.001" in 5" to 10"
from **William Ramsey, Oakland, CA**

Dear Guy,

In 1965 I started to work for a styrofoam cup manufacturer, and was suddenly faced with more precision taper turning than any average toolmaker would get in ten lifetimes! After a couple of days fuming, an engineer named Doug Bell and I came up with something that made getting and holding tapers to ±0.001" in 5"-10" easy, if you have a taper attachment.

You will need: trig tables or a calculator with trig (for tangents)
      a micrometer setting standard between 5 and 10" long
      a 1"-travel direct reading indicator

The procedure is as follows:

Adjust the taper attachment somewhere close to the desired angle.

Assuming the use of a 5" standard, set up a carriage stop about 2-1/2" to the left of where the taper turning attachment's clamp goes past center.

Move the carriage to the right far enough to catch the 5" standard between the stop and the carriage.

Now comes the tough part: mount your indicator on the carriage, with the axis of its plunger movement oriented square across the long axis of the lathe; on the top slide, arrange something for its plunger to bear against (or rig them the other way around, if you prefer), and set the indicator to high or low zero, depending on which way your taper goes.

15

Move the carriage to the right again, and bring it back up to the mike standard to make sure your indicator still reads zero.

Remove the standard and run the carriage to the left until it hits the stop.

As you move the carriage toward the headstock, the taper attachment will cause the topslide to move (in, or out) relative to the carriage.

Compare your indicator reading to 5 times the tangent of the angle (per side) you want to cut.

When the taper attachment slide is adjusted to give you a reading of exactly 5x the tangent of the desired angle, go ahead and cut the taper - you will be right on the money.

I made lots of 32 oz. shake molds - which require matching male and female parts - to produce cups with a wall thickness of 0.045", over 8-1/4" and smaller, over the next 5-1/2 years.

I would be most interested to hear if anyone knows of this method being used before August 1965. So far as I know, Doug Bell and I originated it. Since then, toolmakers have carried it to Ontario, Stuttgart, Glasgow, Fisherman's Bend, Jo'burg, Allahabad, Bombay, Sao Paulo, and Leningrad.

Bill

(I love it! And thanks to Bill's willingness to share this slick idea with the guys, it'll be much more widely known now. If anyone knows of an earlier use of this method, let me know, and I'll pass it on to Bill. GBL.)

and finally.....

A LOW-TECH, HIGH-PRECISION FIXTURE FOR SETTING THE COMPOUND REST FOR CUTTING TAPERS
based, in part, with permission, on an item in
AMERICAN MACHINIST & Automated Manufacturing Magazine
for September 1987, page 167.

In **TMBR#2**, at the bottom of page 89, I said that I had once cut a batch of #2MT shanks with the topslide of my lathe slewed around to the appropriate angle. That is only half the story. The unembellished truth is this:

Several years ago I needed to make about half a dozen #2MT shanks for a project I was working on with my friend Roy Bickerstaffe. I procrastinated over the job for several days. Finally the time came when there was no choice - they had to be done, and done NOW.

The only thing to do was to grab the job by the face and do it. I slacked the screws that secure the Super 7's topslide, and slewed it to what looked about right, and cut the first taper. Now you're not going to believe this, guys, but when I got the surface finish somewhere near ok, and tried it in my #2MT sleeve gage, it fit!! I couldn't have got it closer if I'd done all the things Fred Jackson speaks of in his letter! I was so tickled with myself that I buzzed Roy to tell him. Roy laughed, and said, in his inimitable Yorkshire accent, "Make a bag full - next time it'll take you an hour to set it right."

In AMERICAN MACHINIST & Automated Manufacturing Magazine for September 1987 at page 167 in the "Practical Ideas for Practical Men" department, Robert E. Sandberg of Johnston, RI, detailed a simple fixture that requires little effort or skill to make, and that allows one to dial the compound rest to the correct setting to produce any desired taper, using a sine bar and gage blocks (space blocks would probably also do fine).

Basically, what Mr. Sandberg said to do is this: Make a "shelf" that can be mounted in the lathe by gripping one part of the shelf in the lathe chuck. Put a sine bar and the right gage blocks on the shelf, and indicate the compound rest parallel to the sine bar. You are now set to turn your taper!

The shelf is approximately as drwg below. The reason for making it so the square bar can be swung through 90° is that as a sine bar goes nearer and past 45°, inaccuracies in the height of the gage block stack begin to have a greater adverse effect on the accuracy of the angle produced by the sine bar.

The welding required to make the shelf could be eliminated by running 3 1/4-28 set screws into the piece of 2"φ CRS to lock the 3/16" plate in its slot in the 2"φ CRS. The 3/16" plate would not need to be ground flat stock; so long as it was nice and flat (all 4 edges machined, not sheared) and free of burrs, I think it would do fine. The piece of 3/4" square ground stock you could get from any of the usual machine shop supply houses, local or mail order; $20 should get you enough to make 2 of them.

Mr. Sandberg's original description took up a 2-1/4 x 2-1/4" square of type beside a drwg 2-1/8 x 4", so I think there is no need of more words from me. But I will add this: the gage block stack height required to produce the 1° 25' 50" angle with respect to centerline for a #2MT is 0.1248266". Digging around in my box of 0.000,1" space blocks, I see that my best bet would be to use the 0.125" space block. With that I'd get an angle of 1° 25' 57" - i.e. I'd be about 0.000,17" fast, hence within the 2 tenths "fast" allowed by the British Standard Mr. Jackson speaks of above. With a set of gage blocks, one could come slightly closer.

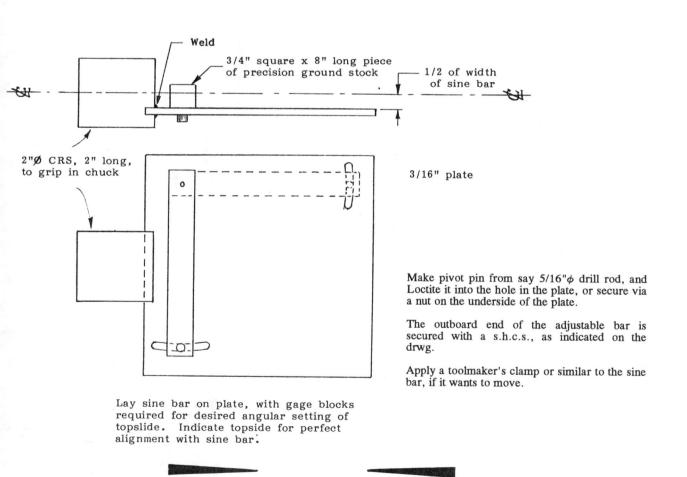

Make pivot pin from say 5/16"φ drill rod, and Loctite it into the hole in the plate, or secure via a nut on the underside of the plate.

The outboard end of the adjustable bar is secured with a s.h.c.s., as indicated on the drwg.

Apply a toolmaker's clamp or similar to the sine bar, if it wants to move.

Lay sine bar on plate, with gage blocks required for desired angular setting of topslide. Indicate topside for perfect alignment with sine bar.

".... it would obsolete my drill press..."
from Kevin Kimball, Boise, ID

Dear Guy,

I recently was shown a tool system which was so impressive that I have been anxious to tell someone who might have an interest in such things.

This info came to me via a travelling salesman. This gentleman, who is retired, and now just travels around the western US selling these tools, gave me quite a sales pitch. As I am a Project Engineer for an industrial contractor, with offices and projects all over the continental US, I am constantly barraged by suppliers, vendors and peddlers attempting to sell me something. Consequently, I was somewhat skeptical of the claims made by this fellow, and found his product literature underwhelming. I must have looked unimpressed, because he told me, "Next to machinists, engineers are the hardest to convince". I didn't have the heart to tell him he was really bucking a stacked deck with me!

This tool system consisted of an anvil, a vise jaw assembly, a drill spindle, a cast "foot" and a V-block. A sixth component, not included, is a length of 1-1/8"ϕ CRS. This is provided by the buyer, since its length is job dependent. The anvil/vise jaw arrangement was very interesting since they could be used together (as a bench vise (4") with 360° rotation in the vertical plane) or separately (as a massive clamp, hickey bar, pipe wrench etc.) However, all this was not nearly as exciting as the drilling set-up.

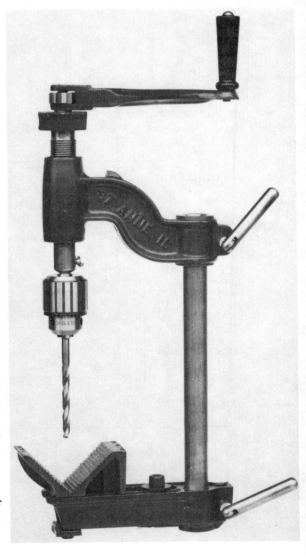

The salesman claimed that with this tool it was possible to drill holes up to 3/4"ϕ through mild steel plate, by hand, without undue effort. Additionally, that with the use of a 3/8" drill motor (low speed), holes (up to 2"ϕ!) in hardened steel were no problem. He topped this off by saying that all this was possible with HSS bits bought at the local hardware or discount store. The thought of drilling a hole in steel plate with a hand cranked drill struck me as so absurd that I nearly told him we were not interested, but he seemed so sincere that I told him to set up a demo and I would watch.

The salesman placed a piece of 3/8" MS plate on the "foot", then slid the spindle down the shaft until the drill bit (1/2"ϕ) came into contact with the plate. The spindle and the "foot" were clamped to the 1-1/8" CRS column via clamping screws and attached handles. The spindle crank was then turned at a leisurely pace, while torque was applied with the free hand to a hexagonal flange of about 3" A/F on the spindle. As the spindle itself is threaded, a constant pressure (the literature says up to 1000 lbs.) is applied to the bit. The hole was drilled through in approximately 30 seconds in this manner. I examined the bit - it was indeed a typical HSS bit, the same brand sold at the Ace Hardware Store here in town.

I was then invited to try this for myself. I did so, and found that in seconds, without exerting myself in the least, I too could drill through this steel plate - there was no trick here.

The salesman then took the handle and its ratchet assembly off, and placed a three-sided drill adapter, and a 3/8" drill motor (Sears) on the spindle, and proceeded to drill holes in all manner of steel pieces, including a 1" square HSS lathe bit, a 1"ϕ end mill (using the V-block), and a bearing race from a piece of construction equipment - a Caterpillar™ loader, I think. This last feat was accomplished using a masonry bit (price $1.96 at Ace) no less.** As a finale, he cut a 1-1/4"ϕ hole through a truck frame (very hard, in my experience) with a hole saw. This cut was lubricated with a constant flood of WD-40.

The package price of this set-up, complete with all components save the 1-1/8"ϕ CRS bar, was $495. This strikes me as a little severe, but I have to say the tools did appear very well made and were certainly effective. This system is so simple in appearance that I believe something similar could be made in one's basement shop without undue effort. The drill assembly in particular was so convenient, versatile, and efficient that I think it would obsolete my drill press, especially since I do not use my press for any precision hole drilling. I can think of few applications where this device would not do as well, and many where it would work and my drill press would not."

Kevin

** I understand that the trick to this (which in no way detracts from the efficacy of this equipment) is that the masonry drill must be new - it cannot have previously been used for drilling rock or concrete. This is not to say that a new bit must be used for every hole - just that one previously used for drilling masonry won't do. GBL

I was very much intrigued by the above letter, plus the leaflet that accompanied it. According to the latter, the Cole drill could be used to drill out broken studs in vehicle wheels or engine blocks, as well as as a gear puller. I couldn't help but think how often we could have used such a tool on my uncle's ranch when I used to work for him as a kid.

In due course I contacted the makers, and ordered one. It is well made. In fact, it is guaranteed for life, and they sure don't tell you to baby it, either. They say that if using the vise as a pipe wrench, to stick a 1-1/8" bar into the hole in the bottom end of the outboard jaw, and put 3 or 4 men to heaving on it if necessary - "... you won't break it."

The whole outfit doesn't carry a fancy finish, just as-cast iron, painted a nice deep green. I took the clamping handles (straight pieces of CRS) off and went over them with a file and Scotch-Brite™ to knock a few minor burrs and sharp edges off them. In due course, I may make a pair of clamp-on ball handles for it.

In looking over the drill portion of the outfit, my conclusion is that it would take some careful work to make a similar piece of equipment in a basement shop, on a one-off basis. Not that it couldn't be done.... but if you put any value on your own time, and/or if you have to pay someone to do the necessary welding for you (if you don't weld), and buy a drill chuck, you'd be better off buying the factory-made unit.

I had a cute idea which I passed on to the makers, namely a bench mounted base, which would enable semi-permanent installation of the drill on a workbench. I think this would be a very useful adjunct to the Cole drill, and the owners seemed to like the idea, but whether or not they actually take up the manufacture of this item remains to be seen at date of writing.

I'd venture to say that a fella could almost pay for the rig with bets won with unsuspecting friends who think it would be impossible to drill, for example, a 1/2"ϕ hole through 2" plate glass with a carbon steel drill, to say nothing of HSS end mills, lathe tool bits, and the like. The Cole drill *will* do these things.

See inside back cover, also. See inside back cover, also. Photo by Mary Dailey Brown, Prairie View, IL, and supplied courtesy of the maker, Midwest Machine Tool, Inc., R.R. 1, Box 328 St. Anne, IL 60964.

AN AID TO SETTING UP WORK ON A FACEPLATE

Robert W. Brown, one of my guys in England, sent me a good idea. While the principle was not invented by him, his version is quite good. The idea concerned a device to ease the job of setting up some types of work on a faceplate. If you mount a faceplate on your lathe, and then try to mount a job on the faceplate, you will need all three hands, and possibly more besides, to hold job, clamps, packing pieces, wrench, other wrench, etc. What to do?

Make a duplicate of your lathe spindle nose, and make its back end a light press fit into a ball bearing of suitable size. Press the newly made part into your bearing and then mount the ball bearing to a base plate - Robert Brown said he used a chunk of 1-1/2" or 2" sq. CRS for this latter purpose.

You may elect to bore, in your baseplate, a recess into which to press fit or Loctite the outer ball race, or you may choose to make some small strap or heel clamps. Anyway, the baseplate should fit into your bench vice; if not, you can attach a gripping block to its underside.

The drwg shows the idea in generalized form. The particular bearing you use will have a big influence on just what yours looks like in the end, so all I will do here is to show the basic idea.

What this gives you is a means of mounting your faceplate solidly over at your workbench in the bench vise, with the working face of the faceplate looking up at the ceiling, and having the faceplate free to rotate just as slick as if it were on the lathe. Now, gravity is working *for* you, instead of against you. You can put the job on the faceplate, pile all the clamping stuff around it, and go to work methodically clamping it down just the way you want it. A dial indicator or indicating pointer (e.g. a surface gage) can be rigged to read out the degree of concentricity of the job as you rotate the faceplate. As my friend Everett Arnes would say, "This is Cadillacin' down hill."

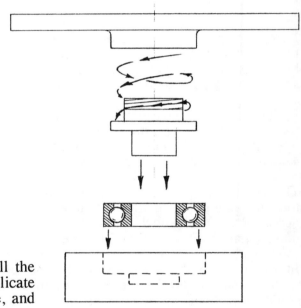

When your set-up is the way you want it, and all the bolts are tight, unscrew the faceplate from the duplicate spindle nose, cart it and the job over to your lathe, and put it in place on the lathe spindle nose.

One addition that would be helpful is some ready means of locking out rotation of the bearing when desired. Such a feature would be handy when you want to take the faceplate off the fixture.

— · —

A Miniature Impact Hammer

Idea: You can use an automatic center punch as a miniature impact hammer. You can make up special tools to go in the business end, to suit the job at hand. This idea ought to be quite useful to gunsmiths, for driving pins in and out, staking pins/screws in place, etc.

FIRM JOINT CALIPERS, CROSSED BELTS and GLORY DAYS
from **Fred Schellmat, Detroit, MI**

Dear Guy,

I'm an old timer who goes back to the days when apprentices came in early in the winter to light fires in oil drums to heat the shops, and one motor drove all the machines through overhead line shafting and belts.

My uncles were both old time machinists working in machine shops and boatyards when boats for the Great Lakes were still being built in Detroit's downriver areas before the turn of the century. As a youngster I remember them telling me about boring out big brass propellor shaft bushings by the light of gas flames piped to the machines.

I always admired their ability to work to close slip fits or press fits with big firm joint calipers, and to work to a couple of thousandths of an inch with a steel scale and a magnifying jeweller's loupe by "splitting the line". They could also "split the line" when shaping contours on a shaper, and do it with reasonable speed.

Some technologically oriented historian/anthropologist should collect these old stories and anecdotes from the few remaining old timers and record them for posterity. They represent invaluable technological history of developments leading to the modern technological age.

Sometimes in my dreams I can still hear the squeak of the crossed belts shifting to the return stroke on the big old gear rack Liberty planers, with big hand forged toolbits peeling off cast iron, and the "clink" when the shavings hit the wooden floor. Those were our "glory days", Guy!

All the best.... Fred

I was over to see Bill Fenton one Saturday afternoon, and although he wasn't feeling his usual self, and should in fact have been in bed, we ended up down in the shop. We got to looking at some of the various firm joint calipers that hang, with many other items, from the rafters in his shop. As you can see in the photos, some of Bill's calipers are a pretty fair size.

One was a pair of moderate size made or bought by Bill's father in Scotland before 1893 (see TMBR, page 191).

Another was a larger pair with a reversible point that gave him both hermaphrodite and inside or outside caliper functions in one tool. These were well made tools, handsome - though not highly finished - that had served their maker and his son for over 100 years and helped both put bread on the table for their families.

Next was another big pair of outside calipers that Bill had made. He told me he'd milled the taper on the two legs as a pair, and had then had the Army blacksmith bend them as a pair. The legs were then separated and re-assembled in the desired configuration, with a heavy rivet. Bill explained that he'd scraped the mating surfaces of the joint, and had made the rivet maybe 5 or 6 thou undersize so that when it swelled up as it was being hammered up, it would not bind in the hole and lock the joint up solid. He pointed out how he had frosted the exposed surfaces for appearance's sake.

"Feel that joint," said Bill. It took a firm hand to move it. "You can caliper something at the far end of a shipyard, and carry them back to the shops on your arm. They won't move from where you set them." These were wonderful, well made tools.

Later, I looked up an item I'd seen in ME* about firm joint calipers, and found a drwg of a somewhat different form of washer that would be a good way to make the joint in a pair of firm joint calipers.

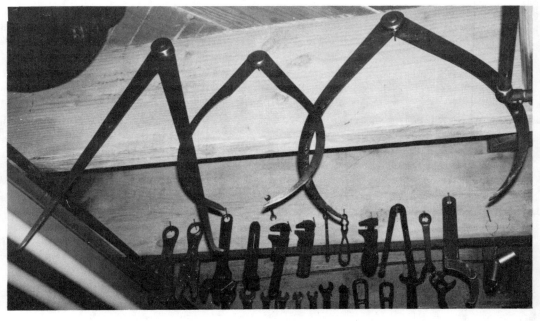

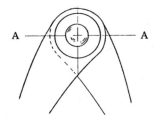

Section through A-A

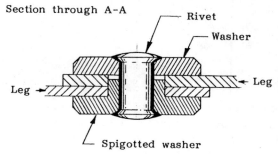

Rivet

Washer

Leg ——▶   ◀—— Leg

Spigotted washer

* By Tubal Cain; May 1989 M.E. page 581

22

SHOP-MADE TAPS

Bill Fenton showed me an unusual tap one day when I visited him. It was unusual for 2 reasons. First, it was about 14" long. Second, it was obviously shop-made, not bought. Bill told me he kept it simply as a memento of its maker, his friend Jack Wood, another long time member of the BCSME, and a superlative model engineer.

The tap was required to cut a 1/2-32 thread, and had been made from a piece of 1/2" drill rod about 3-1/2" long. The back end of the tap was turned down to 5/16"ϕ and cross pinned in a 5/16" hole in the end of a longer piece of 1/2" CRS.

Bill said Jack had made it to reach up to the front end of a model locomotive boiler, to tap a bushing previously silver soldered into the boiler's front tubeplate.

This tap struck me as an object lesson in just how easily a special tap can be made. Jack had simply center drilled the ends of the drill rod, turned a short taper at the working end so as to bring the tip down to the root diameter of the thread, and then screwcut the desired thread, before milling* 3 flutes on the rod.   * see next page

Bill pointed out to me that Jack had draw filed the reliefs behind the flutes.

Now I have seen the term "draw filed" used incorrectly so often that whenever I see it I want to ask the user if he even knows what the term means, let alone does he know how to do it. Bill, of course, knows everything there *is* to know about filing (see **TMBR#1**, page 4), so I asked him instead why he had used that term in this particular case, when drawfiling is not the approach I would have used to file in the reliefs back of the cutting edges. Bill explained something that had not previously occurred to me:

> By drawfiling - that is to say, by handling the file somewhat as one might a drawknife or spokeshave - you have greater control of exactly where the file is cutting, because you are handling it by its long axis, typically with your hands several inches apart. Thus, you can tip the file by very small amounts with great control, to move the cutting action closer and closer to the edge of the flute. Think about it. Very interesting, no?

The tap was then heat treated in the usual way. (I don't know what Jack Wood's "usual way" was, but you know mine, from **TMBR#1**, page 45: coat with soap to prevent scale formation, heat 'til it won't attract a magnet, plunge straight down into cold water or brine, reheat, and temper by color.)

The tempering color appropriate for a tap depends on the duty to which it will be put. If it is of small diameter, and to be used in brass, one might let it down to a brown color (i.e. relatively softer than say a lathe tool bit). If it is of a larger diameter, and/or to be used several times, (i.e. long life of the cutting edges is wanted) then I would leave it harder - say straw color.

The above is here mostly as an encouragement to any hsm who might have reason to make a tap - it doesn't take a magician, and the same thing could be done for any size tap you are likely to want. Bill has made many taps in exactly the same way. Two of these are shown in the photo at left, the smaller one is a 3/16"-16 tpi Acme thread tap.

Jack Wood's method of extending his 1/2-40 boiler tap is worth tucking away for future reference also - you might have reason to extend a tap to reach a hole that would be inaccessible to a standard tap.

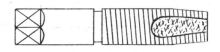

* Taps can be fluted by milling a flute with an end mill to give a radial face, the same way one might flute other types of cutters. However, you can instead simply file 3 tapered flats on the tapered and screwcut tap blank - see drwg at right. Although obviously not the most efficient form of tap, such a tap will cut, as witness the following...

One of BB's pals has a chainsaw engine converted for use as a model aircraft powerplant. The original aluminum drive washer (that's what it's called, but it ain't thin like your average garden variety washer) on it had woofed the cookie in an unscheduled landing, so BB made a new one for him. The main hole on this item requires to be tapped full length of the part, but the tap (5/16-18, I think, but it doesn't matter) was too short to finish the job. I suggested he file 3 flats on an allen head capscrew of the same size, so it would cut, and use that. He did, and it worked fine. He said it took him all of 5 minutes.

A Note on the Incremental Cut Method of Ball Turning
as detailed in TMBR#1

I've had several customers phone me because of difficulty in regard to the ball turning procedure detailed in **TMBR#1**, page 72-79. In every case, the point of puzzlement is the same: *Where does that second value for "C/S" (0.003) in the table at page 79 come from?*

The answer is, "Just pick a number out of thin air," because *it is an arbitrarily selected value.* You pick what seemeth good to thee as a reasonable first infeed for the cross slide, and you calculate the next number (L/S) based on it, and then pick another reasonable infeed value, and so on.

You may also know that we brought out a small book called **"Tables & Instructions for Ball & Radius Generation"** in 1990. This is a 3-part, 104-page book. **Part 1** explains the basic method, which requires nothing more than a parting tool in the lathe to make ball handles, ball end mill blanks, etc. **Part 2** explains how to generate male and female radii with a ball end mill, or the side of an end mill, in a vertical mill. Say you wanted to machine a 1/4" radius on the corner of a piece of material. If you have a 1/4, 5/16, 3/8", 1/2" or other size of ball end mill at hand, you can do it. You can also do it with the side of an end mill. And it's fast, easy, and accurate. **Part 3** consists of over 70 tables - one per page so they're uncluttered - for generating both balls and radii from 1/64" to 1" radius. There's a memory aid on the inside back cover, so you can flip to it if you need to, and then go to work from the table that suits your job at the moment. It's just 4" x 6", so it fits in your tool box, and it's coil bound, so it'll lie flat in use.

This little book is somewhat of a sleeper for us, because a lot of guys don't realize how slick this method of producing spherical shapes works. There's a part in the control system of the P-40 Warhawk that looks like a ball with a piece of round bar poked through it. Bill Mahuson, one of my guys in Webster, NY wrote to tell me he'd used "the method" in making a new one for a P-40 that was being restored, and he said it worked out real fine.

A GRAVITY-OPERATED PANIC SHUT-OFF
BUTTON FOR A MILL-DRILL

Bill Fenton has a small mill-drill type milling machine which he bought 20 years ago when he retired. On it sits a 5" milling vise which he almost NEVER removes. With this outfit Bill has made the Donkey Engine outfit described elsewhere herein, also a 1-1/2"-to-the-foot Climax logging locomotive, a log-carrying railroad car to the same scale, and a great many smaller projects.

On many visits to Bill's shop, I've noticed the slick-working panic button on his mill's on/off switch, and have thought to myself, "I oughta put that in my next book - the guys'd love it". I was over to see Bill about something or other one day when I had my camera with me, and so I snapped a couple of pictures.

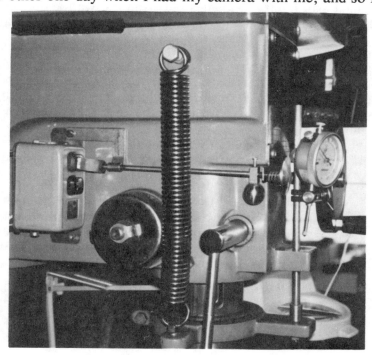

The first photo shows the mechanism "at rest". The push button is up front, just to the left of the dial indicator body; it is lightly spring loaded, and it plus the push rod and the foot over by the actual on/off switch are kept lined up on the OFF button by the weight of a 3/4"ϕ ball soldered to a 1/2" sq. x 1" long piece of CRS, visible just to the left of the spring and the front bracket. This allows Bill to shut off the machine by just reaching up and giving the button a little pat.

As is also clear in the above photo, the push rod travels in 2 simple sheet metal mounting brackets, one of which is screwed to the front of the machine, and the other to the side of the machine. The hole in the side bracket would have been formed by rolling the end of the material much as I show in the drwgs about making a hinge elsewhere in this book. None of the parts involved are at all hard to make, and the net result is a definite improvement over the factory arrangement, particuarly if a panic stop is called for.

In the second photo, Bill has turned the operating button counter clockwise about 1/8th of a turn, so that the foot at the other end of the push rod is now ready to stomp on the start button.

Turn and push and the machine comes on. Slap and it's off.

Larry May, another BCSME member, made this up for Bill.

SHOP LAMPS

If you can't see a job, you can't do much with it. If you do close work, you need lots of light on it so you can see what you're doing. "Luxo" type lamps are good in some cases, but they vary in quality - the good ones cost plenty and the cheap ones don't seem to last.

I was in Vic Baldwin's shop one day, and saw there the mother of all Luxo lamps. Vic had made it about twice normal size. The reach was maybe 6 feet from the base to max. extension of the light. Building one would not be difficult if a fella had access to suitable light-weight square steel tubing or similar. For same, if you can't get it locally, check out Small Parts, Inc., and/or other similar suppliers. You could also use solid CRS - round or square - but the weight might get out of hand pretty quick.

If you look in Popular Science Magazine for October 1981, at page 141 you will find a simple, jointed arm lamp that comes out from the wall about 30 or 40 inches, in a plane parallel to the floor. It is well detailed in one drwg the size of a playing card, and looks like it'd be quite effective.

In another shop, I saw a jointed arm lamp made by welding about a 2"ϕ washer to the end of a 3' piece of 3/8"ϕ CRS bar. The welding was done so that it was on one side of the washer only. An "elbow"-type joint was then made by assembling 2 arm/washer weldments together with a bolt, and - if I recall correctly, a piece of fairly heavy leather between, plus washers and a nut or 2. If I were doing it, I think I'd be inclined to try upping the number of washers to 3 or 4, with leather ones of the same size between. Why? I suspect that a multi-layer joint of that sort would have nice frictional properties - if so, you'd probably end up with a very smooth-working lamp arm.

> Making the leather washers would be easy: punch or drill the hole (probably about 1/2"ϕ) in a piece of leather, and then sandwich it between 2 of the washers you're going to use, via a bolt and nut through the sandwich. Now use a razor sharp knife to cut off all the surplus leather sticking out past the steel washers, and there's your washer, nice as you could ever want.

Now all that is about enough at the conjectural level on the subject of shop lamps. What follows next is a neat item that is not at all in the realm of conjecture, and that contains ideas within ideas. I hope you like it.....

"A Most Revolutionary Lamp Base"
or
A Tangle-Free Spin-Around Luxo-Lamp Base
from **Robert W. Wolfe, Foster City, CA**

Dear Guy:

I've made a number of "civic improvements" in my shop that make good use of stock parts, rather than parts made (machined) entirely from scratch. One of these is a mounting base for a swing arm lamp (often known as a "Luxo Lamp"), that allows the lamp to be spun around and around (at "hand operated", not "motor driven" speeds!) without tangling up the electrical cord that feeds it.

My lamp base has 3 machined surfaces. It takes about half an hour and maybe $5 or so of stock electrical stuff to make.

It might bear mention here that I had previously made several rotating electrical contact (slip ring) arrangements from scratch, and I was now in want of several more. A couple of friends had been after me to make up some for them also, but I kept putting it off - the fussy machining involved hardly seemed worth the time and effort required, and the item, when done, although useful, is not out where you can look at it from time to time and take pleasure in remembering a job well done.

The first photo below shows one of three "work islands**" I have in my shop: with all equipment aboard, plus some overhang, it takes up less than 4 square feet of floor space. It includes a grinder to the left, drill grinder to the right, buffing wheels in the front, and on the back (opposite side) a floppy abrasive disk and a wire wheel. On this kind of set-up, a lamp with a regular factory-installed electrical cord is a nuisance, if not a disaster. What you really want - and what I have - is a lamp that can be spun through 360° as many times as you please, with no effect on the cord.

** I make these "islands" by casting a 18 x 18 x 6" block of concrete (weight, ±150 lbs), with 1" water pipe legs cast right into the concrete to support the top. To render the island readily moveable, I install casters on one side and stabilizing screws in the corners opposite the casters, to take up differences in the floor. These "islands" are stable, solid, and very handy.

The second photo (next page) shows my prototype lamp base of this type, and the third photo shows the assembled spindle arrangement for the lamp.

Any hsm can readily make this type of rotating lamp base.... BUT first, a word of CAUTION: this electrical brush/collector ring set-up should only be used in applications where the required current-carrying capacity is on the order of 5 to 6 amps. *(Watts/Volts = Amps.)* It should also be noted that this is also a high friction device, and should therefore only be operated at very low RPM, and only from time to time.

It's fine for feeding juice to a Luxo lamp, and you may come up with other application ideas, as I have myself. But don't build one into something that will overload it (mechanically or electrically) and cause a house fire or other difficulties.

(NOTE: if the use of an "unapproved" electrical device lead to a fire that damaged your house, most insurance companies would refuse a claim, or at least try very hard to do so. So be warned, be smart, and watch it!! GBL)

NOTE: See the WARNING NOTE at the bottom of page 33.

The idea for a rotating electrical pickup/lamp mount came to me one day as I was working at my workbench, and had in my hands a plug and jack that I was mounting on some electronic equipment. Suddenly I began to look at the plug and jack from more of a mechanical than electrical perspective...

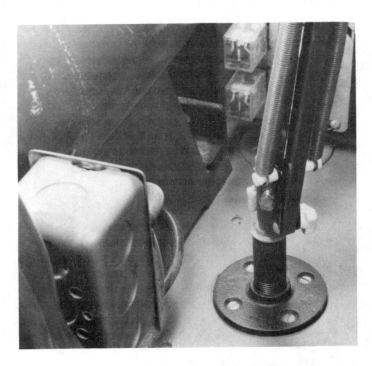 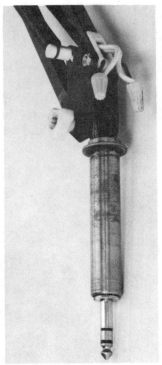

I measured the diameter of the plug and found it was about 1/2". Well, that's about the ID of 3/8" water pipe, and I had a piece in the scrap box. A quick test fit of the plug to the pipe and I was off to the lathe. In a matter of 30 minutes or so I had a working prototype mounted on a scrap board and clamped to my workbench.

Pleased with the result, I made up a nicer one, with finely polished pedestal, turned base flange, reamed and polished spindle bearings, etc. It looked great when I mounted it in its permanent spot, but.... it was a total failure! Every time I turned on a machine, the vibration caused the lamp to do the darndest doggie-dance you've ever seen. I tried various materials and methods to build some drag into the system, and eventually found what I consider to be the best arrangement - or at least one that would be very hard to beat.

If you like the idea so far, here's how to go about making up one of these.....

Revolutionary Tangle-Free Spin-Around Luxo-Lamp Bases:

NOTE: See the WARNING NOTE at the bottom of page 33.

1.   Use 3/8" galvanized water pipe running inside 1/2" galvanized water pipe for the Spindle and Pedestal, respectively. The Spindle bearing wants to be a nice, close, free-running fit without binding anywhere in its rotation, yet it should stay put once the lamp is moved into position and adjusted, regardless of vibrations (e.g. from a motor) that may be introduced into the table on which it is mounted.

I have made several dozen of these Lamp Bases, and of all the materials I have tried, the 1/2" pipe, galvanized on the inside, plus proper (simple) machining of the Spindle, as detailed below, produces the highest level of "stay-putness". If there was such a thing as an "index of stay-putness," I believe the mat'ls I suggest would read out at about 9.9 on a scale of 1 to 10.

2. Note that the inside diameter of standard weight 3/8" water pipe is usually a few thousands under 1/2", whereas galvanized pipe in the same nominal size is a few thou over 1/2"ID. This slight oversize condition is preferable and saves some otherwise necessary machining at the base of the Spindle.

3. Next is the choice of a pipe flange for the base of the Pedestal. You don't want a pipe flange of the kind normally found in the neighborhood hardware store - the kind that looks like it was cast before the beginning of the Industrial Revolution, and dipped in a lukewarm galvanizing vat. If you know of an old time hardware store - a REAL hardware store - you should be able to get some very nicely formed and undipped pipe flanges. If you can't, or if the idea of exposed threads on the pipe offends your tender sensibilities, it's no big job to turn a base flange, and drill and bore it for a press fit (or use Loctite) on an unthreaded portion of 1/2" water pipe that will form the upright element of the Pedestal.

4. The first operation is to pass a 1/2" drill into the ends of the 3/8" water pipe and a 5/8" drill into the 1/2" water pipe. This is done to knock any high spots off the galvanizing or the pipe seam. If using the threaded arrangement between the Pedestal and its Base, be sure that the inside diameter of the threaded end of the piece used for the Pedestal is full size**.

   ** (The inside diameter of 1/2" galvanized pipe is typically a few thou over 5/8", except possibly over the peak of the seam. The actual ID will depend on the thickness of the galvanizing and variation in manufacture. Passing the 5/8" drill through the pipe only knocks off any high spots from the seam. You may think that by knocking any high spots off with a 5/8" drill, the effective ID of the pipe is now 5/8", but it turns out that the effective ID is greater than 5/8" except for the lip at the end, which will be exactly 5/8". We want to open up the end of the pipe to the full actual ID of the particular piece of pipe we are using, so that there will not be a small lip at the end of the pipe to catch the turned Spindle that otherwise fits nicely for the balance of the length of the Pedestal pipe. No doubt some will use a sharp tapered reamer to flare out the crushed area caused by the manufacturer's cut-to-length operation, while others will just give it a lick and a slam.)

5. The Pedestal may be any height desired, but it must be tall enough to accommodate a Spindle at least 2-3/4" long. The Spindle must be no less than 2-3/4" long, so that it can accommodate the 1-1/4" length of the lamp tang, plus the 1-1/2" length of the plug from Radio Shack. For example, if the surface to which the lamp is to be mounted is 3/4" thick, then the Pedestal must be at least 2" long. With this consideration in mind, cut the Pedestal to the desired length and face the top end.

6. The next step is to turn down the outside diameter of the Spindle, and note that it wants to be *turned* to size. The final product wants to be a cleanly turned part, but in no way polished, so lock up your files temporarily. In order to get the highest index of stay-putness, the Spindle's bearing surface wants to be sized to the inside of the Pedestal by turning down the Spindle for a close, free running, no-bind fit. (See TMBR#1, pages 12-15, for tips on how to remove that last half thou.) Also, when making test fits, be sure to clean off all the chips, to avoid any false indications of binding.

7. The length of the turned portion of the Spindle will be the height of the Pedestal, plus the thickness of the table top to which it is going to be mounted, plus 1/16" for the thickness of the washer which will be soldered against the shoulder at the top of the Spindle, plus 1/8" to allow for the wall thickness of the switch box on the underside of the table top, and a hair to spare.

8. After turning the Spindle bearing, shorten to just over finished OAL (measure from the bottom end, and do the cutting/machining at top end) leaving a shoulder about 1/8" long; then face off the end cleanly. Next, bore out a 7/16" SAE washer to a slip fit over the turned portion of the Spindle. Solder the washer against the small shoulder at the top end of the Spindle. The washer thus forms a thrust bearing surface to ride on the top of the Pedestal pipe.

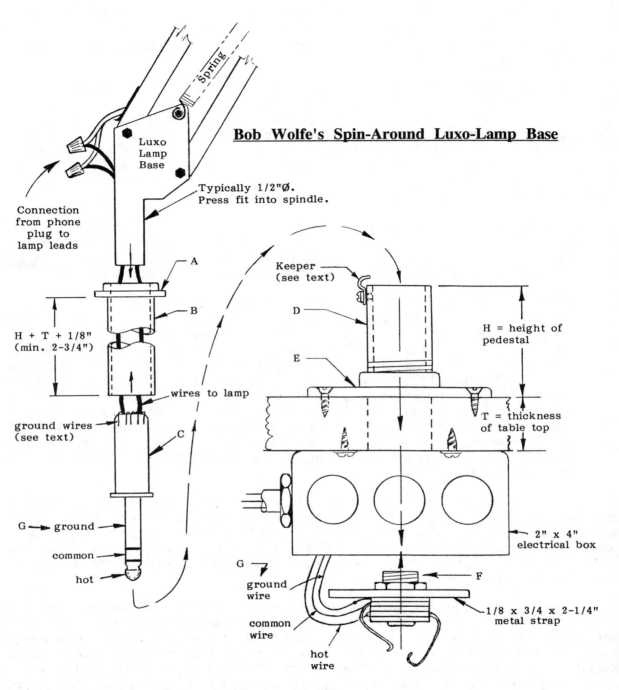

Bob Wolfe's Spin-Around Luxo-Lamp Base

Spring

Luxo Lamp Base

Connection from phone plug to lamp leads

Typically 1/2"Ø. Press fit into spindle.

A

B

H + T + 1/8" (min. 2-3/4")

wires to lamp

ground wires (see text)

C

G → ground

common

hot

Keeper (see text)

D

E

H = height of pedestal

T = thickness of table top

2" x 4" electrical box

G

ground wire

common wire

hot wire

F

1/8 x 3/4 x 2-1/4" metal strap

A = 7/16" SAE washer with its hole opened up to be a slip fit over the turned OD of "B", and soldered against shoulder on "B".

B = Spindle; 3/8" galvanized water pipe, turned down to be a close smooth fit in Pedestal "D"

C = 3-conductor 1/4" phone plug, press fit up into "B"

D = Pedestal; 1/2" Galvanized water pipe

E = Pipe flange

F = 1/4" open circuit phone jack

G = see comments in text re grounding

30

9.   The final mechanical part to make is a small Keeper that mounts near the top of the Pedestal. This Keeper extends up and over the top of the washer on the Spindle and protects against inadvertently lifting the lamp up and out of the Pedestal. The Keeper can be fashioned from any relatively light material - I use 20 gauge sheet steel cut about 3/8" x 1". Attach the Keeper to the Pedestal with a #4-40 machine screw short enough to not foul the inside wall of the Pedestal.

I was puzzled by the need for a Keeper. If the lamp tang is to be oriented vertically - which is most likely - one would have to swing the lamp very hard or roughly to make it lift up 2 or 3" and fly across the room, no? So I asked Bob about it, and here's his reply:

"I have not found a Keeper necessary on lamps mounted with their tangs oriented verti- cally, as in the photos, but as an extra safety precaution, I have put one on every unit I've made, since I don't know how others may use or treat the device. It's difficult to plan for all eventualities, but as an example, I have a small utility table on which I mounted a swing arm lamp on one of these "Wolfe" bases. I recall one day catching myself reaching for the base of the lamp when I wanted to move the table a few inches. It is only necessary to raise the lamp about 1/2" up in the Pedestal to release the plug from the jack, and if the lamp cord (to the jack) is plugged in, the sparks will fly!!** The Keeper thus provides that little added bit of safety which is, for the most part, unnecessary."

** Which is to say, you will cause a short, which is infinitely better not to do. Therefore DON'T pull the plug out of the jack, deliberately or inadver- tently, when the cord that feeds juice to the electrical box under the lamp base is plugged in. NOTE: See the WARNING NOTE at the bottom of page 33.

Now for the electrical portion...

(NOTE: All of the swing arm lamps that I've worked with are wired with two-conductor wire. That means there is no ground wire. When making one of these revolving lamp mounts, I think it is worthwhile to add a ground wire, and I have done so on all the lamps I have modified in this way. RWW)

10.   We need a 1/4" 3-conductor phone plug, (Radio Shack Part #274-139), and a 1/4" jack (Radio Shack Part #274-312). The plug has a plastic cover with a slight taper to it - small at the top end, and larger where it screws onto the body. If you are using galvanized pipe for the spindle, a test fit of the plug into the base of the Spindle should show the plastic cover on the plug just starting to press fit when the plastic cover is about half its length into the Spindle. If you are using some other material, such as black iron pipe or tubing, for the Spindle, it will be necessary to ease the inside diameter to about 0.505/0.507" for a distance of 1-1/2".

11.   Begin the wiring by cutting off the excess wire from the lamp, leaving a pigtail about 2" long extending from the base of the lamp. Set the lamp aside for now. Cut the molded plug off the other end of the wire just cut from the lamp, and discard the molded plug.** Unscrew the plastic sleeve from the phone plug. Inside you will find 3 lugs, each of a different length. The longest one (considerably longer than the other two) is the strain relief lug; we will use it for a ground wire.

** I got all worried about cuttin' the molded plug off the end of the factory-supplied lamp wire and junkin' it, thinkin' that we would want it when we go to plug the lamp in. (And you may.) Bob explained that his approach is to junk the molded plug, and then hard-wire the lamp into the wiring on the "island" where it is mounted. If you do it Bob's way - which I think is good - you won't be "plugging it in" somewhere to feed it the juice, so you don't need the molded plug. Bob says he typically wires a double

receptacle to the "island", and provides it with an extension cord long enough to reach one or another outlet in his shop, no matter where he moves the "island". The lamp is hard wired into the table's private electrical system, rather than being plugged into either hole of the double receptacle.

Bob also points out that, as already noted, a ground wire should be added to the setup, and therefore even if you did want a "plug-in" arrangement, it would require a 3-conductor cord and plug. GBL

12.   Now, at one end of what remains of the lamp wire, stagger (i.e. cut) the two conductors to match the lengths of the 2 shorter lugs in the phone plug. Strip off about 1/16" of insulation from each wire, and solder the bared wire ends to the two staggered lugs of the phone plug. You might think that 1/16" of exposed wire is not enough, but we want the insulation on the wires to come right up to the lugs in the phone plug, and what happens is this: as the wire is heated in the process of soldering, the insulation will shrink back slightly, exposing 3/32 to 1/8" of bare wire. Push the wire end up onto the lug by a similar amount during the soldering operation. Slick, eh?

13.   Now, crawl up to the opposite end of the lamp wire, remove an inch of insulation from one of the two conductor wires, and twist the strands together lightly - not tightly, just lightly. Cut off this inch of bare wire and solder it to the extreme end of the strain relief lug in the phone plug. This bare wire will be the ground connection to the Spindle.

14.   Now, re-install the plastic cover on the phone plug so that the lamp wires, and the inch of bare wire, extend through the hole in the top end of the plastic cover. (We will connect the other two wires to the lamp wires at Step 18 below.)

15.   With the plastic cover screwed tightly on the phone plug, untwist the bare wire just recently soldered to the strain relief lug, and bend each strand down over the outside of the straight portion of the plastic cover, spacing them evenly around the perimeter. Make sure each strand is independent and does not cross over another strand. Trim all of the stands so that they extend down the side of the cover no more than 1/8" along the straight portion of the plug cover.

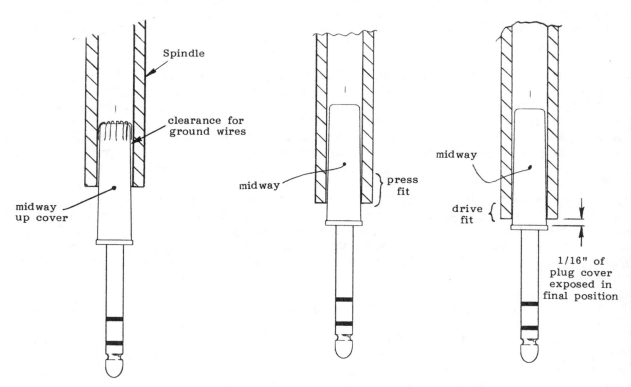

16.   Feed the insulated wire up through the Spindle, and then test fit the phone plug into the end of the Spindle. The bare (ground) wires bent back over the end of the plug cover should be a slip fit into the end of the Spindle. If the fit turns out to be too tight, reduce the amount of perimeter that the ground wires use up around the plug cover. If it is still a tight fit, reduce the number of strands making up the ground connection. When a slip fit of the ground wires is attained, drive the plug into the Spindle until only about 1/16" of the plastic cover still protrudes from the end of the Spindle. This should require some hefty driving. However, do not hammer on the tip of the plug; use a sleeve of some sort, such as a proper size socket wrench, and drive directly on the body of the plug.

17.   Feed the insulated wire leading up from the phone plug up through the round tang on the bottom end of the lamp, up into the base of the lamp, and out the back of the base alongside the 2" pigtail you left when you cut off the wire at the base of the lamp arm. Press fit the tang of the lamp into the top end of the Spindle. You may have to spread the tang slightly in order to get a good tight press fit.

18.   Cut the wires leading up from the phone plug to a length that matches the pigtail ends, and pair up the 4 wire ends with small wire nuts.

19.   Drill a 7/8" hole through the table where the lamp is to be mounted. Mount the Pedestal Flange Base directly over the hole. Next, knock out the slug in the center of the back of a 2x4" electrical box. Position the box on the underside of the table with this hole aligned with the hole through the table, and screw it solidly in place.

20.   Cut a piece of 1/8" x 3/4" strap - steel or aluminum - to a length of 2-1/4". In the center of this piece, drill an appropriate size hole to mount the phone jack (i.e. the other piece you bought at Radio Shack). Be sure the jack is solidly mounted to the strap, using a lock washer. The strap allows the jack arrangement to float freely but prevents the jack from turning more than a few degrees when the lamp is rotated.

21.   Connect the wires leading to the electrical box mounted under the table to the phone jack. Study the jack carefully to make sure that the wires are connected to the correct lugs on the jack. If the box is not grounded, be sure to run the ground wire to the box as well as to the jack. If solid wire is used, run the extension to the box, then add two inches of stranded wire to the ends of the solid wires, in order to make sure that the jack on its mounting strap floats freely inside the box.

22.   Drop the Spindle into the Pedestal and then press the jack all the way up onto the phone plug. Screw a blank cover plate onto the electrical box. Apply the little Keeper to the top of the Pedestal, switch on, and ENJOY.

Post Script: I don't know the actual ampere capacity of the phone jack set-up under discussion above. Although I may be overly conservative in rating my set-up at only 6 amps, I know that it will handle up to 6 amps, and I suspect it may be able to handle 9 or 10 amps without any problem. It depends on the application and conditions.

Bob Wolfe

Added to the 4th Printing: **PLEASE READ THIS WARNING NOTE**

Some readers have expressed concerns regarding the electrical safety of Mr. Wolfe's idea, or have reported to me that they know of electrical fires originating from similar arrangements, although none from units built according to the information contained in this book. When this first came to my attention, I called Mr. Wolfe and asked him if he had ever had any such problems. He said that some of his rotating phone jack electrical pick-up units had then been in service for at least 11 years, with no problems of any type. He also dismantled the first one he ever made, examined it for any evidence of problems such as overheating, and called me back to say there was no sign of ANY problems with it.

Use your own cautious judgment in how you make use of this idea, test what you build, DO NOT omit the "Keeper" (see pages 30/31), and consult an electrical engineer or qualified electrician if in any doubt.

Also added to the 4th Printing:

33   It is also specifically recommended NOT *TO USE* the electrical cord caddy design shown on pages 34 and 35 of previous printings of this book.

THE PILOT WHEEL ENGRAVER'S VISE
a nice example of the machinist's art

I attended the annual Firearms Engraver's Guild of America (FEGA) show in Reno in January of '97. The first thing you get at these do's is a name tag that can be read from 3 or 4 feet away, which turned out to be a good thing. Besides being stopped and spoken to enthusiastically by several people who said they had my books, I met a number of people whose names I've seen in the **Gun Digest** since I was a kid. At one point I turned around to find myself standing about 3 feet from Al Biesen, the legendary gunsmith from Spokane, WA. I introduced myself ("I used to have one of your rifles. I bought it second hand....."); he turned out to be very much regular people, friendly, and very interesting to talk to.

Of course there were lots of fine custom guns, and acres of engraving at the Show, but one of the neatest things I saw was what must surely be the Rolls Royce of engraving vises. Dubbed The Pilot Wheel Engraver's Vise by its maker, John Madole (4869 Old Columbia Road, Campbellsville, KY 42718; phone: 502-789-0088), it is as nice an example of the machinist's art as you are ever likely to see.

An engraver needs to be able to set his work at many different orientations (tilt), and he needs to be able to rotate the work while at any desired orientation, to execute scrolls and similar. The Pilot Wheel Vise accomplishes these 2 functions in a slightly different manner than a traditional "engraver's ball". The latter typically sits in a leather donut, and, being ball-shaped, can be tilted in any direction in the donut. The top half of the ball rotates on a shaft extending into the bottom half.

In the Pilot Wheel Vise there is no separate donut base. The Vise sits on its own 7-1/2"ϕ flat cast iron (or bronze) base. The ball and vise can be tilted to any desired orientation when you loosen the Pilot Wheel by bunting it counterclockwise with the heel of your hand. When the vise is oriented the way you want it, you simply bunt the Pilot Wheel in the opposite direction with your hand, and go back to your engraving. The lock-up of the ball by the spin of the Pilot Wheel is very positive.

The vise *per se* plugs into a hole in the top of the ball, and rotates under the control of the engraver's hand. The vise is tightened on the workpiece by a T-handled hex socket key which also operates the rotation lock screw in the ball.

The vise jaws can be moved to either of two positions on their nuts, to provide for holding larger and smaller workpieces - changing from one jaw position to the other involves removing and re-inserting four 1/4-20 socket head cap screws. The vise jaws are brass lined, to prevent marring work gripped therein. These jaw linings can be replaced with shop-made special jaws if required.

Every part of the Pilot Wheel Vise is machined all over, including the cast base. Fit and finish are impeccable. The ball is cast iron, first rough then finish machined, and finally ground and lapped to a *very* high degree of sphericity. The ball seat in the base, and the bronze locking ring above the spoked Pilot Wheel, are machined and lapped to an excellent fit with the ball.

The spokes on the Pilot Wheel feature an unusual type of "rope" knurling which can be seen in one of the photos on the facing page. Mr. Madole explained to me how to do this type of knurling - watch for full details in **TMBR#4!**

The hand engraved nameplate on the Pilot Wheel Vise is a work of art in itself; its apparent perfection stands up to the closest examination under a magnifying glass, as does the knurling on the Pilot Wheel spokes.

Here, as I said, is as fine an example of the machinist's art as you are ever likely to see. Everything about it says, very quietly, that the person who made it cared about what he was doing.

"..... a nice example of the machinist's art ....."

The Pilot Wheel Engraver's Vise

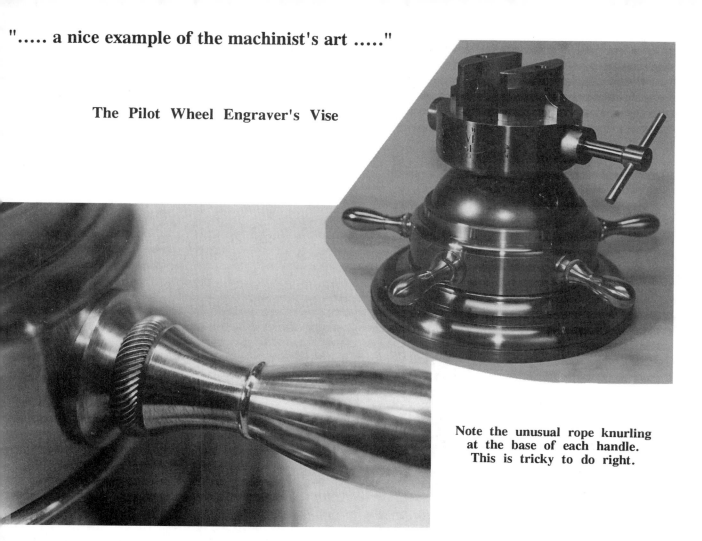

Note the unusual rope knurling
at the base of each handle.
This is tricky to do right.

The hand engraved
nameplate looks
even better under a
magnifying glass.

BILL FENTON'S JEWELRY

Sometimes I find something has passed right under my nose a number of times without catching my attention. When it does eventually click into focus, I'm amazed it took me so long to notice it. Such was the case with Bill Fenton's jewelry.

"Jewelry?!" you snort. "Who cares about jewelry? What we want is interesting metalworking info!" ..... Yeah, I know, but wait'll you see Bill's jewelry...

About 1973 Bill built a steam operated 2-drum donkey engine, and a wood fired boiler to steam it, and a little water tank to go with it. All this stuff was built at a scale of 3/4" to the foot, or 1/16th full size. When Bill got to a certain point in his project, he had to make a skid to put it all on, so he went down to visit the guys at Opsal Steel, an old line Vancouver outfit that makes heavy duty blocks, swivels, hooks, etc. - the stuff you find wherever wire rope is used.

Bill took his donkey engine with him, and - as you might imagine - the guys at Opsal were tickled to death with it.   When Bill asked them if they had any info on the general design of a full size donkey engine skid, they dug out an old catalog and flopped it open to a drwg giving complete details of a skid based on a pair of fir or spruce logs bucked off to an OAL of 65', 7'ϕ at one end, 5'ϕ at the other.   The cross timbers were 12' long, hewn to about 24" square, and set into notches in the skid logs.

They gave Bill a copy of the drwg, plus a copy of an old (1947) Opsal catalog, and off he went. A day or two later, Bill was poking about up in the bush just north of Vancouver, looking for a couple of nice straight dead fir trees, each of which would yield a piece of wood 4' long, about 5"ϕ at one end and 3-3/4"ϕ at the other. Pretty soon Bill had his whole outfit - engine, boiler, water tank and all - mounted on the most authentic lookin' little skid you ever saw, right down to a corrugated sheet iron roof over the donkey engine.

What he had, at that point, was a working model of the type of portable steam powered winching machinery once used in spar tree logging operations all along the west coast of North America from Alaska to California, (and in the Louisiana swamps, too, I believe).

But to be useful, a donkey outfit has to be teamed up with a spar tree and a mess of cables - some of them as thick as a man's wrist - plus blocks of various sizes. So Bill started whittlin' out scaled down blocks, and other related gear, collectively known in the loggers' argot as "the rigging" or "the jewelry" - the cables, hooks, swivels, chokers, tongs, etc. used to extract merchantable timber from a tract of forest via high lead logging methods.

(Maybe I'd better explain what I mean by a spar tree: a large tree in a convenient location is selected as the tree from which to rig cables to make a temporary aerial ropeway with which to bring out the logs from the surrounding forest.

The rigger - top dog in the pecking order on a logging show - will climb this tree, aided by spurs and a climbing belt. He goes up maybe 100 or 120 feet (or less, if the tree is not that tall), cutting off any limbs he encounters on the way up, and then proceeds to "top" the tree. When the top of the tree tips over and falls away, the remaining trunk sways through an arc of maybe 40 feet or so, with the rigger ridin' it out like a fly on a blade of grass. Once the tree settles down again, the rigger will climb up on the flat top of the not-yet-rigged spar tree, and - depending upon where the story is being told - will either (a) eat his lunch, or (b) have a wee-wee before commencing the work of bringing up a light cable and block, which he uses to hoist up heavier tackle, and with that, yet heavier tackle. When he's done, a number of large to very large blocks (the largest is called the Bull Block) are cabled to the top of the spar tree, and all the various other cables, guy lines, and so on are in place, and stretched out to, and very solidly anchored to, various large stumps near and far. When it's all done to the rigger's satisfaction (and he will be particular about the entire job, because not only his reputation, but the lives of most of the crew depend upon it being done right), down he comes. And when these guys want to come down off a spar tree in a hurry, if you've never seen it done before, you may just about faint - they drop 10 or 15 feet at a time, flipping the climbing belt away from the tree and dropping, digging their spurs in and dropping again. I've seen a rigger come down off a spar tree a hundred feet high in a matter of a few seconds.)

Now you know what a spar tree is.

Bill sets up his spar tree logging outfit on a couple of 4 x 8' sheets of plywood set end to end. For the stumps to which he will run his guy wires, he bolts down pieces of tree trunk maybe 2"ϕ x 1-1/2" tall, from the underside of the plywood. The "logs" to be brought out of the "forest" are pieces of small tree trunks maybe 1-1/2"ϕ x 12" long.

Bill has run his spar tree and donkey engine outfit at big public fairs like the annual Pacific National Exhibition (PNE). He fires the boiler with finger sized pieces of wood, and once he gets up a good head of steam, he hauls in several "turns" of logs with the aid of a crew recruited from among the crowd that invariably gathers.

(Bill got me to be his "whistle punk" one time, at the PNE in 1975. He had a cardboard sign about 12" x 18", with various whistle signals written on it in black felt pen -

I was supposed to blow the donkey engine's little steam whistle to signal the "loggers" out in the "woods" what was about to happen next: pick up the turn, tighten the haulback line, pull in the turn, etc. I guess I got kinda tangled up, what with havin' 2 left feet at the best of times, and Bill barkin' signals like "yo!" and "yo!yo!yo!" at me, to indicate what he was about to do, and all this with a crowd of a couple of hundred people watching!

So here I was, all tangled up in the whistle cord, so to speak, and - understandably enough - Bill began to get a little exasperated with me.

"What's the matter with you?" he bellowed. "It's all right there on the sign!"

"But I can't read, Bill," I said apologetically. Cracked the whole crowd up but good. Even Bill had to smile. And thus I escaped the sort of beating that I'm sure would have been administered on a real logging show to any whistle punk even half as inept as me.)

So that's the background about Bill's spar tree and steam donkey outfit. And I've seen it many a time, sitting right next to the bandsaw in his shop. Stewart Marshall and I visited Bill one afternoon, and in due course we got to looking at the donkey engine, and then Bill pulled out a handful of blocks, chokers, chains, swivels, etc. from a little cardboard box that sits on the front end of the skid, ".... and this is what they call 'the jewelry'..." he explained with a chuckle.

Suddenly, Bill's "jewelry" snapped into focus for me - here was a "model" that was not only as cute as a bug's ear, but that would be interesting and appealing, and at the same time would have a certain degree of utility.

(NOTE: Whenever you make something with which to lift things, you should test it with enough of an overload that you are CERTAIN it can safely handle the load you intend to lift with it. More on that below.)

Some of the blocks used in spar tree logging are huge - some weigh over a ton. Bill's are all made to scale, from 1/4" aluminum plate & CRS mostly, with 1-1/4" to 2"ϕ sheaves running on shouldered pins in bronze bushings. The chains and shackles are made from various sizes of nails. The swivels work like parts out of a Swiss watch. The chokers are as functional and smooth working as full sized ones. The logging tongs are made from a couple of 1/8" Allen keys, forged to shape just like the ones they make at Opsal.

So, if you happen to have a need for some really good looking hardware to pull your living room drapes, or to make an adjustable

| Pass Line | Tail | Haulback | Bull |
| Block | Block | Block | Block |

floor lamp, cable-suspended easy chair, or other similarly practical item, here are some ideas on how to make some that will surely fill the bill.

I borrowed Bill's most exquisite pieces of jewelry, and took them back to our shop, where the boys laid them out on one of the big benches in the main R&D room, and we analyzed them in minute detail... Which is to say I dismantled them on the dining room table, with pen, pencil, paper, and my 6" dial caliper at hand. Thus was spent a most enjoyable morning. Thereafter, I repaired to my drafting board to devise an easy way to lay out a scale block shell with dividers etc. on a piece of 1/4" aluminum plate. By following my drwgs, you can make an easy meal of the layout work, and then drill, saw, file and machine to your layout lines. (If you want to make several blocks of a given size, I think it would be more practical to make a pattern, and make castings, or have them made - see elsewhere herein for some ideas on backyard foundry operations.) More on methods below.

If you decide to make up a set of these little fellas and put them on display in your shop or in the billiard room, they'll be much admired by all who see them. But if you plan to use them for any sort of real, albeit small, hoisting jobs, do test them first at about double the intended working load. This is normal industrial practice...

I spent the summer of '66 working for Lister Bolt & Chain, an outfit that made forge-welded boom chains, deck chains, anchor chain, and other custom industrial ironwork. (Lister has moved since, but was then located only a stone's throw from the Opsal plant, and I would occasionally be sent over there to pick up a set of load binders, which Opsal stocked, but which we did not. [And yes, there are a pair of little Fenton-built load binders!])

Every so often we would get an order for several sets of heavy lifting slings, to be made up from cable and chain. A typical order might be for 5 slings, and would include instructions worded more or less as follows: "Test one sling to destruction. Test remaining 4 to twice working load of 5 tons....". The sling tested to destruction had to sustain more than 5 times the normal working load. The other 4 would then be tested to 10 tons, and certificated with a permanently attached metal tag. To carry out such tests, we had a large hydraulic cylinder in a concrete pit, fully instrumented, and capable of exerting a pull on the order of 200 tons, as I recall.

You might get a kick out of knowing what a big chain looks like when tested to destruction: its links will be pulled out from looking as at "A" at right, to a shape about as at "B", and at some point, one link will fail. The links which remain intact are so deformed by the test load that what was once a chain is now as rigid as a crowbar.

Let's have a look at some of Bill's jewelry, and then at how one might go about making similar equipment.

The Bull Block
The sheave in Bill's Bull Block is 2.015"ϕ, and has six fairly heavy spokes created by drilling, sawing and filing out 6 triangular holes. If you want to follow suit, whip up a big drwg, say 3 or 4 times full size, and lay out whatever seemeth good to thine eye. You may want to omit spokes altogether, or you may prefer to keep the "spoking out" work - as clockmakers call it - to a minimum, but want some sort of a pattern of spokes, in which case consider drilling 7 holes, each 0.365"ϕ, on a 1.150"ϕ pitch circle.

Why seven holes?
    In an article in FWW about making a treadled wood lathe almost entirely from wood, complete with a 9-spoke flywheel about 24"ϕ, I noted the comment that an uneven number of spokes in a wheel is "...more interesting, balanced, and pleasing to the eye..." than an even number of spokes would be in the same wheel.

I suspect this is true, so why not here also? If you're making a set of blocks such as these for the fun of it, 7 holes is no tougher to do than 6, and might look nicer, somehow.

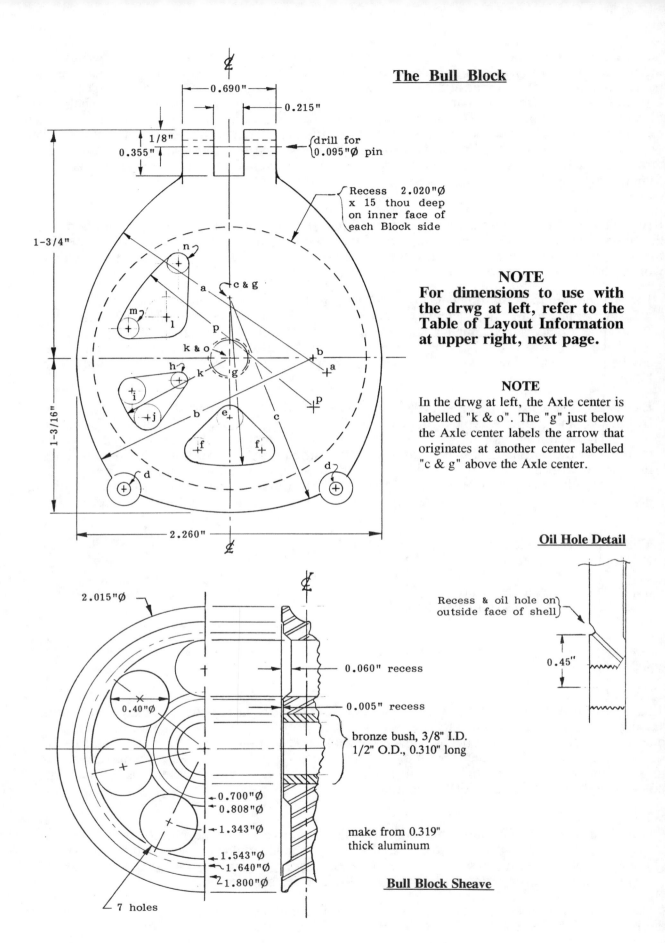

The Bull Block

0.690"

0.215"

1/8"
0.355"

drill for
0.095"Ø pin

Recess 2.020"Ø
x 15 thou deep
on inner face of
each Block side

1-3/4"

1-3/16"

2.260"

NOTE
For dimensions to use with the drwg at left, refer to the Table of Layout Information at upper right, next page.

NOTE
In the drwg at left, the Axle center is labelled "k & o". The "g" just below the Axle center labels the arrow that originates at another center labelled "c & g" above the Axle center.

Oil Hole Detail

Recess & oil hole on
outside face of shell

0.45"

2.015"Ø

0.40"Ø

0.060" recess

0.005" recess

bronze bush, 3/8" I.D.
1/2" O.D., 0.310" long

0.700"Ø
0.808"Ø
1.343"Ø
1.543"Ø
1.640"Ø
1.800"Ø

make from 0.319"
thick aluminum

7 holes

Bull Block Sheave

Table of Layout Information
for the
BULL BLOCK SIDES
(make from 1/4" aluminum plate)

Axle

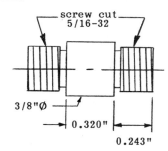

screw cut
5/16-32

3/8"Ø

0.320"

0.243"

Axle as drawn should be suitable for both Bull Block and Haulback Block, but check with other parts. (By an oversight, my original drwgs do not show details of the Bull Block Axle.)

	X	Y	Radius	Diameter	
(a)	0.719"*	-0.113"	1.850"	-	arcs a & b meet on the horizontal ₡.
(b)	0.618"*	0	1.750"	-	
(c)	0	0.464"	1.656"	-	
(d)	0.781"*	-1.000"	1/8"**	0.100"Ø drill, spot face 1/4"Ø, to leave a lug 0.165" thick (**outside of main block profile)	
(e)	0	-0.460"	-	3/16"Ø drill	
(f)	0.234"*	-0.710"	-	3/16"Ø drill, 2 places	
(g)	0	+0.464"	1.288"	gives part of bottom aerating hole	
(h)	0.367"*	-0.172"	-	1/8"Ø drill	gives corner holes for lower left & right aerating holes
(i)	0.717"*	-1/4"	-	0.0200" drill	
(j)	0.600"*	-0.453"	-	-	
(k)	0	0	0.862"	gives outer arc of lower left & right aerating holes (may give a slightly imperfect fit, but it'll be close enough)	
(l)	0.428"*	5/16"	-	5/16"Ø drill	
(m)	0.744"*	0.227"	-	drill 0.160"Ø	
(n)	0.375"*	0.727"	-		
(o)	0	0	-	drill 0.282"Ø, tap 5/16-32 (or similar) for axle, after layout is complete.	
(p)	0.618"*	-3/8"	1.580"	gives outer arc of upper left & right aerating holes	

* Where a dimension in col. X is followed by an * (e.g. 0.234"*) it means that *two* centers are required, one at X = +0.234", and one at X = -0.234", both on the same "Y" co-ordinate, so you can lay out the same feature on both sides of the main ₡.

Tie Bolts (2 required)

Make from 4-48 redi-rod or equal, 0.825" long; plus two 0.170" a/f x 0.092" tall hex nuts per bolt, plus a spacer sleeve 3/16"φ x 0.300" long.

Shackle

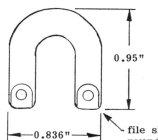

0.95"

0.836"

file slight flats & round ends as shown

Shackle Pin

0.095"Ø

0.170"

0.85"

forge nail head and cross drill #60

cross drill #60 for 0.029"Ø wire loop retaining ring, or make a small spring pin (ξ) to fit.

41

The Haulback Block

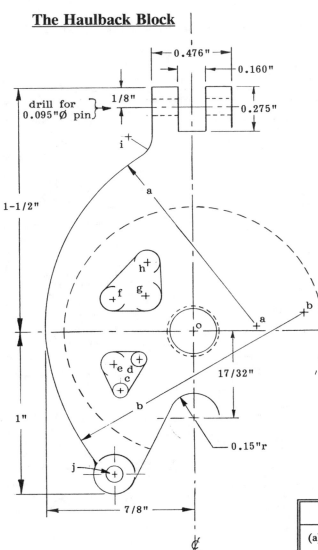

- 0.476"
- 0.160"
- 0.275"
- 1/8"
- drill for 0.095"Ø pin
- 1-1/2"
- 1"
- 7/8"
- 17/32"
- 0.15"r

Haulback Block Details

SHEAVE
Brass bushed aluminum, 1.532"Ø x 0.311" thick, with six 0.228"Ø holes equally spaced on a 0.915"Ø PC. The mouths of these holes are heavily chamfered.

OIL HOLE
The oil hole is a little simpler than on the Bull Block, and comes thru at about 45°.

AXLE
Drill rod, 3/8"Ø x 0.320" long, w/ 5/16-32 x 0.243" long screwcut portion at each end.

SPACERS
3/16"Ø x 0.285" long, plus tie bolts & nuts as for Bull Block.

SHACKLE
Make from a 3/16"Ø nail or similar - 0.85" tall, 0.740" across the opening. Otherwise about the same as for the Bull Block.

Attached to the shackle on the Haulback Block was another shackle, made from a 3/16"Ø nail, 1-3/16" tall, 0.690" across the open end, c/w with a 0.150"Ø x 0.8" pin, the latter cross drilled for a cotter pin, and having a 0.210"Ø x 1/16" head.

Table of Layout Information
for the
HAULBACK BLOCK SIDES
(make from 1/4" aluminum plate)

	X	Y	Radius	Diameter	
(a)	0.375"*	0.031"	1.25"	arcs (a)&(b) meet on the horizonal ₡.	
(b)	0.656"*	0.113"	1.513"	-	
(c)	0.433"*	-0.360"	-	3/32"Ø drill	corners of lower aerating holes
(d)	0.320"*	-0.168"	-	3/32"Ø drill	
(e)	15/32"*	-0.200"	-	0.152"Ø drill	
(f)	15/32"*	+0.200"	-	0.152"Ø drill	
(g)	0.280"*	+0.219"	-	3/16"Ø drill	
(h)	0.267"*	+0.425"	-	0.152"Ø drill	
(i)	0.375"*	+1.196"	-	0.288"Ø drill	
(j)	15/32"*	-7/8"	1/8"**	0.100Ø drill, and spot face 1/4"Ø to leave a lug about 0.165" thick.	
(o)	0	0	-	drill 0.282"Ø, tap 5/16-32, for axle, after layout is complete.	

* Where a dimension in col. X is followed by an * (e.g. 0.433"*) it means that *two* centers are required, one at X = +0.433", and one at X = -0.433", both on the same "Y" coordinate, so you can lay out the same feature on both sides of the main ₡.

** Outside of main block profile.

The Haulback Block

Bill's next block down in size is called a Haulback Block. It's about 2-9/16" overall height, compared to the 2-7/8" OA height of the Bull Block. You can see the similarities from the photo, but you can also see that they're not identical.

Center-to-center distance for the two pins at the bottom of the block is about 15/16" and a hair. Where the shackle attaches, the shells are 0.476" compared to the 0.690" dimension on the Bull Block.

The axle oiling hole is a little simpler, and comes through at about a 45° angle. The recess on the inner side of the block sides would be 5 thou over Sheave ϕ, and 15 to 20 thou deep - this info is not on the drwgs, nor can I find it in my rough drwgs at the time of finalizing this material.

The Tail Block

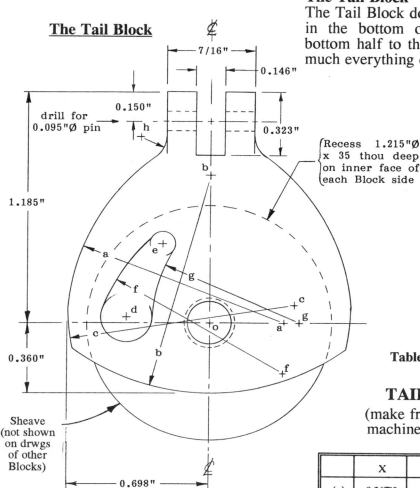

7/16"

0.146"

0.150"

drill for
0.095"Ø pin

h

0.323"

Recess 1.215"Ø
x 35 thou deep
on inner face of
each Block side

1.185"

0.360"

0.698"

Sheave
(not shown
on drwgs
of other
Blocks)

The Tail Block

The Tail Block doesn't have spacers and tie bolts in the bottom of the shell, since there's no bottom half to the shells. The drwgs tell pretty much everything else you'd need to know.

Table of Layout Information
for the
TAIL BLOCK SIDES

(make from 1/4" aluminum plate
machined down to 0.221" thick)

	X	Y	Radius	Diameter	
(a)	0.367"*	0	1.067"	-	arcs (a)&(c) intersect @ a point approx. 0.075" above the horizontal ₵.
(b)	0	+0.755"	1.115"	-	
(c)	0.420"*	+0.093"	1.115"	-	
(d)	0.410"*	+0.031"	-	0.250"Ø drill	
(e)	0.232"*	+0.406"	-	1/8"Ø drill	
(f)	0.363"*	-0.250"	0.950"	**	
(g)	0.440"*	0	0.725"	-	
(h)	0.344"*	+0.958"	1/8" radius	-	
(o)	0	0	-	drill 7/23"Ø, and tap 1/4-32 (or similar) for axle, after layout work is complete.	

* Where a dimension in col. X is followed by an * (e.g. 0.367"*) it means that *two* centers are required, one at X = +0.367", and one at X = -0.367", both on the same "Y" coordinate, so you can lay out the same feature on both sides of the main ₵.

** This drawing was made with (f) 0.256" below the horizontal ₵. Arc f did not fair into hole "E" as well as in my original drawing. Suggest you use y = -0.250", so the curves will fair in better, or expect to do some blending with a file. Of course, much also depends upon the overall accuracy of your layout, and upon how fussy you wish to be.

Tail Block Details

SHEAVE
1.200"Ø x 0.239" thick, with four 0.206"Ø holes equally spaced on a 0.785"Ø PC. Sheave Bushing = 0.4375" OD.

AXLE
Drill rod, 3/8"Ø x 0.246", with 1/4-32 x 0.192" long ends; OAL = 0.630".

This Block is of a different design from the others and does not have spacers, tie bolts and nuts at the bottom end.

SHACKLE
Make from a 0.140"Ø nail, 0.765" tall. Shackle pin is 0.095"Ø x 0.695" OAL; cross drill at one end for a wire ring or cotter pin, and forge flat at the other end.

43

The Pass Line Block

The Pass Line Block is the smallest. Bill said he didn't "aerate" this one, although it would be in full size practice. (The holes in the shells of the full size blocks are for weight savings: in a big bull block there are 10 holes, each probably big enough to stick your head through, which would mean a saving of several hundred pounds of material without sacrificing strength where needed.)

The Pass Line Block is not illustrated. If you want a block smaller than a Tail Block, just make one like the Tail Block but at say 0.75 or 0.8 times the dimensions given for the Tail Block.

Multi-Sheave Blocks

Although Opsal does not make full size blocks of the foregoing type in multiple-sheave form, you could, with little difficulty, modify any of these blocks to incorporate 2 or 3 sheaves. For each extra sheave, you would need a separator plate of say 1/16" CRS, having about the same profile as the outer shells, reamed to suit the axle, and match

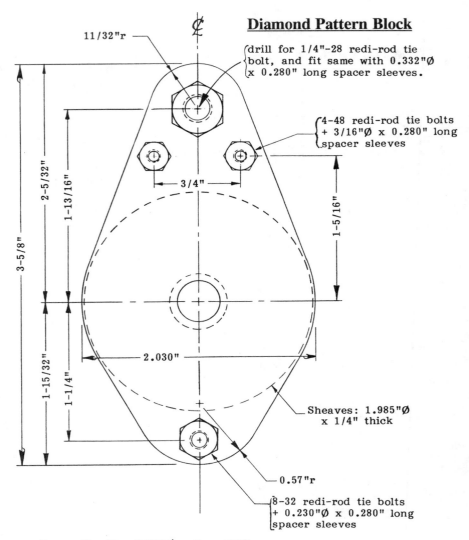

Diamond Pattern Block

drill for 1/4"-28 redi-rod tie bolt, and fit same with 0.332"Ø x 0.280" long spacer sleeves.

4-48 redi-rod tie bolts + 3/16"Ø x 0.280" long spacer sleeves

Sheaves: 1.985"Ø x 1/4" thick

8-32 redi-rod tie bolts + 0.230"Ø x 0.280" long spacer sleeves

Upper Shackle: 5/16"Ø nail or CRS
Lower Shackle: 0.150"Ø or heavier

Side Plates: 1/16" CRS or similar
Sheaves: 1.985"Ø x 1/4" thick.

Axle Pin: 3/8"Ø, running in 1/2"
O.D. bronze bushing

If for serious work, enlarge all over

Drawn from Opsal Drwg #1028 of
03-21-'74, for their #1488 Block.

Overall length of Opsal #1488 is 18-3/16"Ø;
O/A width is 10-1/8". Sheave is 9-15/16"Ø.

drilled at the bottom also for spacers/bolts. Connecting the separator plate(s) at the top would take a little thinking to get them to look right.

Opsal does make multi-sheaved blocks of a type generally referred to as "diamond pattern" blocks. When I went in to see the people at Opsal Steel, they were more than gracious: besides treating me like I was doing them a favor, rather than being a nuisance, they gave me not only a tour of their entire plant, but also a copy of their shop drawings for the diamond pattern multi-sheave blocks found in their current catalog under the heading "Heavy Duty Hoisting Blocks." (They're in the 1947 catalog, too!) The drwgs at right above give details to let you cook up something similar.

If you want to make a pair of multi-sheave blocks of this type, start by cutting a number of pieces of 1/16" or 1/8" CRS plate roughly to size. Drill/ream each piece for the axle shaft, and then stack them up on a pin (or make a jig) to drill the holes for the 3 assembly bolts. Then stack up the plates for a given block on another jig pin a hair shorter than their combined thickness, clamp the pieces together with service bolts (i.e. bolts to be used in the course of making the parts, not in the completed item) in the 3 assembly holes, and mill the outer profile of the several plates for that block at one time.

Do the same for the plates for the next block, and when these are all done, make up your sheaves and bushes, make spacers and studs to suit, thread the ends of the studs to take little hex nuts, and put 'em all together.

Just how to thread the line through a pair of multi-sheave blocks is a whole study in itself. The Navy (any navy) is into this kind of stuff, and if you do some digging at your local library, you will undoubtedly come up with enough info to make your eyes bulge.

Now, having dissected Bill's little blocks, I'm not sayin' anything outta turn when I say they ain't perfect. But their little imperfections - the aerating holes are not identical from one side of a block to the other, for example - give them character and personality: every one is "one-of-a-kind"! If you decide to make a set, don't try to make them perfect - lay 'em out nice, drill the starting holes at the corners of the aerating holes, saw out the waste material, and file 'em up so they look right, file in your radii, etc., but don't spend endless hours trying to get everything just PERFECT.

The sides, or shells, of Bill's blocks show traces of what may have been black paint. Opsal blocks - real ones - are cast, machined, and then painted a sort of blue-green, with red sheaves. After you mill and file the shells to final form, the nicest finish would probably be a glass bead blast job. You could paint or anodize over this if you wanted to.

Bill's blocks don't look new. They show file marks - or more probably abrasive paper marks - also numerous small nicks and dings, quite appropriate for "jewelry" that's been in service 20 years. You can be sure the real ones collect their fair share of nicks and dings, too.

Bill has a little piece of fine steel cable shackled to his Bull Block. An old logger in the crowd stepped up to speak to Bill after one of his demos at the PNE one year. He pointed out that Bill's outfit lacked a "safety strap" on the Bull Block. This is basically a piece of steel cable with an eye at each end; it is looped over the sky line, and hooked onto the clevis at the top end of the Bull Block. If one of the spar tree guy lines snaps, or other disaster befalls the spar tree, the safety strap keeps the Bull Block and other assorted bits of jewelry from crashing down on the donkey puncher or the hook chaser below.

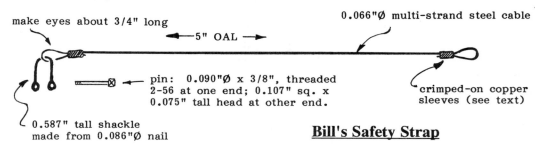

make eyes about 3/4" long

←—5" OAL —→

0.066"Ø multi-strand steel cable

pin: 0.090"Ø x 3/8", threaded 2-56 at one end; 0.107" sq. x 0.075" tall head at other end.

crimped-on copper sleeves (see text)

0.587" tall shackle made from 0.086"Ø nail

Bill's Safety Strap

Now I hope ta tell ya, guys, Bill don't run no chicken crap outfit - that very night he went home and made a safety strap for his Bull Block.

How would you make them?
To make a pair of block sides for one of these blocks (if I were not using castings), I'd be inclined to go at the job thus:

Saw out a pair of blanks from 1/4" aluminum plate, each being a rectangle slightly over the finished overall dimensions of a block side. (High accuracy on the blank sizes is not required, unless you elect to do your layout work on the surface plate, as below).

Locate and drill/ream a service hole at the axle hole location.

Mount each block side in the 4-jaw chuck to machine the recess on the inner face, or do it in the mill with a boring head while set up to drill the service hole.

Then, bolt a pair of block sides together via their service holes, grip the work by its edges in the milling vise, and move the milling machine table so as to put the axle center directly under the milling machine spindle centerline. Zero out the feedscrew dials, and then move to various co-ordinates and drill corner holes for all the "aerating holes", as Bill calls them.

Alternatively, *start by machining the blanks to an accurate rectangle of the exact OA dimensions of a block side, and then* do the layout work with the aid of height gage and surface plate. Once the layout work is done, you can either centerpunch each hole location, or locate over each hole center with a center finder, and drill the holes. Since you have to make 2 block sides for each block you want, you get to use each setting of the measuring tools twice. If you are making several identical blocks, then the layout work will go even more quickly, (although castings would begin to make even more sense).

Make up templates in scrap galvanized sheet metal to permit scribing the arcs that define the block side's outer profile and the few curved portions of the aerating holes. Provide each of these templates with a straight edge that coincides with the vertical centerline of the block side, and a hole - very carefully located - to register with the service hole at the axle hole location.

The Bull Block has one arc (repeated on both sides of the centerline) where the central "register hole" will coincide with the center of the arc. On all the other templates, the register hole will NOT coincide with the center of the arc the template is to draw.

Use a register pin to position each template in turn on the service hole, and scribe the various arcs. Turn the templates over to get arcs right and left of centerline.

Then saw, mill and file each block side to finish profile.

Finally, mill and drill the ears for the shackle, drill out the service hole to tapping size and tap same for the screwcut ends of the Axle.

As can be seen in one of the photos, the shells are nicely rounded over. This can be done (with some effort) with a file, or a lot faster on a belt sander. Here again, the use of castings would reduce the amount of work involved.

A feature of Opsal blocks is the recessing of the sheave into the block sides. (This is doubtless not exclusive to Opsal blocks; what I'm trying to explain here is why they do it.) The minimal width of the opening thus presented between the sheave and the sides of the block denies entry to much of the debris (branches, mud, etc.) that must get dragged through such blocks by the cable, and also prevents smaller cables that may be attached to the main cable from slipping in there and jamming up everything. Such events are what drives riggers to the brink of palpable insanity.

Shackles
Most of Bill's shackles and shackle pins were made from common nails. If you make a set of blocks like these, don't decide to put a serious (i.e. valuable) load on them without giving them a severe test first. Better quality steel than that found in some nails might not be a bad idea,

although some - such as TREE ISLAND brand nails (made here in Vancouver) - are pretty good stuff.

We'll get into nails, their quality, and their strength from an engineering point of view, in a few minutes.

All the smaller shackles were turned down from larger diameter stock, with lumps slightly larger than 0.086"ϕ left at each end. These lumps appear to have been cold forged down to a flat lobe about 0.086" thick, and then cross drilled, with one of the two holes then being tapped. Bill made several, all the same, and then bent them into a U shape in a simple jig in his hydraulic press. From the way Bill told me how the holes in the ends of each shackle blank had come out right in line from the bending jig, it was obvious how much pleasure he'd gotten from just this one aspect of the whole project originally, and again in recalling it. (We'll see some ideas on a shackle bending jig in a few pages, so stay tuned.)

Bill made the chain links for his "tag line" (another part of "the jewelry") from nails, the ends of each link being soft soldered together.   There are a dozen little shackles on the tag line alone, all identical, jig bent, c/w square headed, threaded pins. I don't see the little chokers and the rest of the tag line as having much utility for most readers, so I have not shown it, but you might get a kick out of the following:

Bill's little choker hooks, which are maybe 3/4" long, are as functional as any choker in the big woods. They require a small back spotfacing job inside when made in this size (the full size ones are cast, and come from the sand pretty much ready to use), so Bill made a little wee cut-

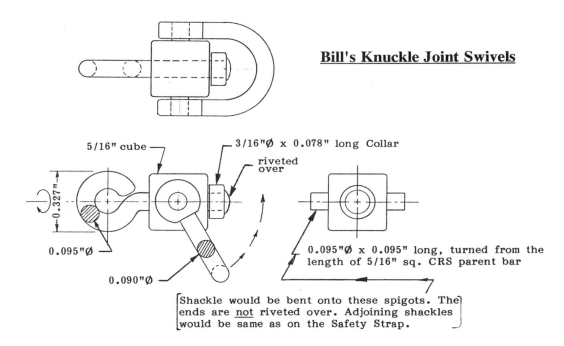

Bill's Knuckle Joint Swivels

5/16" cube

3/16"Ø x 0.078" long Collar

riveted over

0.327"

0.095"Ø

0.090"Ø

0.095"Ø x 0.095" long, turned from the length of 5/16" sq. CRS parent bar

[Shackle would be bent onto these spigots. The ends are not riveted over. Adjoining shackles would be same as on the Safety Strap.]

ter with the teeth filed to cut in the "up" direction, hardened it, and did the necessary on each choker in short order. Bill gets a kick out of showing such cutters as this to visitors to his shop - and well he might, for they are beautifully made, and each one did, in it's day, the job it was made to do.

Like I said above, I don't see the chokers and chain rigging as being useful/safe/practical, but I will do a drwg of the little knuckle joint swivels, because they work like parts out of a Swiss watch, and they might be useful somewhere. There are 5 of them, all alike, on Bill's tag line. (An Opsal-built swivel for 1-3/4"ϕ cable weighs about 75 lbs.)

The body or knuckle for one of these swivels would be made by chucking a piece of 5/16" sq. CRS in the 4-jaw and centering it up nice with the aid of a parting tool blade in the toolpost, touching it to each face of the material in turn, and noting the cross slide feedscrew dial reading at each face. Adjust opposite chuck jaws until the readings are the same on opposite faces. Then, face the material, turn down to 0.095"φ spigot for 0.095" and move towards the chuck 5/16" plus the thickness of the parting tool blade, turn another 0.095"φ x 0.095" spigot, and part off. Shift the piece of 5/16 CRS in the chuck, and do everything above all over again, for as many swivels as you plan to make. Here again, the shackles are made from nails, jig bent.

HOW GOOD ARE NAILS?

You might wonder, as I did, about using nails as a source of material for the shackles, pins and such in these blocks. I phoned the Tree Island people mentioned above, and learned a few interesting things about nails.

First, using good quality nails for this purpose (that is to say, *"model"* blocks) is not down there on a level with sleepin' in yer underwear. Nails are usually made from hot rolled 1008/1010 to 1018 steel - i.e. material about like ordinary CRS for quality and carbon content. The black mill scale typical of HRMS is removed in an acid dip at an early stage in the manufacturing process, and the subsequent cold working of the steel is what gives the finished nail stiffness (which makes it resistant to bending). [When I think of all the soft nails people have unloaded on me! On the other hand, practice makes perfect - Margie says she's never seen anyone who can drive bent nails as well as I can.] The tensile strength of a decent nail may be on the order of 120,000 psi, which is not bad at all - cold rolled 1008 has a tensile strength of about 48,000 psi, while that of 1018 CRS would be about 58,000 psi.

Now that stiffness is going to work *against* you somewhat when you're bending up a shackle, but the thing to do is to make a little jig, and cold bend the shackle blank in this jig, with the aid of your vise or a small arbor press. (We'll look at such a jig in a few pages.) On the other hand, that same stiffness is working *for* you when you use a nail as a pin in one of these little shackles, or to pin a gooseneck shackle to the top (ears) of these little blocks.

How to tell if you're looking at good quality nails in a store? Look at a handful of them. Good nails have a nice, clean, uniform appearance, sharp, well formed points, well centered heads, and they'll be generally free of burrs and warts. In short, a good nail looks good in its various details, 'cause somebody saw to it that they made a good job of making it. Cheap nails won't look like good ones.

If you find a source of good nails, pull out your dial calipers, measure different sizes, and then buy some of each size you need to make all your little forged parts.

I got to wondering just what sort of a load a little shackle like the ones Bill put on his jewelry could take. How to figure that out falls into the realm of the science of Strength of Materials, a subject that professional engineers must master. I ain't one, so I buzzed the Mechanical Engineering Department at the University of British Columbia and spoke to Dr. Bruce Dunwoody.

Dr. Dunwoody told me how to approach the calculations and what formulae to use. I took a crack at it, but I didn't feel confident that I had done everything right, so I buzzed him back. If I were to write down my attempt and send it over to him, would he go over it and tell me what I was doing right or wrong, if anything? Yes, he would, so I did and he did. What follows is the result of yet another example of my tireless pursuit of new stuff for your fun and enjoyment.

Basically, I had 2 questions:

1) What load will the pin in a shackle stand?

2) What load would be required to tear the eyes out of a shackle?

Let us deal with these 2 problems in the order stated above.

To determine the load a shackle pin can sustain, we have to calculate 2 things:

1) Bending Moment, or Moment of Inertia, which is given by the formula

$$I = \frac{d^4\pi}{64}$$ *FORMULA 1*

where   I = Bending Moment, and
             d = pin diameter in inches,

and

2) Strength of the pin, which we will call S, and which is given by the formula

$$\textit{Strength of the pin} = S = \frac{fld}{2I}$$ *FORMULA 2*

where   f = 1/2 of the load on the shackle, in lbs.
             l = unsupported length of the pin, in inches
             d and I as above.

I got tripped up on l, thinking it would be the distance between the eyes of the shackle. Dr. Dunwoody straightened me out:- it is 1/2 of the pin length between the eyes, as shown in the drwg below.

This is so because the load is passed from the shackle through the pin to a chain link or cable eye which would be nominally centered on the pin. Thus, 1/2 of the total load on the shackle is carried through each half of the pin, and the unsupported length of pin is approx. 1/2 of the shackle opening.

A factor of safety of 5 would be appropriate to use, to determine the minimum material strength required to sustain a given load.

Opsal's #66 High Lead Shackle, with a safe working load of 26 tons, is supplied with a pin 1-1/2" in diameter, and a 1-5/8" spread between the eyes of the shackle.

Therefore...

$$I = \frac{d^4\pi}{64} = \frac{1.5^4\pi}{64} = \frac{5.0625\pi}{64} = 0.2485 \text{ inches}^4$$ *FORMULA 3*

and

$$S = \frac{fld}{2I} = \frac{(\frac{26 \text{ tons}}{2}) \times (2000\frac{lb}{ton}) \times (\frac{1.625''}{2}) \times 1.5}{2 \times 0.2485}$$ *FORMULA 4*

$$= \frac{13 \times 2000 \times 0.8125 \times 1.5}{0.497} = \frac{31,687.5}{0.497} = 63,757.5 \text{ psi}$$

Multiplying this by the safety factor of 5, we would get

63,757.5 x 5 = 318,788 psi, or say 320,000 psi

This means that the pin would need to be made from a steel alloy having a yield strength exceeding 318,000 psi in order to insure against it bending (permanently) at the center of the pin. Dr. Dunwoody made this additional comment: "Of course, this analysis is very rudimentary. Other considerations such as plastic flow and a finite (i.e. measurable, rather than the theoretical point or line) width of contact between pin and wire loop could strongly affect the actual load at which observable bending occurred."

Now if we consider a small shackle with a pin of 0.150"ϕ and an unsupported length of 1/4" (= 1/2" divided by 2), and if we were using a nail having a tensile strength of 120,000 psi, what load might it sustain?

$$I = \frac{d^4\pi}{64} = \frac{(0.15)^4\pi}{64} = \frac{\pi \times (0.000,506)}{64} \qquad \text{using FORMULA 1}$$

$$= \frac{0.001,590}{64} = 0.000,025 \; inches^4$$

$$S = \frac{fld}{2I} \qquad \text{and FORMULA 2 again}$$

where S = 120,000
       l = 1/4"
       d = 0.15"

Again using a safety factor of 5, and rearranging the above equation to solve for "f", we would have ...

$$f = \frac{2Is}{ld5} = \frac{2 \times 0.000,025 \times 120,000}{0.25 \times 0.15 \times 5}$$

$$= \frac{15}{0.0375 \times 5} = 80 \; lbs \; \longleftarrow \qquad = 1/2 \; of \; sustainable \; load$$

This means that you could hang a weight of about 160 lbs. on the shackle without bending the pin. However, the shackle eye might tear out... which brings us back to the second question posed above, namely:

2) What load would be required to tear the eye out of a shackle?

The eye might tear out as in the drwg at right.

For this mode of failure:

$$\textbf{Strength} = S = \frac{f}{t(D - d)}$$

where S = strength of the pin
       t = thickness of the eye.
       D = diameter of the eye
       d = pin diameter

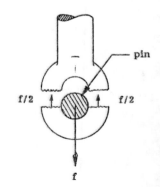

Let's stick to the little shackles made from 120,000 psi. nails. Bill's shackles in fact did not have an eye formed exactly as above, but let's say D = 1/4"ϕ, d = 0.100"ϕ, and t = 3/32"...
Then

$$S = \frac{f}{\frac{3}{32}(0.25 - 0.100)} = \frac{f}{\frac{3}{32}(0.150)}$$

$$f = (120,000)(\frac{3}{32})(0.15) = 1688 \; lbs$$

Therefore, the eye is very unlikely to tear out - much more likely that the pin would bend long before the shackle eyes would fail.

There is a 3rd problem, concerning the pins that join some of the gooseneck shackles to the tops of the block shells, as shown in the drwg at left below.

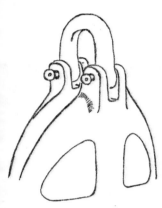

Typically, in Bill's blocks, the pin diameter is 0.095". Either the "eye" at the top of the block shells would tear out (unlikely), or the eye in the gooseneck shackle would tear out, or the pins would shear off much as if they were being cut in a bench shear, where a piece of wire or similar is put into a hole in two plates, and then one plate pivots (slides) across the other, shearing off the wire. This is the most likely sort of failure: double shear. As Dr. Dunwoody explained to me: ".... the two pins would have to shear simultaneously in 4 locations (two places on each pin) for the shackle to separate from the block."

The total load required to do this would be:

$$Load = 4 \times shear \; strength \times \frac{\pi}{4} \times (pin \; \emptyset)^2 = 4 \times 120,000 \; p.s.i. \times \frac{\pi}{4} \times (0.095'')^2$$

$$= \frac{4 \times 120,000 \times \pi}{4} \times 0.0090 = 120,000 \times \pi \times 0.009$$

$$= 4 \times 120,000 \; p.s.i. \times \frac{\pi}{4} \times 0.095^2$$

$$= 3402 \; lbs, \; or \; say \; 3400lbs$$

Conclusion: Even if the load required to shear these pins was only half of that amount, the pins in some of the other shackles in the rigging would fail first.

I don't know whether you found the foregoing interesting or not. If you read this far, presumably you did, to at least some degree. I have about exhausted my knowledge and understanding of this subject in the above, and I may even have made a slip or two, but I think not. The science of Strength of Materials owes much to a man called Timoshenko. No, he was not an Irishman by the name of Tim O'Shenko. He was a Russian engineer, Stephen P. Timoshenko, born about 1878; he moved to the U.S. some years after the Russian Revolution. Timoshenko was a very clever man, sometimes subtle, sometimes not, as you will see if you care to read his autobiography, "As I Remember". He went back to the University of Kiev (in the Ukraine) in 1958 and found the current crop of engineering students still using a book of problems he had written while a professor of engineering there in 1908. Like I said, you might enjoy the book.

Having said all the above, I will reiterate my caution given earlier in this section: if you make a set of these blocks, with 0.095"ϕ pins made from finishing nails as their weakest part, DON'T decide to lower your mother-in-law's $38,000 grand piano (or herself) from the window of her 40th storey penthouse apartment with them for something to do on your day off. If you have serious work to do with blocks, get serious blocks, ones well suited to the job at hand.

Harken to this:

P'ssst! Wanna know where to buy some real nice blocks for moderately serious work? Get thee to a boating supply store, and there see the variety of yacht fittings that are offered. Some of it is real cute - Harken Yacht Fittings, of Pewaukee, Wisconsin, makes some very nice looking, well designed stuff. I made up two triple sheaved block-and-tackle outfits one time, in two sizes, from Harken blocks, one having a safe working load of about 750 lbs, the other about double that. I gave the smaller one to my brother as a game hoist, as it was very compact and light - ideal to carry in a hunting pack, and what a boon if you nail a big buck! The bigger one resides in my own shop, where its main duty is re-installing the vise on my B'port. (Never mind laughin'. That vise weighs 75 lbs, and has hardened steel alignment keys on its underside. You think I want to wrassle that suckah up onto the table of my B'port and shivvy it around until the keys drop into the middle T-slot?! No thank you. My way, I can set it down on the table of the mill like it was a kleenex, and what's more, I'll still be able to do so when I'm 90, if I should live so long, and can still remember what the vise is for!)

Now, let me see, what wuz I talkin' about? Oh yeah - my big block and tackle... Handy rig to have around. Fella visited here one day. Parked his car in our (moderately steep) driveway. When he was ready to leave, he let the brake off before he took his car out of "Park", and it jammed that way, with so much weight on the little pawl or whatever it is that does its business in the transmission. What to do? No sweat. Out came the old Harken tackle, which we rigged to a short post at the top of the driveway. We took up the slack, and then put a good pull on the line, and his tranny popped out of "park" as easy as shuckin' an oyster.

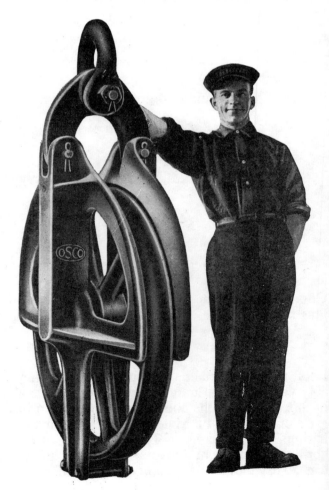

"Try hangin' this'un
on a nail, Lautard!"

52

WIRE BENDING JIGS

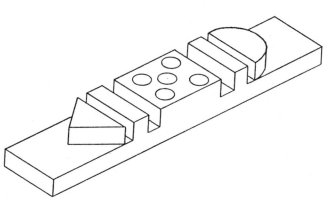

There are times when one wants to make a controlled, or relatively controlled, bend in a piece of wire. Various aids to this end can be had, or made. I have seen a simple cast aluminum jig as drwg below. It could easily be duplicated in short order in any hsm's shop, screwed to a block of wood, and thereafter would be found useful for a variety of jobs. A piece of wire can be placed in any of the grooves, and bent, or can be bent around pins placed in the holes drilled in the top face. These things sell for very little money, but a die cast aluminum one won't stand up as well as one made from CRS.

A Shackle Bending Jig

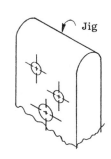

I came across a neat idea about bending a shackle out of wire. This is essentially a "U" with a bent eye at each end. What the fella showed to do - see drwgs at left - was to bend the material over a former, which has three holes drilled in it. That forms the "U" part of the shackle. Then, the wire is bent through 90° along the length of the former. Then - and this is the clever part - you insert into the former the pin around which the eye is bent! The eye section is bent, with excess material angling out over the base of the eye. The excess is then cut off, and the eye hammered into a single plane. One might also want to weld the eye shut - see below.

The foregoing appeals to me no end. This idea came from an item in ME for July 7, 1955. The author of the article is given only as "A.E.U.". He quite correctly said that to look well when finished, work of this type must be done with bending jigs etc., to avoid having hammer and plier marks all over each piece. "AEU" also showed a jig for forming links for chain. While I wouldn't say it wouldn't work, how he proposed to use it was not made clear in the article.

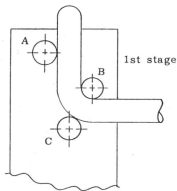

1st stage

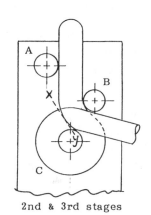

2nd & 3rd stages

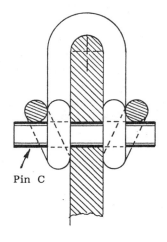

Pin C

1st stage: Bend wire through about 180° over the jig; then bend both ends 90° to the right around pin B.

2nd stage: Insert Pin C in its hole and wrap the working part of the wire about 375° clockwise around Pin C.

3rd stage: Cut off excess wire on line X-Y and cold forge the eyes flat against the jig body; solder or weld eyes after removal from the jig.

Pin C should be a loose fit in its hole in the jig.

I happen to be somewhat familiar with full size chain-making practice, and if I were going to make up some "model" chain, I would wind up a coil of whatever I was going to use for link stock around some sort of a mandrel whose shape would reflect (although not necessarily "mirror") the final inside form of the links I wanted. I would cut single turns off this coil, and close each link with further careful bending/hammering. Finally I would weld each link shut with that miniature oxyacetylene welding torch described elsewhere herein.

This basically is how chain was made here in Vancouver 25 years ago. I'm speaking here of deck chains, log boom chains and similar, with links maybe 8 or 10" long or more, formed from 1", 1-1/4, or 1-1/2"ϕ steel bar.

Today, hot rolled steel bar is fed into a Swedish-built chain making machine which cuts it to a length just right for one link. This blank is heated by an induction current to red heat in a matter of seconds, and then bent around a former block, hooked into the link ahead of it, closed, electrically welded, and trimmed of burrs. The "chain makers" are attendants only - they watch the machine, and clear any jams. In my day, the chain maker stood next to a gas forge, and "fire welded" each link by hand, assisted by a power hammer fitted with dies to suit the particular size of chain he was making. It was hot work in summer. There was then not one man in a thousand who could stand it, and probably far fewer would care to today.

Now, lest I begin to sound like an old man reminiscing in his coffee mug about how tough things used to be, let me say that it was tough... but I only worked in the warehouse. I occasionally got over into the blacksmith shops where chain and other products were made, and I was a most keen observer. I learned a little - just a very little - about bending and forming steel. Those men were masters of their craft, and to watch them work was fascinating.

(NOTE: The term "former" here does not mean something that has been referred to earlier rather than later; it means a form around or over which a piece of metal is bent to produce a desired shape in the workpiece.)

The Duo-Mite Bender

I got a little Duo-Mite metal bending tool quite a while back, and it has recently come in handy - up to now I'd made the odd hook or whatever on it, but nothing serious..... No, that ain't quite true - I did make some paper clips on it one time.

"Paper clips?! Are you nuts, Lautard?"

Well, no, it ain't exactly like you think, guys. These were oversize paper clips, about 2" wide, and 5" long, bent up from 0.150"ϕ wire. I made the first one just to see what the bender could do. Then I made a couple more, and sent one to Tim as a gag. I later made a run of about a dozen - Margaret found a use for some of them at her desk. Then I had the idea that they might make a good advertising gimmick. A friend in computer sales took 30 of them to hand out to customers.

I saw an item on the TV news one night about some school kids who were short of money to go on a trip they'd been planning. I contacted their teacher, and told her I had an idea for a fund raiser: I would show the kids how to make and sell these big paper clips in a big local shopping mall. (I don't know if it would work, but I think people would buy one for a modest price, if they saw the kids making them, and felt that the money was going for a good cause.) But the teacher said, "Oh we don't want to EARN the money, we want somebody to GIVE it to us."

The conversation got pretty brief after that.

The big paper clip looks about as at the top of the next page. If you want to make some, go ahead. (But don't quit your day job right away!)

If you make a bunch of them using the stops that come with the Duo-Mite, and if you use nice straight wire, it goes pretty fast. Adjusting each paper clip after bending is the most challenging part.

More recently, I saw a book called "Slocom & Boterman's New Book of Puzzles". Along with many wooden "burr" type puzzles that a basement machinist could quite easily make in metal, and end up with something really unusual, this book included several bent wire puzzles. I suspect that if a person wanted a bent wire project to use as a kids' fund raising project, a wire puzzle would - much as I hate to admit it - go over better than oversized paper clips.

An Over-sized Paper Clip

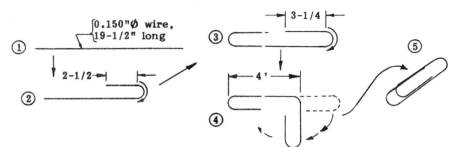

1. Start with a 0.150"φ wire blank 19-1/2" long.

2. Bend through 180° around a 1.166"φ circle die, starting at a point 2-1/2" from one end.

3. Change to a 1-1/4"φ circle die, and bend the other end, starting at a point 3-1/4" from the end.

4. Switch back to the 1.166"φ circle die. Start the 3rd bend at a point 4" from the first bend.

Adjust by judicious clouting after completion.

Another (practical) use for the Bender came up about 2 years ago. My friend Terry buzzed me to ask how he might make some hooks to hang cooking pots from an overhead pot rack in his kitchen. And what to make them from? I told him where to get some stainless steel wire, and when he got it, to bring it over to my place "... I've got a bender that will do it as slick as a whistle!"

Terry was tickled pink with the hooks we made. His 3/16"φ stainless steel wire was maybe a little heavy - we busted the head off one of the 2 button head socket cap screws that anchor the side plates onto the sides of the Duo-Mite Bender's clamping jaw, but that was easily fixed: we pulled out all 4 screws (2 each side) and replaced them with regular 10-32 shcs, which are a little meatier under the head.

   (I think what led to the screw failure was the fact that the screws holding that one side plate may have gotten a little loose, without me noticing it. One screw head snapped off utterly, and the other one's head was nearly off when I removed it. The Bender did fine with the rest of the work, once we reassembled it with the new screws, and tightened them up. But bending that 3/16" stainless was no easy job.)

A few days later, I got a phone call from the guy who'd made the overhead bar Terry's pot hooks were to go on. He's into making stuff for restaurant kitchens. Would I like to make him some pot hooks? No, but I told him where to get a Duo-Mite Bender, so he could make them for himself. [Oxford General Industries, Union City Road, Route 68, Prospect, CT 06712; phone (203) 758-4467]

Right after I hung up the phone, it struck me that making pot hooks might be a whole lot more lucrative than oversize paper clips.

If I want a hook of a special shape to hang something on in the shop I will dig the Duo-Mite Bender out, stick it in the bench vise, set it up, and have my hook made faster (and better lookin') than if I'd tried to bend it by hammering the wire* around a piece of round bar held in the vise. If you have uses like that for a bender, look into the Duo-Mite. It will bend m.s. round bar up to 5/16"φ, and can also bend flat bar up to about 1/8 x 1" the easy way - and with a factory-made accessory, 1/8 x 3/4" material can be bent the hard way (i.e. edgewise), too.

* (footnote from previous page) I have a roll of 0.150"ϕ iron wire I got at a local old time hardware store; I made the big paper clips out of it, and find it just about right for making hooks to hang stuff on in the shop.

You don't have to buy a bender... Roy Moungovan, in his book, "Shop Savvy," shows a shop-made bender that would do many jobs the average hsm might come up with. Mostly his arrangement involves the use of socket wrenches as dies around which to bend material.

Speaking of socket wrenches, maybe this is as good a place as any to bring up something else you should know about.... Walt loaned me a wad of back issues of **Sport Aviation Magazine**, and thus introduced me to the writings of Tony Bengalis. Tony writes DIY material on home-built aircraft in Sport Aviation on a regular basis. Many of his past columns have been collected into book form - "Firewall Forward" is one of these. Tony has a considerable ability to show his readers how to carry out a very wide array of work, some of it in metal, often with only makeshift tooling. For example, he shows how to make firewall grommet covers. These could be roughly described as a diamond shaped piece of sheet metal with screw holes at each end and a shallow dished central portion to contain a rubber grommet. The grommet cover holds the rubber grommet against the firewall. Cables, hoses or wires would go through these grommets and through the firewall. Tony shows how to produce these at home, using a hammer, tinsnips, a drill and a couple of socket wrenches in a vise, to a standard that would satisfy the fussiest aircraft builder.

While you may have no interest in homebuilt aircraft, or even grommet covers, you would find it hard to learn nothing from Tony's articles, and indeed from the many other writers who contribute interesting material. In the issues Walt loaned me, there was a serial detailing the conversion (or overhaul?) of a Volkswagen engine to aircraft standards - I found this in particular very interesting.

(I suspect that any hsm who made himself known to homebuilt aircraft enthusiasts in his area would immediately have many friends, and possibly some paying customers if he wanted them. Something to think about, maybe.)

A WIRE HOOP BENDER

If someone you know (wife, daughter, friend) wants to make lampshades, a rig for rolling wire into rings can be made quite easily. (This idea was described in ME for June 28, 1945, page 605, and is included here with the Editor's permission.)

The first item needed is a piece of 1/2" plywood about 8" x 10". Turn up 3 steel rollers, each about 2"ϕ, with a groove in the rim of each about right for the wire to be used. Mount two of these rollers 2 or 3" apart on the plywood in such a way that they can turn freely. Make a slot through the board, perpendicular to and centered between the two fixed rolls. Rig the third roller, which is turned by a hand crank, so it can be slid along in this slot. All 3 rollers should lie in a single plane, parallel with the face of the plywood.

Run a piece of wire of the correct length (theoretically, 3.14 x desired ϕ of ring) into the rollers - the 2 fixed rollers on the outside of the ring-to-be, and the adjustable roller on the inside. As you crank the adjustable roller, the wire will pass through the rolls, taking on a curved shape as it does so. Adjust the sliding roll closer to the fixed rollers and repeat until you get a fully formed ring. More than one pass may be required at a given setting. Once correctly set for the required bend, duplicate rings can be quickly produced.

TIP: You may find it better to cut the wire blank a little over the minimum length, so the bend doesn't peter out at the ends of the wire before you want it to. There will then be some overlap, but you can cut this off to final size, and then solder or weld the ends of the ring together.

USE OF APPLIANCE MOTORS FOR SHOP DRIVES
from SSgt. John E. McClain, FPO NY

Dear Guy,

I called you about 2 months ago and we talked about using appliance motors to drive shop equipment.

Such motors are reversible, often have two speeds, and are cheap. Where a single motor is not powerful enough for a particular application, you can put two of them face to face, so to speak, and run them in opposite directions - that way, whatever machine they are belted to thinks they are both going the same way, and will have the full power of the dual drive. The trick, of course, is in hooking them up to run like this, but it's not hard to do.

Right after we talked, the situation in Kuwait heated up, and I haven't had a chance to send you the wiring diagram until now. Right now I'm sitting on a ship off the coast of Saudi Arabia, so I'm doing this from memory. The diagram should be fairly easy to follow, but a few explanations should help.

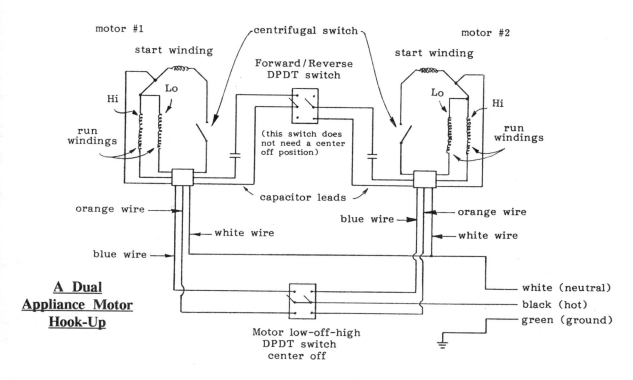

The three wires on the bottom right represent the power supply cord. The switches I used are standard toggles mounted in a regular wall switch box with a blank cover drilled for them. The bottom one needs a center position for off; the top one doesn't, but the same type will work. The switches need at least a 10 to 15 amp rating - the motors should draw about 6 - 8 amps apiece under full load.

The frame my motors are mounted on is fabricated of steel angle. It and the whole lathe are grounded through the green lead on my cord. If wood or other insulating mat'l is used for the motor mounting frame, make sure both motors are grounded, for safety.

You also want to make sure the cord is wired to the plug with correct polarity and the outlet likewise. If they are incorrect, the on-off switch would be on the neutral wire instead of the hot wire, and the motor windings would always have voltage on them. This goes for all machines, not just home wired ones.

A 15 amp breaker controlled circuit should work without a problem, but if you have fuses (instead of switch breakers) it will take at least a 20 amp circuit to handle starting current. The way this is wired, only one motor's starter windings are on at any time, so you don't want to start up with a major load on (e.g. a tool bit already in the cut), but that's probably true in most cases anyway.

The motors I've seen and used all have a rectangular box on the end opposite the shaft, which contains the centrifugal switch with all the wires coming out of it. The start capacitor should have one lead running to it also - this lead should just pull off. So pull it off, and then run a wire from the tab in the box (from which you just pulled the lead) to the Forward/Reverse switch.

The capacitor lead you pulled off goes to the other pole of that side of the switch. Same goes for the other motor.

When the switch is right, you will have 6 terminals on it, only 4 of which will have wires on them. The 2 empty terminals will be kitty-corner. All 6 terminals of the on-off switch are utilized. The center two are jumpered together and connected to the hot (black) wire of the power cord.

I drive my lathe this way. The large double pulley on the countershaft came from the front of a Chevy engine. The step pulley was fabricated from disks of plywood cut on a bandsaw and drilled in the center for the shaft and 3 holes drilled in each for 3 tie rods (3/8" redi-rod) to hold them together. The discs are glued and dropped on the rods which also tie the Chevy pulley to them. The shaft is just slipped in the center and held in place by the excess glue. This works fine because there's no torque on the center shaft - the torque is transferred through the tie rods.

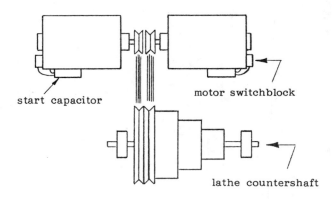

start capacitor

motor switchblock

lathe countershaft

The whole assembly was put in pillow blocks on a temporary frame, and driven by one motor so that the cone pulleys could be turned true using a large file, the end of which had been ground to serve as a wood turning chisel.

In making one of these installations to drive a machine, the motor drive belts want to be new and identical. I'm using a (used) inside-out timing belt for a flat belt (I don't know what it's from - just something with overhead cams). The steps of the cone pulley need to be crowned - I got mine ok by eye.

John

SPOKED HANDWHEELS

Tim and I have batted back and forth the idea of making a spindle crank for the Super 7. Instead of a crank, it occurs to me to use [or make something like] (a) the 3-armed speed handle off my Kurt vise, or, (b) maybe even better, about a 5"ϕ handwheel with spokes set into the rim like a ship's wheel. I once read a letter in ME from a guy who had thus modified the tailstock barrel handwheel of his lathe, and he said he found it real nice to use.

ANOTHER USE FOR A MAGNET

Here's a one liner:- A strong flat magnet can sometimes be used as a stop on the end of your milling machine vise. Oops.. went onto two.

A Lathe Spindle Nose Thread Modification
to give a self-cleaning thread
from **Jobie Spencer, Tuscola, IL**

In early '92 I had a letter from Jobie Spencer, some of whose ideas you've seen in **TMBR#2**. He'd bought himself a big old Monarch lathe (none finer) and was in the process of fixing it all up. Towards the end of the letter he told me a couple of neat ideas, as follows:

"For lathes with a threaded spindle, it often happens that a chip will lodge between the threads while screwing on a chuck. You're supposed to keep those threads cleaned off, but sometimes a chip will sneak in anyhow. Sometimes a chip will fall out of a hole or opening inside the chuck when the chuck is being installed. However it happens, it's a nuisance.

"A solution which helps to alleviate the problem is to use a die grinder or Dremel tool to grind back the first thread on the end of the spindle to a square end. Grind it all the way back to a full thread, so as to eliminate the beginning taper of the thread.

"This makes the leading thread similar to the teeth on a tap, and it will sweep away or scrape chips or particles ahead of it with the square end, instead of allowing them to become wedged between the threads. I have fixed my 10" South Bend this way, and have had no more problems with chips in the threads since. "

Jobie

Another chap, **Robert Brown, of Chester, England**, suggested much the same thing. Quoting directly from his letter,

"....I had a chuck spin off a few years ago, leaving a slight mark on the end thread of the mandrel nose, so whilst making a backplate in the Fall of '91, I screwed up my courage to do something about it. I'd read the idea in a reprint of some material from a very old issue of *American Machinist Magazine*. The idea is to remove the first thread on the mandrel nose for about 1/4", half a turn from the end (right where the damage happened to be, on mine), taking it right down to the root, and leaving the sides of the notch square. It's a bit of a fiddly job to do neatly - I ground up a tiny cross-cut cold chisel, and finished off with a flat needle file. The notch then acts like a thread cleaner

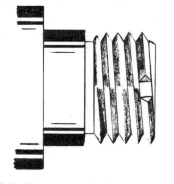 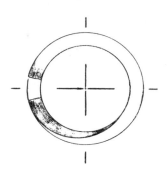

every time you screw a chuck on, and believe me, it is very efficient! I screwed my 4-jaw on and off 3 times, and each time the notch was full of stuff - and I thought the chuck thread was clean to start with! ..."

Robert

The above idea, which differs slightly between the two write-ups, makes a lot of sense. I usually "chase" the thread in the chuck with a toothbrush before I put a chuck on my Super 7. I also wipe the nose of the lathe spindle thoroughly with toilet paper, and then give it a squirt of 30 wt. engine oil when I'm about to screw a chuck, face plate, or whatever onto the spindle.
GBL

Also from **Jobie Spencer's letter**: "... Another problem can arise with the above mentioned South Bend, and other lathes with flat drive belts, especially endless fiberglass belts. The former owner of my lathe had one installed, and to do so, the headstock has to be disassembled, which is why I haven't done something with it (the belt).

"The pulleys get worn slippery and the belt becomes slippery as well. If you get the slightest bit of oil on the pulleys, the belt will pop off when starting up a heavy inertia load such as a chuck. The belt will also jump off when starting up a collet assembly like the Jacobs Rubber-flex unit, which uses an inertia type handwheel for closing, and a lot of the types of work that one would be doing with such a collet outfit is of a size that needs to be run at top speed.

"My friend Rupert Evans, former industrial arts teacher and retired Dean of the University of Illinois College of Education, owns a lathe with a belt such as mine, but it has a thin leather lining cemented to the inside. Fiberglass belts are nice because they don't stretch. Rupert says to keep the tension on them at all times. But fiberglass also gets slippery, if you get any oil on it. The leather lining absorbs the oil, and still provides a grabby surface. In fact, the oil seems to help it - keeps the leather from drying out.

"But I discovered another trick to use on fiberglass belts: toothpaste - with or without fluoride! It makes things tacky enough so the belt will grab, and a little goes a long way.

"I didn't originate this idea myself. I read about it somewhere as a way of stopping fan belts from squeaking. While we're on the subject of how to stop belts from slipping, here's another one: I was having trouble with my trip hammer belt slipping. An old blacksmith told me to put some honey or molasses on it. it works, too - for a while."

Jobie

An untried conjecture from GBL: Take the pulleys off, plug the bores with cotton waste or similar, and then sand blast the pulleys' working faces with coarse sand.

~ ∾∾∾ ~

A MINIATURE 4-JAW CHUCK FOR SMALL WORK

It's sometimes nice to have scaled-down workholding devices for holding small parts. On the next page are dimensioned drawings of the various parts of a small 4-jaw chuck off a watchmaker's lathe. If the size shown doesn't suit you, there's no reason not to scale it up to a size that does suit - the same general design should work well up to about 4"ϕ.

On the little chuck I measured, the mounting shank was a stub arbor with the same profile as the collets the lathe was designed to take. You could use instead perhaps a 1/2"ϕ stub maybe 2" long, it and the chuck body being screwcut say 7/16-20 or similar.

Naturally, this being a 4-jaw chuck, you are likely to be running less than perfectly balanced set-ups in it, so you should not over-rev it, but I will warn you just in case: don't over-rev such a chuck!

The adjusting screws in the chuck had a left hand thread, which gives "conventional" movement to the jaw upon turning the screw (clockwise moves it in, counterclockwise moves it out). If you were going to make yours this way, you'd need to make or buy a left hand tap.

The major parts of such a chuck would do well to be hardened (or case hardened) for wear resistance and smooth action. A cutter would be required to machine the "T-slot" for the collar on the adjusting screw. Good fits, and careful work all around.

NOTE: One thing that is **not shown on my drwgs** *is that there must be holes right through the chuck body from front to back, one such in each slot, to let the 4 jaws go into place, and to let the screws that suck them into place come in from the back...*

The little drwg at bottom right in my drwgs shows the way the gripping faces of each jaw were machined in the original.

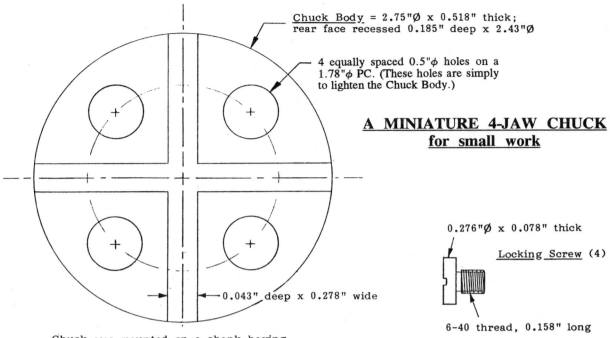

Chuck Body = 2.75"Ø x 0.518" thick; rear face recessed 0.185" deep x 2.43"Ø

4 equally spaced 0.5"φ holes on a 1.78"φ PC. (These holes are simply to lighten the Chuck Body.)

A MINIATURE 4-JAW CHUCK
for small work

0.276"Ø x 0.078" thick

Locking Screw (4)

6-40 thread, 0.158" long

0.043" deep x 0.278" wide

Chuck was mounted on a shank having the profile of a watchmaker's lathe collet. You could make a shank to fit your lathe spindle nose socket.

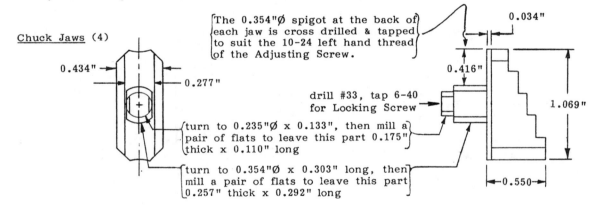

Chuck Jaws (4)

The 0.354"Ø spigot at the back of each jaw is cross drilled & tapped to suit the 10-24 left hand thread of the Adjusting Screw.

0.434"

0.277"

0.034"

0.416"

drill #33, tap 6-40 for Locking Screw

1.069"

turn to 0.235"Ø x 0.133", then mill a pair of flats to leave this part 0.175" thick x 0.110" long

turn to 0.354"Ø x 0.303" long, then mill a pair of flats to leave this part 0.257" thick x 0.292" long

0.550

Adjusting Screw (4)

Slotted Plate Washer (4)

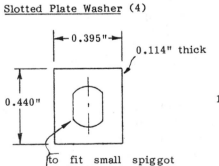

0.395"

0.114" thick

0.440"

to fit small spiggot on back of jaw, with maybe 5 thou play on vertical axis

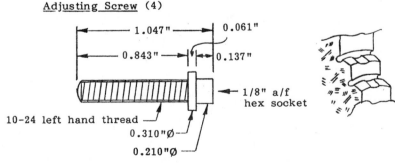

1.047"

0.061"

0.843"

0.137"

1/8" a/f hex socket

10-24 left hand thread

0.310"Ø

0.210"Ø

NOTE: The Jaws go into the slots in the face of the Chuck body from the front; the Slotted Plate Washers & Locking Screws go in from the rear. Also required - but not shown on the drwg - are 4 radial holes in the Chuck Body; these take the Jaw Adjusting Screws.

SAFE SPEEDS FOR LATHE CHUCKS

I had a call one day from one of my guys suggesting that I include herein something on safe speeds for various sizes of lathe chucks. He told me he'd once worked with a guy who'd seen a chuck break up. The way it happened was something like this: a fella had some sort of job in about an 18"ϕ chuck, on a lathe of a size to match, and had everything perkin' along nicely. Somebody came along and told him to speed it up. So he did. To about as fast as the lathe would go - maybe 15 or 1800 rpm. When he touched the tool to the work again, the whole chuck just came apart. The guy who'd seen it said it was just like a bomb goin' off, and I can well believe it.

The two lathe chucks I use on my Myford Super 7 are by Pratt-Burnerd, and they're marked "made especially for Myford Series 7 lathes". I got them with my first lathe, an ML7, which had a top speed of about 1100 rpm. When I got my Super 7, I switched the chucks, as well as some other accessories, over onto the new lathe.

The Super 7 has a top speed of about 2500 rpm. I got to wondering one day what about if that was too fast for the 6" 4-jaw? I mean, if you set the belts for top speed, and eased the clutch out, that suckah winds up like a turbine. If the chuck decided to come unglued, you'd certainly notice it, at least until some of the pieces started sinkin' into your forehead.

The more I thought about it, the more it worried me. So I decided to write to Myford's about it. Back came a very courteous letter saying fear not, the chucks you have can be run safely at the Super 7's top speed.

Still, the suggestion that some info along these lines would be worthwhile seemed like a good one. I hunted up addresses for Buck, Cushman, and Pratt-Burnerd, and sent off letters to them all. No reply, so I phoned them. Buck sent along a catalog which included the info tabled below. Still no reply from the other fellas, so I buzzed Pratt again, and got to talk to a guy called Herbie.

Herbie said the reason he hadn't written back yet was that he hadn't found out anything that was worth telling me. It seems the further he went into the matter, the more unable anybody was to provide solid info. It all depends upon the type of chuck (2-jaw, 3-jaw, 4-jaw, 6-jaw) and whether it's made of steel, semi-steel, or cast iron, and whether the screws screw into the backplate from the front of the chuck, or do they screw into the chuck body from the back side of the chuck? There just doesn't seem to be a simple answer to the question, nor a formula into which you can plug numbers, and get good answers. Herbie said it had been mighty interesting, but he really didn't have much to tell me.

One thing he or one of the other fellas had said was that it is more usual for the workpiece to break up than the chuck, or that the workpiece breaks up first, and takes the chuck with it, when a safe speed is exceeded. I told Herbie I'd got some info from Buck, in their catalog, and he said that any such info therein would be about as good as could be had, and could be taken as Gospel.

I'd prefer to modify that just a hair: it should more likely be taken as a good guide for the types of chucks it covers.

The Buck catalog says first off, never run a chuck except when gripping a workpiece. (i.e. don't spin it up empty). I don't know why this is so, and I've run mine up to full speed empty more than once with no ill effects, but if they say not to, they must have a reason, so I will follow it from now on.

Also note that a chuck loses some of it's gripping force at high speed (which would probably explain why workpieces come out of chucks at high speed, and destroy the chuck when they do).

Also, Buck says to contact them for further info on particular applications of their chucks - their experience would be extremely wide and varied, and it is at their customers' disposal. Which is no doubt true of the other chuck makers as well. Safe speeds any maker will quote you will no doubt be based on test-to-destruction data, plus a margin of safety. Naturally, that the chuck is used properly, and not used if damaged, is the user's responsibility.

So take the info below as a guideline only, be a little conservative, and if in doubt, write to the maker of your chucks, and ask for safe speed info on that particular model. And do a test of your own, but stand well out of the line of fire when you do.

SIZE	APPROXIMATE SAFE SPEED	
	for steel-bodied	
	3-jaw chucks	6-jaw chucks
6"ϕ	3400 rpm	2500 rpm
8"	3100	2200
10"	2700	1900
12"	2200	1500
15"	1800	1300

NOTE: The above information **applies to specific Buck chucks**. It does **not** apply to, or cover, all makes and types of chucks of these sizes, by any maker, in any material. Cast iron bodied chucks will **not** be safe at the speeds shown above.

CHUCK MOUNTING
or
How to fit an unmachined backplate to a lathe and its chuck

Machining a chuck backplate to fit your (threaded) lathe spindle nose is sometimes necessary, and may strike some as a rather intimidating task. It doesn't need to be, if you know how to go about it. Let's look at the basic way, and a couple or 3 variations.

The worst case (in one sense), and at the same time about the only one for which you might need any instructions, is if you have to make a backplate from an unmachined casting. This requires the least outlay of money, but obviously takes more time. If a standard backplate is not available to fit your lathe spindle nose thread, order a backplate casting of a suitable size for the chuck you want to mount. (Most mail order machine shop supply houses carry unmachined chuck back plates.)

While you're waiting for the backplate casting to arrive, measure your lathe spindle nose thoroughly, and then make 2 plug gages. Make one exactly the same ϕ as the register on your lathe spindle nose. Make the other a little longer, and of a ϕ exactly equal to the root ϕ of the thread on an existing backplate.

Set these aside, and buy a duplicate of your lathe spindle nose, or make one from a 3 or 4" long slug of say CRS or HRMS. This item is an important and useful piece of shop tooling, and should be treated accordingly. If you can buy a factory-made spindle nose adaptor for your lathe, it may be worth the cost. Mine is a duplicate of the nose on my Myford Super 7, and has a #2MT shank sticking out the back. This allows me to mount my chucks on the tailstock, or in the #2MT socket of my 6" geared RT, plus it serves me as an undoubtable gage, should I ever need to make a backplate or other spindle nose fitting for my lathe.

When your backplate casting arrives, go over it with an old file (i.e. not your best one) to knock off the warts and pimples. Then, chuck the backplate casting by the rim, with its back side out, and machine it all over (except for the OD and the front face, which is of course presently looking at the lathe spindle nose). In the boss, bore a hole the exact size of your root-diameter plug, as deep as required - probably right through the casting. Enlarge what is for now the tailstock side of this hole to just exactly accept the register-diameter gage, and to the right depth for same. Chamfer the outer working face of the boss, and screwcut an internal thread to a nice fit on the spindle nose duplicate. When you have a fit, pull 'er off, yank your chuck off the lathe, and try your new backplate in the spindle nose - it should go on as slick as a greased pig. (If it doesn't, you're dead! Well, not quite - there are ways to re-coup even that situation.)

Now, with the part-machined backplate screwed home on the spindle nose, machine the OD and front face to suit the chuck...

Myford's owner's manual for the Super 7 says to make the chuck back plate "a light tap fit" in the recess of the chuck it is to carry (assuming it is a recessed back chuck). I would suggest half a thou under chuck recess ϕ.

Fig. 1

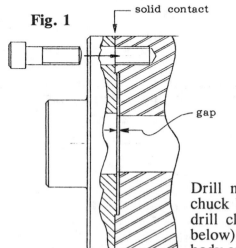

— solid contact

— gap

If the chuck being mounted is a 3-jaw, Myford says to have contact as in Fig. 1, left, whereas if a 4-jaw independent chuck is being fitted, do it as in Fig. 2. For the case of a 6" 4-jaw, they recommend sinking the hole for the spindle nose deeper into the backplate and recessing the backplate into the back of the chuck somewhat, to minimize overhang. [My 6" 4-jaw chuck, made for Myford Series 7 lathes by Pratt-Burnerd, is bored and threaded right in the back of the chuck body casting itself - the chuck is not backplate mounted. This reduces overhang even further.]

Drill mounting bolt holes in the backplate to match those in the chuck body. If the mounting bolt holes in the chuck are tapped, drill clearance size holes (typically 1/64" over bolt size, but see below) in the backplate. If the bolts are to pass through the chuck body and screw into tapped holes in the backplate, drill tapping size holes in the backplate.

Fig. 2

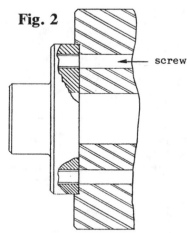

— screw

solid contact on face
and rim of backplate

Naturally, it is imperative to de-burr the holes and all else so that the chuck seats nicely onto the backplate as intended.

It is also necessary to allow for clearance around the bolts in the clearance drilled part, during the marking out process. Where the backplate is to be tapped (more likely to be the case on a 4-jaw), a simple way to locate the holes in the backplate is to make up a center punch that is a nice fit in the chuck body's tapped holes, and use these holes as an alignment guide for the centerpunch. Drill and tap one hole in the backplate, put a bolt in place and tighten it lightly to lock the backplate, and then centerpunch the other 3 holes. Dismantle, and drill and tap the backplate for the last 3 screws, de-burr all, and reassemble.

OK, that's the basics, to which Myford adds 2 further points:

(1) When tightening the mounting bolts between a chuck and its backplate, go around all 3 or 4 bolts several times, tightening them all evenly and gradually in succession. (Same idea as torquing down the bolts on an engine head.)

2) *Always* clean and oil the spindle nose threads and register before screwing anything (chuck, faceplate, etc.) onto the spindle nose.

Variations on the above:
Stewart Marshall told me of one trick he uses when possible in fitting a backplate to a 3-jaw chuck. He deepens the recess in the chuck body, if enough "meat" is available, and then machines the backplate to incorporate a spigot to enter this recess almost to full depth (say 0.010" shy) and makes it maybe 20 or 30 thou under the ϕ of the recess in the chuck body. He then fits 3 setscrews - maybe 10-24, or more likely 1/4-20 if for a larger chuck - in the chuck body, radially, and enlarges the clearance holes in the backplate to maybe 20/30 thou oversize also. Next, he mounts the chuck to the backplate, screws the backplate onto the lathe spindle nose, and uses the 3 set screws to shift the body into concentricity with the backplate. When he has it lookin' good, he nips the bolts up just a little, and then puts something (e.g. a piece of 1/2"ϕ drill rod) in the chuck, and indicates it. If it doesn't run true, he will use the 3 setscrews to shift the chuck body around on the backplate until it does run true, whereupon he tightens the chuck mounting screws.

By this means, a 3-jaw of less than "precision" class can be made to perform very well on work of a given diameter, although you will likely have to re-adjust it for work of different diameters.

One writer in ME years ago suggested what would be basically the opposite arrangement to Fig. 1 above - i.e solid contact between the inner recess of the chuck and the spigot on the backplate, and a gap of maybe 50 thou between the chuck and backplate in the outer portion where I have drawn it in full contact. His rationale was that if the chuck and backplate become stuck together by dirt and/or rust over the years, you can put something in the gap to pry them apart. It's an idea, but I think what is obviously wanted is to have the solid contact where the screws come through to join the chuck and backplate. If they're snugged up hard with a gap between the two, there is going to be some distortion, and everything you ever hoped for goes out the window.

One other idea is to drill a series of indexing holes in the rim or rear face of the backplate. One fella wrote in M.E. that he'd put 3 tommy bar holes in the rim of his chuck backplate, and many a time since had been as happy as a lark for having done so. If I were going to do something like that, I'd drill 24 or 60 holes in the rim. If I wanted something to roll the chuck around with, I'd use the chuck key.

Interestingly, neither Machinery's Handbook or the American Machinist's H'book (my copies, anyway) have anything to say about chuck backplate fitting.

More On Flat Lapping

As a result of reading my account of the lapping activities of the old gagemaker in **TMBR#2**, several readers wrote asking where to get lapping plates, lapping abrasives, and related supplies. Listed below are two outfits which will send you a catalog on request, and between them, can supply everything you would need.

For **cast iron lapping plates**: Challenge Precision Surface Equipment, 1433 Fulton Street, Grand Haven MI 49417; phone: (616) 842-8300.

For all **hand lapping equipment, abrasives**, etc.: Speedfam-Spitfire Machine Tool Group, 509 N. Third Ave., Des Plaines, IL 60016-1196; phone: 1-800-423-8855; or try (708) 803-3200. (With this outfit, ask for their product info on hand lapping, abrasives, etc., as these items are not shown in ther general catalog.)

Do-All also sells granite lapping plates, and no doubt there are other makers offering similar wares - see Thomas' Register, at your local library.

MORE IDEAS ON SHOP-MADE REAMERS, DRILLING PLASTIC, ETC.
From **Larry Balchen, Nordic Metalworks, Jonesport, ME**

Dear Guy,

Leo Fortin, of Taunton, MA, although now retired, was a toolmaker almost without compare. I worked with him for some time, learning about extrusion dies for multi-conductor electric cable. Here's something I learned from Leo, and that ties in with your info in **TMBR#1** about shop-made reamers:

If you grind, and then stone, a three-faceted pyramidal point on the end of a hardened pin - dowel pin, mold ejection pin, or a HSS drill blank - the result is a pretty effective reamer. The 3 flats can be ground freehand or with some type of fixture.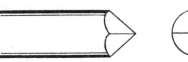

There is also what is sometimes referred to as a drill reamer. This is simply a regular twist drill with a slight radius stoned where the margin (the twist) meets the conical cutting lip. In many cases, this type of reamer will give a better finish than a conventional factory made reamer. The accuracy you can get is also pretty astonishing, plus it's a lot cheaper than buying reamers. Again, it's used just like a regular reamer.

A good tip for drilling plastic (Plexiglass, etc.) is to use a drill that is ground the same way as for brass, i.e. with a slightly flat face on the cutting lips. This prevents the drill from hogging in or grabbing, and prevents the plastic from shattering as the drill exits a through-hole. It eliminates the need to back up the piece being drilled, too.

"Vanishing oil" (from The Man-Gill Chemical Corp, 23000 St. Clair Avenue, Cleveland, OH 44117; phone: (216) 486-5300) will give you a really nice finish on plastic. It's especially nice when tapping clear plastic like Plexiglass where the finished threads will be visible. Try it, you'll like it. (See also 'Dead Dogs and Other Matters,' elsewhere herein. GBL) **

I found your idea about oiling a file interesting, and I like the way it works. However, it might also bear mention that when filing, one should resist the temptation to run a finger over the area being filed, as the oil from one's fingers is just what it takes to make a file slip. This is more pronounced on a flat surface than an edge. Try it and you'll see what I mean.

Larry

** When Larry first wrote to me about this, he referred to it as "'Vanishing Cream', from Magnus". Well, I figured it must be some sort of cosmetic you'd put on yer face to make yer spots 'n wrinkles disappear, so next time I was in one of the big department stores at our local shopping center, I went to the cosmetics department, picked out the prettiest girl there, and asked her if I could talk to her for a minute. She smiled and said yes, so I explained that I was writing a book, and one of my American customers had suggested the use of vanishing cream for tapping holes in Plexiglass. 'Course I had to back up and explain to her what tapping meant.... she wasn't entirely convinced at first that I was playin' straight with her ('specially when I told her the title of the book was "The Masochist's 3rd Bedside Reader"!), but once we got past that, she turned out to be more than just a pretty face - she knew all about Vanishing Cream. She said vanishing cream was an old product, but that while you might have been able to get it a long time ago, it wasn't sold here any more. She agreed it might still be available in the US. I said ok, I would check back with my customer, and find out where he buys it. She seemed a little disappointed that her profoundest thoughts weren't going to get into my book, but I thanked her - sure enough she'd done her best to help. And she sure was purty.

So I buzzed Larry.... Turns out it's not vanishing cream, but Vanishing Oil, and it's sold in 15 gallon kegs and 55 gallon drums. When I explained the above to Larry, he roared with laughter, and said to put it in the book.   He also enlarged upon his original comments somewhat, saying that he had at one time worked at a place where they made batteries of all types, including prototypes, which had to look good, too. He said that tapped holes done with Vanishing Oil do not show the frosty appearance one would otherwise expect to see in a tapped hole in Plexiglass. On one occasion he'd had to machine cavities for about 30 cells into a block of Plexiglass about 5" x 10" x 48" long, and when it was done (with both the external flycutting and the internal work aided by Vanishing Oil) you could see right through the block down the full length!

Smallest quantity sold is 5 gallons; it's also used for light stamping and deep drawing operations, and evaporates 100% in 1 hour, hence the name.

Best place to look for it is at a place that sells Plexiglass, or makes things from it. If they don't already know about Vanishing Oil, you'll be doing them a favor by telling them about it. They can buy a 5 gallon pail, and I expect they'd be pleased to give you a small quantity in return for you having put them onto it.

How to Machine the Edge Bevels on a Gib Strip
without special tooling
from Clarence Jones, Kettering, OH

Holding a gib strip the size of your finger and 1/8" thick so that you can bevel the edges may tax your ingenuity, but here's a way to do it that saves having to make any special tooling at all.

1. Machine the female dovetail.

2. Mill the gib strip slightly over final width.

3. Place the gib strip in the female dovetail and lock it there by sliding an adjustable parallel in beside it.

4. Tap the adjustable parallel with a brass hammer to jam it tight.

5. Mill the first bevel on the gib strip with an end mill.

adjustable parallel

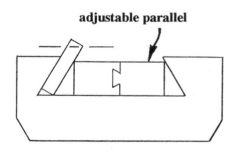

6. Rotate the gib strip 180° on its long axis and put a shim about 10 thou or so thick under it (visible in 2nd drwg if you look closely; GBL), secure as before, and machine the other side.

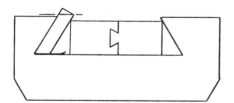

To machine seats for the gib screw tips, put a small center cutting end mill (i.e. a slot drill) in an electric drill and feed it into the holes while the gib strip is jammed in place as above.

As Clarence says, not elegant, but it'll work.

MAGNETIC GOODIES and QUICK CHANGE VISE JAWS

I had a call one day from **Steve Cannon, of Chesnee, SC.** He told me he'd made a "Haralson Hose End" (TMBR#2, page 148), and now all his friends wanted them. He asked me where he could buy a tap to cut that thread, instead of having to screwcut it. I wasn't able to tell him, but we ended up having an interesting conversation.

Steve is a tool & die maker by trade, but has wound up ramrodding a company that imports magnetic products from Germany. A while after we talked, along came a package of magnetic stuff - little magnetic discs with cup hooks attached, ditto with little eyebolts instead, also, same-size metal discs with nuthin' on 'em but paint on one side, and double-sided foam core adhesive tape on the other side. And a strip of metal rolled up in a coil, similarly backed with foam core adhesive tape. Cutest of all were some magnetic discs with little clips like very short clothespins attached to them.

"Don't get this stuff near your pacemaker," Steve warned me. I don't wear a pacemaker, so I was ok on that score, but it took me a while to tumble to the purpose of all this stuff. Finally it hit me.

Suppose you want to put a (re-)moveable magnetic hanger on some non-magnetic surface - a plaster wall, maybe, or the bottom edge of an overhead floor joist. If you stick one of those adhesive-backed metal discs or a piece of the metal strip to the beam or wall, you can then stick a magnetic disc, complete with cup hook or eyebolt, up there any time you like. Or stick a couple of those magnet/clothespin things up, and poof! a shop drawing hanger! (Of course, this is not so.. so.. so..... *elegant* as Tim Smith's shop drawing hanger (TMBR#2, page 145/6), but it is eminently practical.)

There was even a metal plate, painted white, about 6" square, with holes in the 4 corners. You could screw this (or similar) to a wall, and stick magnetic discs anywhere on it. And let me tell you, those magnets are really powerful!

Steve later sent me a nice little wooden box containing several magnetic probes from 60 thou up to 0.435"ϕ. The first use that comes to mind for these is for the removal of swarf or tap cuttings from holes.

Steve says he also has magnetic tape with an adhesive on one side. What immediately came to my mind was.... **vise jaw liners!** With this stuff stuck to a piece of cork, or copper, or whatever, you could stick 'em in place and have 'em STAY there! Magnetic backed vise jaw liners could be useful for both bench vises and milling vises. (There would be instances where having magnetic material in the vise would be a nuisance, but if the workpiece was non-ferrous, this wouldn't be the case.)

My main bench vise is a 3-1/2" Wilton, mounted on a Schlegel vise lift. BB fell in love with the latter item, but when we tried to get him one, the source named in **TMBR#1** had evaporated. Then I got a catalog from Blue Ridge Machine Tools, and lo & behold, they had 'em. BB immediately ordered one, and in due course put his very nice Starrett #923-1/2 bench vise on it.

My current set of jaw protectors, made several years ago from copper water pipe split and flattened out, were getting a little long in the tooth, so I dug out a set of jaw pads I'd bought from Brownells a good while back but hadn't used yet. These consist of two 4" x 5" pieces of 1/4" aluminum plate, with a 2" x 4" x 3/8" pad of "elastomer" bonded to each of them. Exactly what the rubbery stuff is I don't know, but you can wash it in soap and water, or gasoline. They can be had with red pads, for wood and plastic, or green ones, for metal. Mine are green.

To put them into use, you make a square cut-out in the lower edge of the aluminum plates, dimensioned to fit the slide or tongue of your vise, and they'll stay put. But the Wilton's slide is round, not rectangular. "No big deal," I told myself. "I'll just bore a hole to suit in each plate, and mill out the bottom, and end up with a U-shaped cut-out."

I made a couple of measurements and tried out, in cardboard, what I thought would suit. Good thing I did - with the 35 thou clearance I'd allowed, it wiggled about like a mouse in a wooden shoe. I cut it down to 5 thou clearance (3 might be even better) and started making chips. With a rubber band around each one and its jaw, low down out of the way, to finish off the job, they fit real nice, and they don't fall out.

But... the Wilton's serrated jaws would soon chew up the back side of the aluminum plates. It didn't take long to make a new pair of smooth jaws from 5/8 x 1" CRS. Much better! Even better would be hardened jaws that wouldn't be marred by gripping even soft iron wire. (Here would be a good item to try casehardening, if you don't want to spring for a set of factory-made smooth jaws.)

Another arrangement I've often thought would be good for a bench vise is some sort of a quick change vise jaw system.

It should be pretty easy to arrange a quick change vise jaw system for a bench vise, because the tolerances and fits could be fairly loose, even to the point of the jaw inserts being completely wiggly until the vise comes up on whatever you want to squeeze.

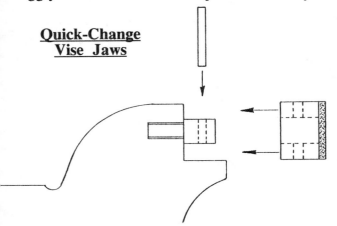

Quick-Change Vise Jaws

I would be inclined to make threaded pins with cylindrical heads, cross drilled as in the drwg at left, to replace the screws that hold the standard jaws in the bench vise. Then, one could make jaw inserts from CRS or aluminum, with holes to match, and with intersecting holes to take retaining pins, which would drop in from the top.

I may be wrong, but I don't think that one would suffer any from having bench vise jaws that are not bolted tightly into place in their recesses in the vise casting. It isn't a milling machine vise, and they'll seat tight enough when you snug the vise up on something.

One could have shop-made jaw inserts faced with leather, copper, brass, lead, lino, canvas drive belting, cork, sheet rubber, or whatever.

The main problem I see with having a QC vise jaw system like this is the time a guy would lose stopping to figure out which set of jaws to use for a particular job. "Should I use the cork jaws? Or maybe the leather ones today? No, I used them yesterday. Maybe I should use the lino ones today..." And so on.

Speaking of using heavy canvas drive belting, I rode down to La Conner, WA on the Harley one day, and while stretching my legs, noticed a little shop called "The La Conner House". On the surface of things, it looked like a card shop - greeting cards, Western blankets, and like that - with a couple of old guns layin' about here & there for atmosphere. But on looking a little closer, I noticed a little bench vise attached to the counter near the cash register. Now obviously you don't need a bench vise to sell greeting cards....

Turned out the owner, Brian Earp, makes black powder rifles right there in the shop, to pass the time between customers, and he does almost everything by hand. Whether he builds them to sell or not, I don't know, but they're nice.

During our brief visit, I noticed his vise jaw protector: a piece of heavy canvas drive belting cut as shown at right. As you can see, when this is bent into a U shape, the rectangular cutout in the middle becomes a notch which fits down over the sliding bar of the open vise. Between the fit of the cutout, and the natural springiness of the belting, it pretty much holds itself in place, while the ends serve as pads for the vise jaws.

heavy web drive belting

A Flexible Pusher for your Milling Machine Vise
from **Bill Webb, Kansas City, MO**

I had a call Christmas Eve '92 from Bill Webb, whose reamer stoning jig is pictured elsewhere herein. In the course of the conversation Bill told me about something I think you'll like a lot.

When faced with the need to hold odd shaped work in the milling vise, Bill uses pushers which he makes as follows:

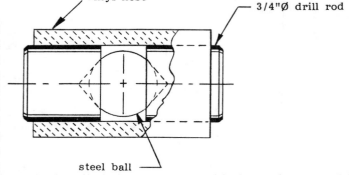

Vinyl hose

3/4"∅ drill rod

He puts a good-sized center drill hole in one end of each of two slugs of about 3/4"ϕ drill rod, having first faced both ends flat. He pushes these into the ends of a short piece of vinyl hose from the hardware store, with a ball from a ball bearing trapped between them. The plugs are a nice fit in the hose. He chucks the whole thing in the 3-jaw and trims the hose ends off clean. This makes an

steel ball

effective pusher, and the 3 pieces involved can be handled as one manageable item, instead of 3 unmanageable pieces. Neat, eh?

See also page 38 in **"Hey Tim..."**, for another type of pusher/workholding aid, that would be more appropriate if the workpiece is more fish-like (slippery) than blocky in shape.

More on Vise Jaw Liners,
and some Horrible Crudities

On Tim Smith's advice, I got a set of "Royal" vise jaw liners. (Travers Tool sells 'em.) They are a polyurethane material molded around a core of thin, perforated sheet steel. Tim said you could hold an egg with them, and they won't mar finished work held in the bench vise. If you grip a screw real hard, the threads will imprint in the polyurethane material, but after a few hours the marks will disappear.

I had some material left over after cutting a set to fit the Wilton bench vise, so when I got my new Bridgeport-type mill in '89, I cut out a pair to fit the 6" Kurt vise I put on it. Then along came a "B'port home-coming present" from Tim: 3 new handles for the machine's knee, table, and saddle cranks. (I only needed one for the table, as the feed crank at the other end was obsoleted by the Lyman power feed unit.) The handles Tim sent had a much nicer, more hand pleasing profile than the original handles.

Having the "Royal" vise jaw liners in the Kurt made it easy to hold the table and saddle cranks to drill out the holes therein and re-tap them to 1/2-13, to suit the new handles. Holding those ball handles would otherwise have been a pain in the neck, as they tapered in every direction. I just put each ball handle in turn in the vise, got it oriented exactly right, and then tightened up the vise until it was completely immobilized. How to do the job without those jaw liners would have been a real head scratcher.

As for fitting the 3rd handle to the knee crank, that was dead easy, but I told Tim afterward in a letter that he'd probably get sick if he knew the crudities to which I stooped to open up the hole in the knee crank (mat'l: cast iron) to tapping size for 1/2-13: I clamped in the bench vise the drill next up in size from the existing hole in the crank, and - cover yer eyes, you purists! - just twisted the crank onto it with my bare hands. The threads in the existing hole helped make it feed onto the drill readily. I repeated the process with bigger drills up to 27/64, after which I tapped the hole 1/2-13. I bet the whole job took me 5 minutes.

And do those new handles ever look & feel *Gerrr-ate!* Thanks, Tim.

More Royal Goodies for the New Mill

I got a set of 4 "Royal" machine mount/leveling pads for the new milling machine. To use these, I had also to get a couple of pieces of black iron, 1/2 x 3" x 25-1/2", and drill them to match the holes in the B'port base casting. Then, after I got the mill where I wanted it to be on the shop floor, I raised the mill, slipped the bars & mounts under, & lowered the mill onto the mounts. This was quite a job, as I was all alone, with only a 24" gooseneck bar to use, and one handed at that, the other hand being busy poking shims into place. Better would have been a 3 or 4' bar (and even better, room to use it!), plus someone to insert shims while I pried the mill up a little at a time. However, Margie was away that day. I just took my time, worked away at it, and in due course I got it done. It's amazing what you can lift with a 24" gooseneck bar.

Another Idea for Removal of Cuttings from Holes

John Fitzpatrick, the boss of Turnmill Machine Company (Troy, MI) sent me a basket ball inflator needle screwed into an air hose end. The needle is just under .080"ϕ, and sticks out about an inch. He said he finds these good for blowing tap cuttings outta little holes and similar jobs - safety considerations not being ignored in the process, of course. Unfortunately, the one John sent did not fit my air compressor hose nozzle, but I like the idea...

A Note on File Selection

A fella goes to the hardware store to buy a file.

"What kind of file?" asks the clerk.

"Anything that I can use to file the point off my parrot's beak. When he eats out of my hand, he bites me, so I'm gonna file the sharp point off his beak."

"You can't do that! A parrot has nerves that go right down to the tip of his beak. If you file the tip off his beak, it'll hurt somethin' awful. It'll drive him nuts! He'll starve to death. That's a terrible idea!"

"Naw, it'll be jus' fine. Just sell me a file."

Couple of days later the guy comes back, wants to buy a shovel.

"I knew it!" storms the clerk. "Your parrot starved to death, didn't he?"

"No he didn't starve to death. He was dead when I took his head out of the vise."

A HANDY DEBURRING TOOL
made from a file
from Tim Smith, Toledo, OH

Tim spotted an interesting item in use at work. As you will see from what he says about it below, every machinist ought to have a couple.

".... Here's a neat idea for a deburring & smoothing tool. All the guys in the shop where I work use them. It's basically just an old 10" hand file with a strip of crocus cloth stretched up one side and down the other, and clamped just ahead of the handle. I made myself two - one for work, and one for around here. Why we never thought of this I don't know, but does it ever work! It leaves a nice smooth finish on parts, especially when the paper wears down a bit. Very easy to make - just 2 pieces of 1/4" CRS or HRMS about 1" square, with a hole

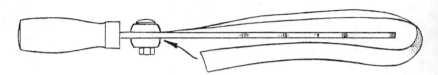

Tuck the ends of a strip of crocus cloth under the clamping blocks, pull it up snug, and tighten the nut down. (Tap the block on the underside, and I think you could dispense with the nut.) Use a 10-32 or 1/4-28 button head cap screw, and bevel off all edges & corners of the clamping blocks.

drilled through for a nut and bolt. The file is easy to drill because as you know they're soft at the tang end. Notch the crocus cloth strip at one end, pull it tight around the file, pinch it between the two plates, tighten the bolt and you're done. Make one - you'll love it. It's a really good idea."

"Knorrostol"

Ever had that sick-in-your-stomach feeling that comes when you open a drawer or a storage box, and find the item inside coated with rust? Well, this is gonna be blunt and brief: Get yourself a tube of "Knorrostol". You won't believe the way it takes rust off steel. Light, recent rust, or old tight rust/stain, it doesn't seem to make any difference - 5 or 6 swipes back and forth with this stuff on a rag and your bare hands, with minimal effort, and the rust comes off!!

You will not believe it can be this easy, but it is. It also works on spotty rust on blued steel. (I wouldn't guarantee it won't have any effect on the bluing, but when I tried it on some blued steel, it took the rust off, and left the blue. Rub some more and no doubt it'd start removing the blue as well.) Supposed to prevent further corrosion, too, tho I can't yet vouch for that.

Look in the Yellow Pages under "machine shop suppliers," and call any outfit that sells precision measuring tools and cutting tools. If they are a distributor for SPI, Inc., they should have it. If not, call SPI headquarters in Los Angeles, (1-213-721-1818) and ask for the name of your nearest distributor.

Not expensive, and good elsewhere than in the workshop, and on other metals besides steel, also. Get some.

A SMALL STOCK STORAGE SCHEME

How do you store short pieces of small material - brass tube, drill rod, etc. - so it won't get bent or mangled, and so you can see what you have and get at it?

Dale Scherbart, a draftsman for the John Deere farm machinery people in Ankeny, Iowa, called me one day to share an idea which I think is good:

Get some 1-1/2"φ plastic water pipe. A 6" length would be enough for a start, but you may want more than one of these racks, in which case you'll need more. To make one rack, mark the pipe off into 2" lengths. Then, while it is still in one piece, cross drill 2 places in each segment, as drwg.

At each hole location, drill 3/8" or 1/2" on the top side of the pipe, and then use a drill just a hair over screw shank size to drill the opposite wall of the pipe.

Cut off and de-burr the 2" long pieces of pipe. Make a nice job of this, because when it's done, the rack is ready to install, and you're going to be using the result for a long time to come, so make it nice, not rough.

To install, screw all 3 in a line, horizontally, onto a wall or the side of a rafter. Put 2 segments close together - say 10" apart, to hold short pieces of mat'l. Mount the 3rd piece further over, so longer pieces can be stored too.

You can see your stock, and fumble through it, in the spaces between the pipe segments.

Also from Dale Scherbart: Moly grease is good stuff, withstands temperatures up to 700°F. Dow Corning makes it, in several formulations: aerosol can, 15% grease, and 30% grease. Best place to start looking for it would be at a bearing supply house.

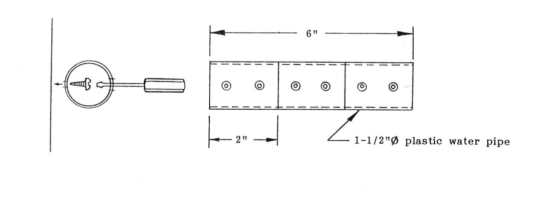

6"

2"

1-1/2"Ø plastic water pipe

Clean Up Time

My shop gets in a mess when I'm workin' in there. Sometimes, if I can't find a tool I just had, when I'm working at the lathe, say, I just take a break, and start putting stuff away until I find it. Once I've found whatever it was I wanted, I go back to work on what I was making. The side effect is that things are neater, which suits me fine.

Knurling on the Straight and Narrow

In September '90 I received a package from **John Dicks, of Saline, MI**. In it was a little rectangular block of aluminum about 1" sq. x 2-1/4" long. It had 6 holes drilled in it to suit various size tap shanks up to 3/8".

The purpose was immediately obvious: put the block over a tapping size hole in a workpiece, drop in the appropriate tap, and start the tap squarely into the hole in the job. (And, as John said, it also helps to prevent an attack of clumsy when removing the smaller taps from their holes, once the threads are in.) Nothing very revolutionary in it, but however simple in concept, it was 1) a gift, and therefore appreciated; and 2) something I should long since have made for myself, but never had, and therefore doubly appreciated. (I did once make a similar item for one size of tap - see TMBR#1, bottom of page 167.)

What was unusual about it (and remember, I said this tapping block was rectangular) was that there was a raised rim around the upper edge, and this rim was.... *knurled*! Oh, gone to sleep in your Bedside Reader, eh? Never mind, we have your full attention now, I think.

I wrote back to John to say thanks, and to ask him to come clean and tell me how he'd done that odd-ball knurling. Pretty soon along comes the following info:

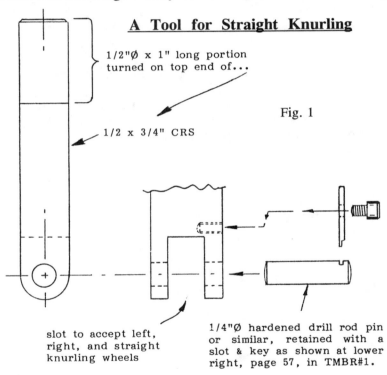

A Tool for Straight Knurling

1/2"Ø x 1" long portion turned on top end of...

1/2 x 3/4" CRS

Fig. 1

slot to accept left, right, and straight knurling wheels

1/4"Ø hardened drill rod pin or similar, retained with a slot & key as shown at lower right, page 57, in TMBR#1.

"The knurling is done with a shop-made tool having a 1/2" square shank about 4" long. This is slotted, and a cross hole reamed for a 1/4"ϕ dowel pin (for the knurling wheel axle) at one end. The other end is turned to 0.500"ϕ for about 1", (see Fig. 1), which lets me use it both in a collet in the vertical mill and in the lathe tool post. I used 1/2" sq. CRS because I had a short piece handy at the time, and anything bigger would not have fit my tool posts at home and at work. Heavier hasn't seemed necessary. A radiused transition where the material goes from square to round might look more elegant, but a square shoulder prevents the tool shank from moving up in a collet in the mill, and it's the quickest way to machine it, too."

(When I came to draw Fig. 1, the idea of a shank 4" OAL, with 3" stickin' out below the collet, for a tool to be subjected to side thrust, bugged me enough that I reduced it to 3" OAL. I then also realized that if you put a 3/4"ϕ x 3/8" wide knurling wheel (which was perhaps not the size John Dicks used, but it's about the most common size) in a 1/2" sq. shank, the side arms of the slot will be only 1/16" thick. My drwg therefore reflects what I feel might be better: I'd make the "wheel end" of it basically like the arms on the caliper type knurling tool detailed in **TMBR#1**, page 54 - 60. The dimensions shown here would still permit the holder to be used in both collet and tool post, but if one wanted less "overhang" below the milling machine collet, it'd be no trick to make the business end from a short piece of 1/2 x 3/4" CRS, and then silver solder it onto the end of a short piece of 1/2"ϕ CRS.GBL)

"By mounting the correct one of 3 wheels, one can produce straight or slanted knurling (coining), or diamond pattern knurling. To do diamond pattern knurling takes two passes, of course.

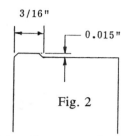

Fig. 2

"To knurl something like the tapping block I sent you, the work is held in the milling machine vise; the tool is mounted in a 1/2" collet in the spindle nose, and lowered onto the work where the knurl is wanted. If a 0.015" high ridge or band has been left where the knurling is wanted, so much the better (see Fig. 2, left). The table is traversed under the knurling wheel, and the knee is raised until an impression of suitable depth is obtained."

John also sent along info about 2 or 3 other items that help him earn his daily bread and put a bit of jam on it.

1) A simple, quickly **adjustable work stop** for the lathe spindle. It works with both chucks and collets. The use of a separate adjusting tool keeps it short. See Fig. 3.

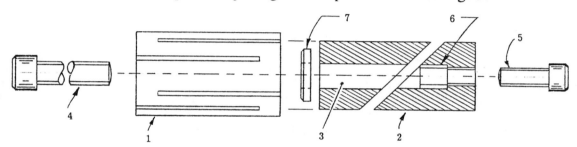

Fig. 3

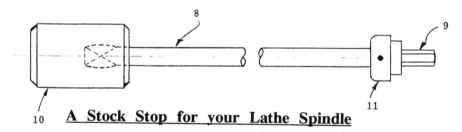

A Stock Stop for your Lathe Spindle

1) Gripping sleeve - made from aluminum tubing of a convenient size to be a slip fit in lathe spindle bore. Slots were cut with a hacksaw.
2) Expander - 6061 aluminum bar.
3) 3/8"∅ clearance hole to allow enough movement to lock up.
4) Expansion screw - 5/16-18 socket head cap screw.
5) Extender stop screw - 1/4-28 shcs (gives a little more length, and can enter the back of a collet).
6) Tap 5/16-18 for (4).
7) Washer - for under head of (4).
8) Adjustment handle - reaches in from outboard end of headstock spindle to adjust (4). Should be long enough to reach the full length of the headstock. (Remove before running the lathe!) Make from seamless tubing, 3/8"∅ CRS, or similar.
9) 1/4 A/F hex key [to suit (4)] silver soldered into end of (8).
10) Round handle - cast or epoxied in place (see text).
11) Alignment guide - made to be a sliding fit in lathe spindle bore, and pinned in place.

NOTE: There is a subtle piece of chicanery incorporated into Fig. 3. Elsewhere herein you will find something about casting machine handle knobs in epoxy, in a shop-made mold. I couldn't resist putting one on this handle. I just swaged the bulge on the end of the rod, and cast the handle on it - all with a few strokes of my pencil.GBL

2) Three pairs of hefty **aluminum vise jaws** for the 6" Kurt vise on his B'port. Each would be made from approx. 3/4" thick aluminum, with drilled & c'bored holes to attach to the main jaws of the vise. Each pair of aluminum jaws shows a variety of holes, pockets, grooves, etc. to accommodate special jobs. These jaws typically start life about 2-1/2" high, and get shorter as they are re-cut etc. John says "...there's not much left of one set - they're about used up, but they don't owe me much now." He said he made one set 4" high for use with a sine bar.

3) **Protective mats** for the exposed areas on either side of the vise on his B'port's table. These were made from plastic seat cover material. A 4" long block of hard plastic that is a tight slip fit in the center T-slot has two holes match drilled, and then tapped, in it from holes in a 1/8 x 1-1/2" x 6" piece of sheet steel, PVC sheet, or similar. The car seat material is dropped on the table, and held in place with the 1-1/2 x 6" item on top, and two #8 screws, with washers, which are run down into the piece of PVC in the T-slot. (Or you could use regular bolts and T-nuts). If you add a thin piece of brass or aluminum to fit up snug against the side of the vise, it'll catch just about all the chips.

By making these covers longer than the table area to be covered, they adapt readily to varying vise locations. The excess portion can be folded under if in the way. Saves wrenches, etc. from marring the table, and keeps chips out of the T-slots.

Tapping of Blind Holes
from **Barrie Boulter, Medicine Hat, Alberta**

The tapping of blind holes (especially in the smaller sizes) is not always an easy job. The chips created by the tap tend to be pushed ahead of the tap into the bottom of the hole. These chips then become packed in the bottom of the hole and prevent the tap from cutting to the full depth. They are also usually very difficult to remove.

An easy solution is simply to fill the hole with wax. When the tap is driven into the hole, the wax is extruded up the flutes of the tap, carrying the chips with it. The hole can be tapped right to the bottom without having to remove the tap to clean out the chips. (Providing you frequently reverse the tap about half a turn to avoid generating long, stringy chips.GBL) As well, the wax seems to lubricate the tap so that no cutting oil is needed.

Filling the holes with wax can be a problem. For larger holes, wax can be melted and poured in, but this would be a nuisance in the middle of a job. To solve this problem, I made myself a gadget to make "wax wires". With a supply of wax wires on hand, one of the correct size can simply be dropped into the hole, and tapping can proceed with no delay.

Wax wires are easily made by extruding them from a tool as shown in the sketch at right. The dimensions of the tool are not important, and it can be made from any suitable piece from the scrap bin. Drill the holes that the wax is extruded through to match the most common sizes of holes tapped in your shop (I used #0, 2, 4, 6, 8, 10 and 1/4").

A mixture of approximately half paraffin wax and half

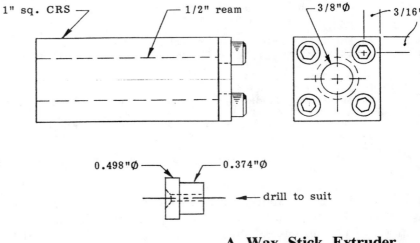

A Wax Stick Extruder

76

beeswax seems to work best for me, as it is not as hard as paraffin wax alone, nor as sticky as beeswax alone. Melt the wax mixture and pour it into a flat tray until the depth of wax is about equal to the reservoir hole(s*) in your extruder die body. When the wax is hardened, remove it from the tray and cut it into strips with a width equal to the thickness. Roll and knead these strips into round bars that will fit into the reservoir hole (just like you rolled modeling clay when you were a kid).

Push a piece of this wax bar into the die body, filling it almost to the top. Pressure from a plug or bolt (in which case the top end of the reservoir hole would need to be tapped to match) will extrude a wax wire of whatever size you want. A large supply of wax wires can be made in a short time, and the problems of tapping blind holes are over. Try it - you'll like it.

* **NOTE**: Barrie's sketch showed several reservoir holes in an oblong metal block, each reservoir ending in a different size extrusion hole. My drwg shows the slightly different approach I would take, which would be to make one wax reservoir, a plunger to suit (not shown), a plate with a hole in it to screw onto the business end, and various (interchangeable) extruder dies. To use, take the plate off, insert the correct size die, screw the plate back on, wad some wax in the top of the reservoir, and push the plunger in after it. You'll get a nice stick of extruded mat'l out the other end. Lotsa fun.

I once watched a tool-and-die-maker-turned-custom-bullet-maker use a shop-built hydraulic press to extrude 0.200"ϕ lead wire from an arrangement similar to, but much heavier than this. It was really quite amazing to see the lead wire come squirting out the bottom of the die! It had to be rolled up on a spool as fast as it came out, to keep it from piling up in a big mess of kinks, and there was a great mighty length of wire came out, compared to the 2"ϕ x 8 or 10" long cast lead slug that went into the top end of the die. GBL

A RECIPE FOR A GOOD CUTTING LUBE
from **Robert J. Molid, Molid Machine and Tool, Dubuque, IA**

Dear Guy,

Here's a good cutting lube recipe I used when I was making a homebuilt aircraft some years ago.

I had to saw wing attachment fittings out of 4140 tubing. I was going through bandsaw blades faster than a Norwegian goes through Lutefisk when my grandad happened by. He told me of an old recipe he used when working in machine shops to drill tough material.

He said to melt four tablespoons of lard and pour it into half a pint of turpentine, and add one teaspoon of powdered graphite. I immediately made some up, and finished sawing all of the remaining fittings with one blade. I have used it many times since and it has always worked.

Be careful when heating the lard, and wear a face shield when pouring the melted lard into the turpentine. After cooling, the mixture has a tendency to coagulate. I don't re-melt it, just shake it up and it still works fine.

Robert

More Uses for Stockholm Tar

And speaking of recipes, here's one from Stewart Marshall. A 50/50 paste of beeswax and Stockholm tar is tops for threaded fittings - shackles, turnbuckles, etc. - exposed to salt water. Stewart uses this stuff on many of the fittings on his little tugboat "Buster". He says it never hardens, and if you were to put it on the threads of a steel shackle pin, say on an anchor, and then left the anchor exposed to salt water for 20 years, you could unscrew the pin like it was brand new.

HOW TO GET BUSTED TAPS OUT OF ALUMINUM

Bob Eaton phoned me one evening and said he'd busted a 2-56 tap off in the aluminum steam chest of a steam engine model he was working on, and how could he get it out?

I told Bob that according to a guy I know who's had the problem, one way is to immerse the job in pure nitric acid*. Some radiator repair shops use nitric acid to clean up brass filler necks on rads, for the sake of appearance. Make sure they give you pure stuff (i.e. stuff right out of the supplier's container, which they haven't been using for their purposes yet).

Watch it - even the fumes from nitric acid will burn you, and also they will cause steel to rust - I'd do the whole thing outdoors from start to finish.

Put the problem workpiece in a glass jar with the acid. Leave it for several hours - 12 to 18 hours may be required. When the acid has done its work, there is no more tap, period. Maybe some rust at the bottom of the glass jar, depending upon the orientation of the hole.

There may be a whitish haze on the aluminum, but this will wash or rub off.

This works for aluminum, not brass or copper.

Try it on a scrap piece of the same material as the job, if you care to, and maybe throw the other end of the tap in there with the test piece.

NOTE: some small taps are stainless steel, and will not respond to this treatment.

> * The use of nitric acid comes up elsewhere herein, in connection with making etched nameplates, etc., so you might want to note this as a source of nitric acid, even if not for tap removal.

"Gunsmith Kinks, Vol. I", has some interesting things to say about broken tap removal, as does Machinery's H'book. The latter suggests the use of an EDM machine, if available, as the best way.... Be nice if a fella had such an animal, tame, that he could keep around the place for this and other purposes.....

MORE ON EDM MACHINE DRAWINGS

In **TMBR#2**, page 150, I mentioned a source of drwgs for an EDM machine you can build at home, although I'm not sure it would be the best route to go - which is to say, there may be better designs available. There was an article on making one in ME back in 1976, for example. One of my guys sent me some preliminary info on one he's made, together with a steel sleeve about 3" long into which he had spark eroded a triangular hole with it. He was willing to supply full details of his machine for use in **TMBR#4**, but you'll be able to get at **another article on building an EDM machine** long before we get underway with that effort - read on...

Shortly before writing the above, my copy of the June/July '93 issue of *Strictly I.C.* Magazine arrived. It was a bumper issue, for several reasons, but here I will cite just one:

At page 3, Editor Washburn says that in the next issue (i.e. August/September '93) he's going to run an article on how to make an effective Electrostatic Discharge Machine. I'd sign up for a year's subscription just to get at that one article. And the color photos of the engines some guys are making will make your eyes bulge!

IMPROVING A MAGNETIC DIAL INDICATOR BASE.

If you inadvertently leave one of the spindle clamping screws on your magnetic indicator base loose, and later pick it up quickly, you may have some of the parts part company with some of the other parts, particularly if the unit has taken part in your shop's particular line of work for a particularly long time, and is somewhat slack in its parts.

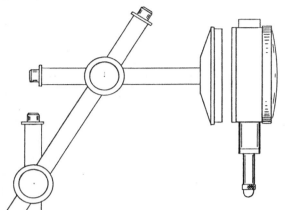

Stan Hill, a self-employed Vancouver machinist, visited me one day, and showed me his simple but effective solution to this problem.

He has drilled and tapped the ends of each spindle of his magnetic base, and fitted each one with a screw and a washer, the washer being of just slightly larger ϕ than the spindle. This prevents the spindle from sliding through the hole in its clamp.

Stan had also replaced the "lug back" on his dial indicator with a plain steel back and had made a mushroom-headed spindle, complete with a magnet, to connect the indicator to the moveable spindle on his magnetic base. This gave the indicator a much greater range of possible movement, because the indicator can be slid about and twisted on the face of the magnet.

A SHOP-MADE ELECTRIC "DIAL INDICATOR"

A while ago I heard about several "indicators" a fella has made for setting up work to run true in the 4-jaw chuck. According to the guy who told me about these things, everybody loves them, and borrows them, and then won't give them back. They have a couple of small batteries in the base, and a light bulb. Electrically, they are made so that when the probe touches the work being set up, the light goes on.

I wasn't told just what the probe was made like, but it might be nothing more than a piece of stiff copper wire; it certainly wouldn't want to be very flexible.

According to the fella who was telling me about this thing, you can set up a job in the 4-jaw with it very quickly. I haven't used one, and I don't see why it would be so very great (unless you were workin' late at night) but like I said up front, the guy who told me about it said everybody who tried it liked it a lot, so it might be something to look into.

Another Finger for the Fingerplate in TMBR#1
from **Edwin L. Dale, Cowpens, SC**

There's always some way to improve on something, and the finger plate project at page 88 in TMBR#1 is no exception. Ed Dale sent along a drwg of a V-type finger he made for his, and which he finds useful. Its wider space between the tips of the fingers gives more room for operations that might need to be carried out there - e.g. if you had to drill 2 holes side by side in a part.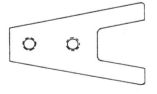

Actually, if you haven't made yourself a fingerplate, you are missing out on a very handy tool that should be in every hsm's toolbox. If you want to make it out of steel, instead of cast iron, there's nothing to stop you. If you want a piece of cast iron from which to make it, you can get same from Small Parts Inc.

PULLING A TEE IN COPPER PIPE
Tee, anyone?

In the first printing of TMBR#2, I asked if anyone could put me onto an article in *Live Steam* on pulling a "Tee" out of the wall of a piece of copper pipe. I'd seen the article, but couldn't relocate it. Sure enough, 2 or 3 fellas soon came back to me with the info I wanted - the article was titled "A Firebox for the Briggs-type Boiler", by R.V. Wood, and appeared in Live Steam for July 1982, page 40. Two guys - **Barrie Boulter** of Medicine Hat, Alberta, and another chap whose identity I'm sorry to say I've lost track of - sent me copies of the article, while **Ken Reynolds**, of Louisville, KY sent literature about similar but not identical commercially-made tooling to do the same job. **David Bezinque**, of Plano, TX sent a xerox of a page from a book called SOFT-TECH, whereon was shown the same rig - it's called an "Aushalser" in Germany, where the tooling is made.

John Maruschak, of Hayward, CA referred me to an item in the Pracitcal Ideas section of *American Machinist Magazine* for December 24, 1962, page 71, and **Bob Johnson**, of Kelowna, B.C. sent me a sketch of tooling he'd made to make a water manifold with 1/2" I.D. copper headers and 3/8" O.D. risers. This method, which I have detailed below from Bob's drwgs, is, I believe, pretty much the same as the 1962 *American Machinist* idea.

While the main purpose of the *LS* article was to show how to make a small boiler from empty propane bottles, the thing in it that had caught my eye was the detailed info the author showed on how to make simple tools to go into a drilled hole in the wall of a piece of copper pipe, and pull a smoothly formed Tee of the desired size right out the side of the pipe, "...to produce," as he said, "a joint that is both strong and of free flowing form, which would not be possible if the riser was just poked into a hole in the header".

The tools are simple, and simple to make. What's more, the Tee's shown in the photos accompanying the article look just as slick as if they'd been put there with a barn full of fancy equipment. In many cases where one might use these tools, a commercially made pipe Tee would probably work as well, but it certainly would not be as impressive, or as much fun.

Since the *LS* article will be readily available to anyone wanting it, via your local library, interlibrary loan, etc., there is no point in presenting the same material here.

Bob Johnson's Tee-Pulling Tooling
(more next page)

Die Block (aluminum or mild steel)

smooth radius

easy fit on pipe O.D.

* = 1" for 1/2" pipe

Idea: One could use such tooling to make tanks for cooling water, washer fluid, fuel, etc. for model engines, antique engines, cars of every sort (an auto restoration shop would be one potential customer), motorcycles, etc.

I once saw a '66 Olds whose owner had made and installed a large cylindrical copper tank just ahead of the firewall to hold extra radiator fluid*. I'm not convinced of the necessity of this addition, as I owned a '67 Olds myself at the time, and never suffered from the lack of extra radiator fluid, but I gotta tell ya, guys, if that copper tank and piping had been polished up nice, you would've swooned. The job was very handsomely done, and I suspect that such an addition, with a polish job sufficient to frost your eyeballs, would up the resale value of an older car, or a "replicar," or whatever, by far more than its cost to install - it'd bring out the magpie in most anybody.

* Walt says what it more likely was was a coolant recovery tank - which most cars now have. If he's right, the plastic tanks they use now are a long step down, in looks, from what that big copper tank could've been!

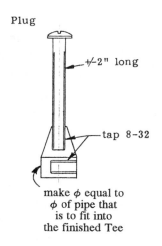

Plug

+/–2" long

tap 8-32

make φ equal to
φ of pipe that
is to fit into
the finished Tee

But why restrict it to radiator fluid? Hot coffee on the road? Extra engine oil? I mean, there's gotta be some excuse to do something like that. Walt says that the "in" thing these days is polished aluminum, not copper, in custom car work, but a couple of articles in the car mags showing some polished copper tanks, etc. would change *that* fatuous notion in a hurry.

Footnote to the above:
I had a call from one of my customers down in northern California one day. He said he had made tooling to pull Tee's at 45° and 90° in pipes from 2 to 6"φ. I would surmise that they required a significant flare in the T juncture, because my man said he'd used the info in my little book **"Tables & Instructions for Ball & Radius Generation"** to make the tooling, and that his employers had then ordered 2 extra copies of the book in case he quit and took his book with him. But here was the thing that really gave me a kick: the customer they were doin' all this work for was Boeing Aircraft Company of Seattle. My, but that boy from Vancouver do git around, don't he?!

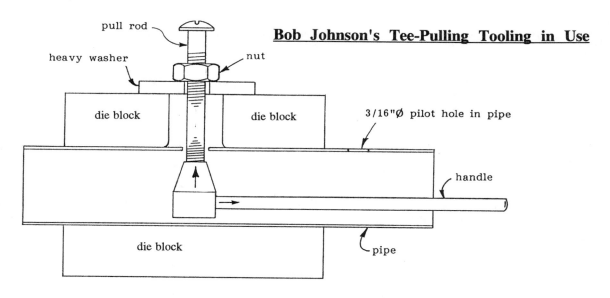

Bob Johnson's Tee-Pulling Tooling in Use

pull rod

heavy washer

nut

die block

die block

3/16"∅ pilot hole in pipe

handle

die block

pipe

Drill pilot holes where Tees are wanted in the pipe. Anneal pipe. Center a pilot hole under the pull-out hole in the die block and snug up the bolts. Screw the handle into the side of the plug and insert plug/handle down the pipe. Screw the pull rod into the top end of the plug and spin the nut down onto the heavy washer. (You will need to prevent the pull-out rod from turning with the nut; I have shown a screw head, but any convenient form will do). Remove the handle when the top of the plug begins to come through the pipe. Pull the plug right through the pipe wall by further tightening of the nut. A little oil or grease on the plug will ease the pulling operation.

"Never mind blowin' yer own horn, Lautard. Tell us about about them commercial tools for doin' the same kind of work... "

Well, ok... If a guy had work for such tools, not only the Aushalser thing, but their tube bending, flaring and expanding tools would be fun to own. The stuff is made in Germany, by a company called Rothenberger, and sold in the USA by Rothenberger USA Inc, (955 Monterey Pass Road, Monterey Park, CA 91754; phone (213) 268-1381).

The unusual items in the line are the Tube Expanding tools and those Aushalsers, or Extractors, as they're unromantically designated in the American catalog.

The Tube Expander goes into the end of a piece of copper, brass, or aluminum pipe, and expands it so the next piece can be slipped into the expanded section. A single soldered joint thus suffices to connect two pieces of pipe, whereas otherwise you'd need a separate fitting and *two* soldered joints. The different sizes of expander heads (covering 5/16" to 1-5/8" OD pipe) are used with a set of handles that would feel like you were using a set of bolt cutters. A bigger multi-ϕ head is also available to deal with pipe up to 4-1/8" OD.)

I would want this type of tooling if I were installing copper pipe plumbing in a house. Where sharp-radius bends were required, you'd need to use regular elbow fittings, but where a larger-radius bend would do, one could make a bend in a length of pipe, cut off the bend, and slip this site-bent elbow into the expanded ends of the pipes to be joined.

But where you want a Tee, anywhere along the length of a run of pipe, that's where the real fun would be! No need to cut the pipe and put in a T-fitting. Just drill a hole in the pipe, slip the Extractor hook into the hole, and twist the hook back out through its own back-up sleeve housing with a ratchet wrench. Slip your branch pipe into the Tee hole, make one soldered joint instead of 3, and you're done.

I can think of all kinds of stuff a guy could make with this equipment. A multi-bar towel rack for the bathroom is one example. Or how about a plant stand in copper pipe? Lotsa opportunities for Aushalser-izing the daylights outta something like that, or a floor lamp or a table lamp.

> TIP: Wipe your soldered joints with a tallow-impregnated rag while still hot, having first done your soldering with the minimum amount of solder possible, and I think you will have a very nice looking end result.

Added later: I was lookin' in the engine compartment of my Explorer, and I'm gettin' some very radical ideas that would shock Ford outta its mind. There's room in there for a copper windshield washer tank (whose contents could readily be heated, too; I've heard that heated washer fluid is much more effective than the same stuff cold, altho how hot it'd be by the time it hit the windshield, if you were under way at the time, I don't know...). Plus I could make a replacement for that hideous plastic coolant recovery tank, too. Hmmmm... Maybe that old fella with his '66 Olds had somethin' after all...

PULLIN' THINGS INTO LINE

In machine tool work, it sometimes comes up that one has to screw one part into another just a little further, so that the alignment is as you want it. How to make it come up to just right?

Screw it in tight, and measure how much farther it has to go, in degrees, to come into proper alignment. Let's say your protractor tells you it needs to roll ahead another 73°, and let's say the thread is 40 tpi. What to do?

One full turn of the thread will advance it 1/40 = 0.025".

To get the part to rotate 73°, it has to screw in 73°/360° x 0.025" = 0.005,069" more to come up right. I'd be inclined to shave 4-1/2 thou off the shoulder, and get the last of it (about 7°) by a little extra torque on the final tightening.

INDICATORS, EDGEFINDERS AND SIMILAR TOPICS

In the letters section of Jan/Feb '91 HSM, there was a letter and several photos from Ed Ellison, of Corona Del Mar, CA. The photo that really caught my eye was #5, which was of a Co-Ax type indicator for use in the spindle of a milling machine...

I wrote to Mr. Ellison, asking if he would be willing to divulge details of the device for inclusion in **TMBR#3**, as I felt it would be of considerable interest to a lot of "the guys".

Ed wrote back in the affirmative, so here it is....

A CO-AXIAL INDICATOR
from **Ed Ellison, Corona Del Mar, CA**

If you own a milling machine, and have a catalog from most any mail order machine shop supply house, you know that for $200, or maybe a little less, you can buy something known as a Blake Co-Ax Indicator.

A "Co-Ax" indicator is nice to have when you need it, but you may not need it very often, and if that is so, you may feel you have better things to do with that particular $200. But you can make one just as good as the commercial item. ("Co-Ax" appears to be a trade name, so to avoid ruffling any feathers, let's use the term "Co/Ax Indicator" here.)

Somebody is going to be wondering: "What's a Co-Axial Indicator?", so let's talk about that first, and then I'll tell you how to make one for yourself.

Suppose you have a job clamped down on your milling machine table, or grabbed in the vise thereon, and the job has either a hole in it, or a round portion sticking straight up - could be anything - a rotary table, a partially machined chuck backplate, a gear, or whatever. Anyway, there it is, and you want to move the mill table so that the center of whatever job or fixture you have clamped to the table is lined up right underneath the spindle centerline.

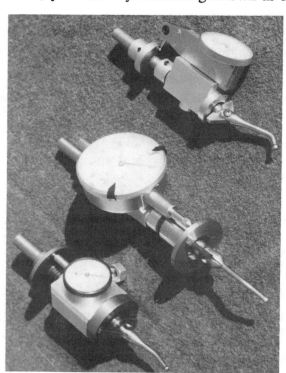

Not all of Ed's Co/Ax Indicators are built exactly as in the GA Drwg which follows.

One way to do it would be to use Osborne's Manoeuvre (TMBR#2, page 159).

Another way would be to mount an indicator in the mill's spindle nose and sweep it around the job until you get a zero runout reading all the way around. The difficulty of actually doing this may only become apparent when you first try it: throughout much of each sweep, the face of the indicator is not looking at you. This makes it somewhat difficult to see what the indicator has to tell you. A mirror can be used to view the indicator, but that is not particularly handy either. There are indicators that have dual faces and dual indicating needles - one on either side - back-to-back, as it were. All of these things are helpful, but....

But what would be really handy would be to be able to chuck, in the spindle nose, some kind of a device with an indicator built right into it, and that will permit the indicator to sit still, facing you, while the spindle rotates, sweeping a contact finger around the job.

This is exactly what the Blake Co-Ax Indicator does, so you can see why it'd be nice to have one.

You can also use this type of indicator to sweep the table of the mill to check that the head is square to the table. Ed says he rarely moved the head of his mill because of the effort involved in dialing it back in. He found that by making a point with about a 45° bend in it for his Co/Ax unit, it would do this job very nicely. See also info on this topic in **"Hey Tim, I gotta tell ya..."**, page 33.

Well, like I said, you can make one for next to nuthin' if you have a suitable indicator on hand. If you don't have a suitable indicator sittin' around, you can buy one for about $25/30, or more if you want one that says Starrett up front, and you can build your Co/Ax unit around it. The General Arrangement Drwg is shown at right.

Most of the parts are straightforward and easy to make, although some of them will bear comment.

But first, let's see how a Co/Ax Indicator actually works. Suppose the Contact Finger (part #7) is in contact with the wall of a bored hole which is slightly offset from the mill's spindle centerline. We switch the mill motor on, first making sure the belts are set for a spindle speed below 500 rpm. (Running it above that speed is unnecessary, and serves only to wear the Co/Ax unit out.)

The dial indicator does not turn with the machine spindle, but the Contact Finger/Rocker/Mounting Shank assembly does, sweeping around the job.

The Upper Flange (#3) is fixed to the Mounting Shank, giving the foot of the dial indicator a fixed reference plane to push against.

Any misalignment between the job and machine spindle centers causes a horizontal movement of the Contact Finger as it sweeps around the job.

The Contact Finger is clamped solidly to the Rocker (part #6); hence horizontal movement of the Contact Finger is, through the linkage between the Rocker and the Mounting Shank (part #1), translated into an upward movement of the Body (#2), Lower Bush (5), and the Indicator (9), which shows up as a movement of the indicator needle.

If the Indicator shows any movement as the spindle is rotated, you simply move the machine's table (and the job) in the X or Y direction to reduce the movements and bring the job into perfect alignment under the spindle. Typically, a machinist familiar with a co/ax indicator can center a job under the spindle nose of his machine in maybe 2 minutes, from the time he turns the mill on.

Okay, I bet that last sentence convinced you it's a good gadget! But you may still not want to pony up $200 to have one in your toolbox against the day of need, so let's look into how to make one.

General Arrangement

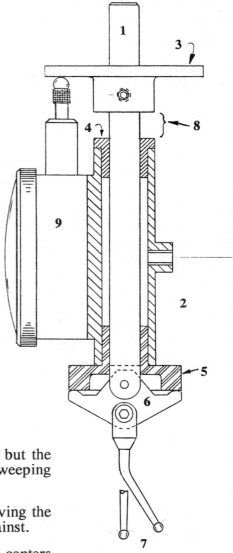

Parts list
1) **Mounting shank**
2) **Body**
3) **Upper Flange**
4) **Upper Bushing**
5) **Lower Bushing**
6) **Rocker**
7) **Contact Finger(s)**
8) **Spring + 2 washers***
9) **Indicator**

* **not shown on drwg**

As indicated in the Parts List on the previous page, there are 8 parts required, plus a dial indicator. Ed suggests using an indicator of 2-3/16" or 2-1/8"ϕ, having 0.000,1" sensitivity, and a plunger range of 0.050". Note that the pointer/rocker/shank linkage of the Co/Ax unit decreases the apparent sensitivity by about 4:1, which is why an indicator having 0.000,1" sensitivity is to be preferred over one reading to 0.001" or half a thou.

Making the parts

1) Mounting Shank
Make from 3/8" drill rod. Length is not critical, so long as about 3/4" sticks up out the top end for chucking purposes. Think about 5" as a start.

Face both ends. Turn one end approximately spherical for nice appearance (you can do this by eye, or with my incremental cut method, or just chamfer at 45° if you don't care what it looks like.)

Move the job over to your mill. Cross drill and slot the spherical end for the Rocker at one setup. Use a slitting saw to cut the slot, centering it as accurately in the material as you can. This calls for "nice work", so do it by-the-numbers - no eyeball stuff here. Use the Rocker as a gage, and aim for a real nice fit with only 0.000,2 to 0.000,4" clearance.

Put a small flat on the upper end of the Mounting Shank for the set screw in the Upper Flange to bear against, so it can't scar the OD of the Shaft. Make the flat as long as required. Moving the Upper Flange up or down gives some adjustment on the spring tension. This screw and the Upper Flange are what hold the whole device together, so make sure the screw does not come loose!

2) Body
Make from a 3/4" square piece of 6061-T6 or 2024-T4 aluminum, or brass, about 2-3/4" long, or in any case, slightly longer than the diameter of the back of the dial indicator you are going to use.

Face ends, chuck in 4-jaw, center up good, and drill/bore/ream a 9/16" hole full length.

Make a backplate from 1/16" brass or aluminum plate, and solder or screw same to the 3/4" square piece, as drawing, making sure the screws (if used) do not enter the 9/16" reamed hole. (The tapped holes for these screws can break into the reamed hole, with no harm, but the screws themselves must not.)

Alternatively, make indicator back and body from a single piece of material 2-3/8 x 2-1/4" x 1". Bore/ream the 9/16" hole, grab in mill vise by the face away from body portion, and mill off sides to leave body, plus integral material for indicator back. Chuck by square "body" portion in your 4-jaw, and machine to the desired thickness and diameter the face that will carry the indicator. Return to the mill to drill indicator mounting holes by-the-numbers or with rotary table, or by careful marking out. Use the original screws from the dial indicator to attach the completed Body to the back of the dial indicator.

Note: The exact form of this part will be governed by the indicator you use. Make the back plate to match your indicator.

3) Upper Flange
Make from bronze or cast iron. Straightforward, but must be a close sliding fit on the Mounting Shank. Setscrew in place on Mounting Shank.

4) Upper Bushing
Straightforward. Make from bronze.

5) Lower Bushing
Ditto - Bronze or cast iron. Machine working face and bore at one setting, or do as described in the next paragraph.

Body

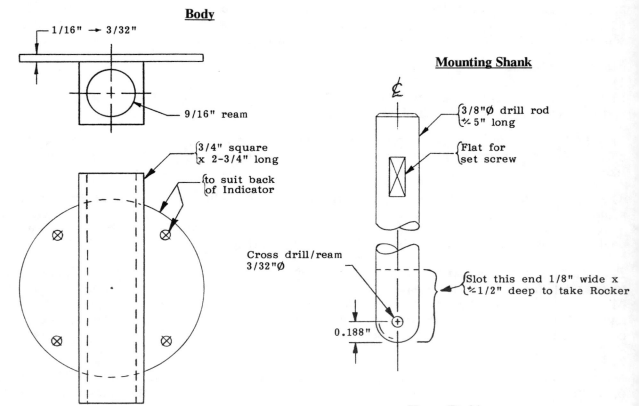

— 1/16" → 3/32"

9/16" ream

{3/4" square
x 2-3/4" long

{to suit back
of Indicator

NOTE: it may prove desirable or necessary to mount the disc that serves as the new indicator back further up the 3/4" sq. piece to give clearance for the non-working end of the plunger - see text.

Mounting Shank

{3/8"Ø drill rod
±5" long

{Flat for
set screw

Cross drill/ream
3/32"Ø

{Slot this end 1/8" wide x
±1/2" deep to take Rocker

0.188"

Upper Bushing

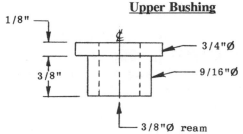

1/8"

3/8"

3/4"Ø

9/16"Ø

3/8"Ø ream

Upper Flange

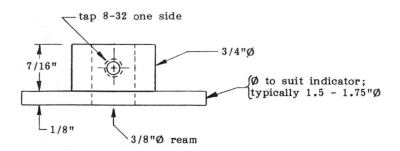

tap 8-32 one side

7/16"

3/4"Ø

{Ø to suit indicator;
typically 1.5 - 1.75"Ø

1/8"

3/8"Ø ream

NOTE: this part should have been drawn the other way up, i.e. having the same orientation it has in the GA drwg.

For accuracy in the finished Co/Ax Indicator, it is important to provide surfaces perpendicular to the axis of the Mounting Shank (part #1), for the Rocker and the dial indicator's contact point to work against. This can be accomplished by boring the hole in the Upper Flange and turning the critical face at one set-up, and doing the same for the Lower Bush/flange. Alternatively, turn a stub mandrel or spud in the lathe, making it of a size such that the parts are a wringing fit onto it, and then mount each part in turn on this spud and skim cut the critical face. Either way, the face will be square to the bore.

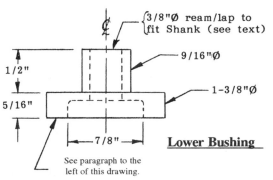

NOTE: These bushings should be a light press fit into the Body, and must be a very good fit on the 3/8" Mounting Shank. Reaming may give an acceptable result. Boring is preferred, with possibly lapping to final size.

6) Rocker

Make from 1/8" x 3/4" ground high carbon flat stock. Lay out the Rocker profile, ream two holes, and mill/file to profile. Polish rocker tips, harden, and re-polish.

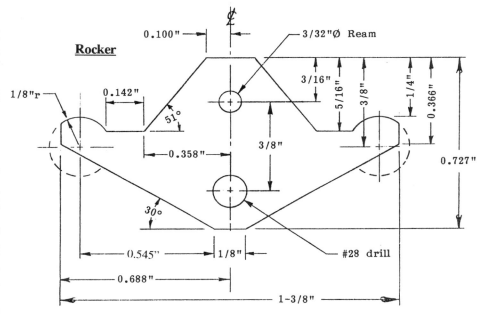

You might wonder how to harden such a part in a basement shop. Follow the steel maker's instructions, if available. If not, just do it the same way you would a cutter. As a general instruction, coat with soap to prevent scale, heat to a point where the part will not attract a magnet, and then dump it into a big soup can of cold water. (Or oil, if using oil hardening stock.)

You might also wonder how to mill the rocker tips to the specified radius so they will be identical... You could do it with the aid of a pair of filing buttons, or just file them to approximate form - they do not need to be absolutely identical.

Simplified forms of the Rocker

There is no need to follow the drwgs slavishly. Ed sent me drwgs for two simplified forms of the Rocker which he said would also do fine, and which would be somewhat easier to make. These are shown at right.

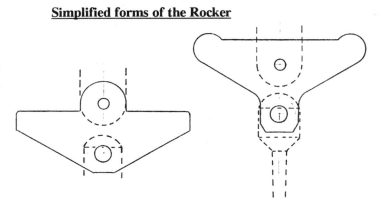

87

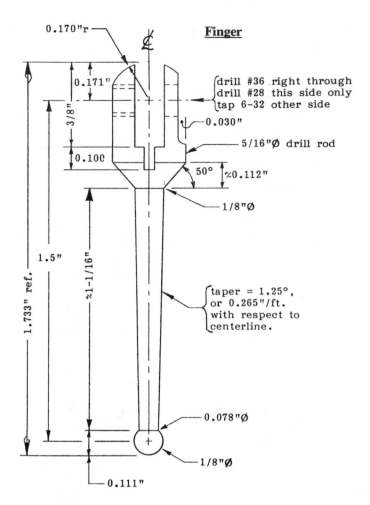

Finger

0.170"r

0.171"

3/8"

0.100

drill #36 right through
drill #28 this side only
tap 6-32 other side

0.030"

5/16"∅ drill rod

50° ≈0.112"

1/8"∅

1.5"

≈1-1/16"

1.733" ref.

taper = 1.25°,
or 0.265"/ft.
with respect to
centerline.

0.078"∅

1/8"∅

0.111"

NOTE: The purpose of the narrow saw cut centered in the main slot that takes the Rocker is to enhance the clamping action of the screw/Finger on the Rocker

7) Contact Finger

Make from 5/16"∅ drill rod. Although the drwgs show both straight and bent forms, the straight one may see little use. It is for reaching into small holes. The bent Contact Finger will do for both round spigots and in holes over about 1/2"∅.

The notes below, and the sequence of drwgs on the facing page, show the method Ed suggests for making the Contact Fingers:

1. Turn 0.156"∅ radius, slot, drill and tap pivot hole.

2. Part off at about 3/16" longer than the final OAL.

3. To keep the slot from collapsing while holding it for turning the small end, cut a small piece of material the same thickness as the Rocker to fit into the slot. It must not stick out beyond the 5/16"∅. With the slot filled, hold the job in a collet if you have one, or do it with extra care in a 3-jaw chuck.

4. Chuck by slotted end, and center drill other end using a #1 (1/8"∅ x 0.040") center drill.

5. Using a small center for support, turn to 0.130"∅, and turn the steep taper at the large end.

6. Use a V-pointed tool ground to cut on either side to neck the Contact Finger down to 0.078"∅ behind the ball tip.

7. Set the topslide over to approximately 1.4° and carefully turn the long taper.

8. Move the V-tool over enough to put its other cutting edge to work on the tailstock side of the ball, and carefully feed in about 1/32". By careful machining, the ball can be roughed out with tailstock center support. After the ball is roughed out, it can be lightly filed to a sphere. Check using a radius gage. It does not have to be perfectly spherical. (For 2 other methods of putting the ball on the Contact Finger shaft, see below.)

9. With a jeweler's saw, cut off the excess piece at the ball end.

10. Bring the ball tip to final form by filing with a light touch, and polish it to a good finish.

11. The spherical tip should be hardened and drawn to a dark straw color at the ball end after bending as drwg.

Steps in Machining the Finger

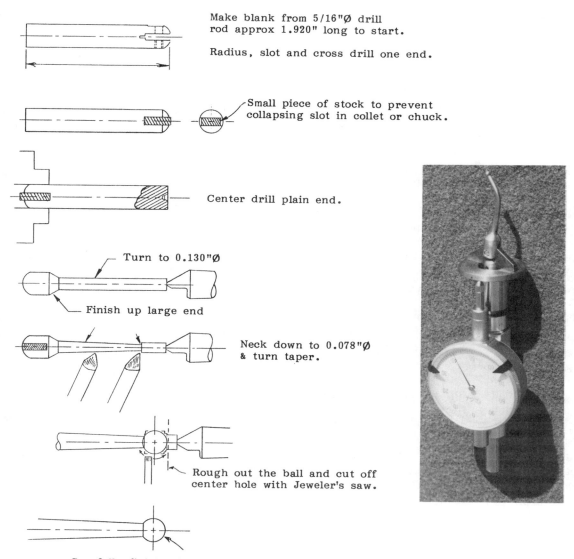

Make blank from 5/16"Ø drill rod approx 1.920" long to start.

Radius, slot and cross drill one end.

Small piece of stock to prevent collapsing slot in collet or chuck.

Center drill plain end.

Turn to 0.130"Ø

Finish up large end

Neck down to 0.078"Ø & turn taper.

Rough out the ball and cut off center hole with Jeweler's saw.

Carefully finish with file and emery cloth.

There may be better ways to do it, but this approach definitely works well. Ed also suggested the following:

".... another, and quite simple, way to put a ball end on the Contact Finger, is as follows. First of all, centerdrill the shaft (per Step 4. above) with a 1/8 x 0.40" center drill. Then turn the shaft as above, but make it only long enough to include the taper down to where the back of the ball would be, and have the small center hole fall at that point.

"Set the shaft vertical in a vise, put a small drop of solder flux in the center hole. Cut a small piece of solder and place it in the hole. Heat with a small torch and melt the solder. Now flux a 1/8"φ ball, and place it on the solder.

"Hold the ball in place with tweezers and re-melt. The ball will drop into the small socket and be stuck firmly in place. You may have to lightly file off a little excess solder. It only takes a very small amount of solder to do the job. Solder is plenty strong for this job. If you use the lead-free solder now used for soldering copper water pipes, it will be very strong.

"Another way of holding the ball for soldering is to center drill a spot in a piece of steel, to hold the ball and keep it from rolling about. Hold the shaft vertically over the ball and use your torch to sweat it to the ball.

"The beauty of this system is that you get a perfect sphere for the end of your Contact Finger, but the soft soldering doesn't anneal the ball."

8) Spring

The Spring is not shown in the drwgs. Make from 0.029"ø music wire, or #12 piano wire. The Spring should have 5 active turns, and a free length of 7/16", with closed ends, ground flat. If I were making it, I would wind about 8 turns of wire onto a 3/8"ø mandrel, starting with a piece of wire at least a couple of feet long. However, from a phone conversation I had with one of my guys one day, I'm beginning to think I still have a lot to learn about spring making. (My springs work, but in winding them as a more or less packed coil on the mandrel in the lathe, and then stretching them until they show the desired number of coils in the desired length "…. you're not developing anything like the full potential….", I was told. Getting nice spring ends with the last coil nice and flat has always been a problem for me, as I suppose it would pretty well have to be, with my system.)

I therefore asked Ed if he had any ideas on making the Spring for the Co/Ax unit. I also asked him what is the difference, if any, between music wire and piano wire. Back came an excellent and complete reply, as follows:

"The spring is not really a very critical item. It's main purpose is to apply pressure downwards, especially when the unit might be used in a horizontal mode. I have even used these units without a spring, with just the weight of the indicator assembly to hold it against the rocker. The spring is mainly, as mentioned, for horizontal use.

"I have a pretty simple way of making compression springs which will give good square ends and nice uniform coils. For this particular spring I used a 25/64"ϕ mandrel. There will be some variation from one man to the next due to the amount of tension each will put on the wire during winding. Don't try to hold the wire in your hand! I make a hardwood clamp shaped much like a toolbit, but having a horizontal hacksaw kerf or slot part way through it lengthwise. This is clamped on the wire by the setscrews which hold toolbits in the toolpost. The clamping can be varied to give a firm restraint on the wire during winding.

"The mandrel is cross drilled for the wire and is chucked in a 3-jaw. Put a 90° bend in the end of the wire to keep it in the mandrel.

"Choose a feed or thread which will give the proper lead, in this case about 0.085" per turn (= about 12 tpi). Set the QC gear box for this, and set the leadscrew rotation to feed the saddle away from the chuck. Put the lathe in slow backgear, and put on a few turns with the halfnuts disengaged, in order to make several tightly packed ("solid") turns to begin with. Then engage the halfnuts and run the lathe long enough to give the desired number of active coils. Then pop the halfnuts to chop the feed and make 2 or 3 more turns with no saddle feed, to give a solid end.

"Cut the spring loose from the mandrel and trim the ends to give one solid turn at each end. On your bench grinder, lightly grind the last coil at each end to give good flats.

"Using this approach, it only takes a few minutes to make a very good spring.

"'Music wire' and 'piano wire' are just 2 names for the same thing. The name piano wire comes from the fact that it used to be used for piano strings.

"It is also not a bad idea to put a nice fitting washer under each end of the Spring to keep it from digging in."

90

9) Indicator

As previously noted, Ed recommends a 2-1/8"ϕ or 2-3/16"ϕ "tenths" indicator having a travel of about 0.050" - any more than this just makes the whole device longer, with no benefit.

(In one of his letters, Ed said he made up one using "... a "Teclock", one of the better Japanese instruments. It has a 1/4" stroke, which is more than enough. At about $28.00, it was relatively inexpensive.... It is necessary to remove the lower cover on the indicator spindle. This does not impair the operation of the indicator, and it keeps the whole unit from becoming too long...")

As noted earlier, the Co/Ax Indicator should be operated at spindle speeds under 500 rpm. Above that is unnecessary, and serves only to wear it out.

Ed closed off his last letter by saying:

"I like nice boxes for precision tools of this sort. One can be made from a block of hardwood milled out to fit the tool. I make a solid lid for it with a nice latch, and give the whole thing a couple of coats of Polyurethane varnish."

And a concluding note: The GA drwg shows how to make provision for attaching a rod to the Body to prevent rotation, but Ed says he does not find he uses this feature very often - usually it can be prevented from rotating by hand.

●

MILLING MACHINE SAFETY
from **Garth H. Spencer, Russellville, AR**

Dear Guy:

In connection with dialing in the head of a vertical mill, you noted in "Hey Tim..." that the worm gearing on the back side of the knuckle casting on a B'port type mill can wear out, over years of use, and that is true. There are a couple of other serious safety points I've never seen mentioned anywhere in writing.

After you have a mill dialed in the way you want it, and the locking bolts are all tight, it is wise to slack off the pressure exerted by both worm gears (for rotation of the head in the vertical plane parallel to the X (table) axis and tilting of the head in the plane parallel to the Y (saddle) axis.)

If pressure is left on the worm gears, it can cause the head to creep, especially during heavy or interrupted cuts. This is especially true of worm gear adjustments, which typically have more mechanical advantage than other types. Table feeds are also subject to this - watch a DRO flicker or change during a cut some time and you will see it happening. The effect is similar to the results of tapping a tight bolt with a hammer while trying to loosen it. It is surprising how many experienced men are not aware that this is a problem.

Also, when dialing the head of a vertical mill, always keep at least one of the locking bolts a bit snug, and keep clear!

A man was killed in a shop near where I worked when a big K&T vertical milling machine head fell. In another shop where I worked, a man had his hand badly crushed for the same reason.

Many who read your books may already know the above, but some will not, so I hope this is helpful.

Garth

It is! There are many safety points about adjusting and operating machine tools of all types. Some points are so obvious they hardly bear mention. Other points are obscure - things can happen that you would never think *could* happen, unless it was pointed out to you. Always think before you act, and always be careful.

SPUDS, BUGS & WIGGLERS

In a letter to ME's "Postbag" section for October 20, 1972, a Mr. N.J.C. Walford of Yeovil (in England) described the use of a shop-made wiggler, or edgefinder. Starrett and others make these, for somewhere around $25 at date of writing, but you might like to make your own.

The 1/2"⌀ mounting shank should be screwcut 1/2-20 or similar, with the 5/8"⌀ body or ball holder portion being bored and screwcut to match. The threaded joint will allow you to effect some adjustment of the spring pressure on the spring loaded shoe that rides on top of the ball.

These 3 parts can be made from brass, steel, or aluminum.

If you have a 7/16"⌀ ball from a ball bearing, anneal it and use it. Otherwise, turn a ball integral with the tapered finger. Absolutely perfect sphericity is not required, and the incremental cut method detailed in **TMBR#1** will work fine. The barrel-shaped tip on the tapered finger should have its maximum OD as close to 0.200,00" as you can get it, and should be hardened, and well polished.

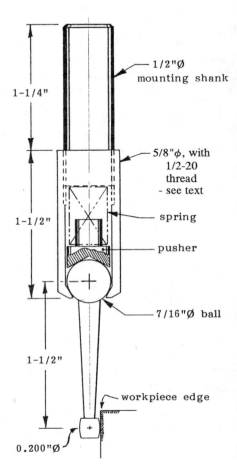

The threaded joint between the shank and body was not shown in the drwg accompanying Mr. Walford's letter, but without it - i.e. if the body and shank were made in one piece - the spring, pusher, and ball would need to be inserted in their hole, and the mouth spun over, or capped with a knurled nut, to retain these parts, and all opportunity for adjustment of the stiffness of the action of the finger would be lost.

"But how does it work?" you may wonder, if you've never seen one of these little brutes used.

Stick the shank in a collet or drill chuck in the spindle of your vertical mill. Switch on at some moderate speed - say 200/500 rpm - while holding the barrel tip between the thumb and next one or 2 fingers as you might if holding an eyedropper full of poison. "Milk" the barrel tip so it is running true - everything spinning on one axis.

Now let's say you have a parallel or some workpiece in the milling vise. Bring the face you want to locate up to the spinning barrel. When contact is made, nothing changes. Go a thou past the point of contact and the tapered finger flicks up and spins above the workpiece, tracing out a conical shape in space, something referred to by mathematicians as a solid of revolution.

Can you figure out why? Think about it for a minute, and then I'll tell you...

If you really want to scramble your marbles, try to develop a formula to describe the path the center of the barrel follows as it goes from spinning all straight down to out-at-an-angle.

I'm told mathematicians love this kind of problem. They engage in never-ending speculations on the comings and goings of some little bug that crawls towards the center of a LP record at so many inches per minute, and then try to describe the path that the bug follows, given that he is going round and round while doggedly boring in towards the center of the LP. If this begins to pale, they introduce slippage, drift, windage, coreolus effects, stops for food, etc.

Now, that was just a distraction.... Have you figured out why the tapered finger flicks up? Bill Fenton demo'd this for me years ago and I never understood why until I was hand writing the first draft of the above, when suddenly it came to me:

When you push the workpiece a thou or so past first contact with the side of the barrel shaped tip of the tapered finger, the barrel is forced off the axis of rotation, which unbalances the whole finger, and - since it is free to swivel at the ball joint - it is flung out and up by centrifugal force, much the same as when the piece of bar began to (excuse me) flail the snot out of the locker next to Phil Lebow's lathe, as reported elsewhere herein.

What's the bent shaft item in a typical "Wiggler" outfit for?

One of my guys asked me the above question, and thus made me think about it sufficiently to come up with an answer. What it's for is holding an indicator. The ball end goes up into the chuck of the main tool, and your indicator clamps to one or other part of the two-diameter shank. This allows you to indicate holes or OD's to test them for concentricity, face run out, establish center distances, check alignment of an edge or surface, etc.

Some Notes on Edge Finders and their Use

Top View of Milling Machine

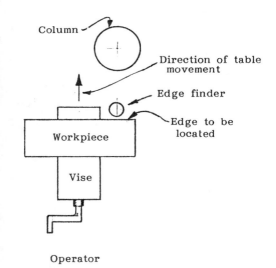

Column

Direction of table movement

Edge finder

Edge to be located

Workpiece

Vise

Operator

Brian King, of Tustin, CA sent me some interesting info on how to use an Edge Finder.

A bird's eye view of a vertical mill, with a job in the vise, and an edge finder being used to locate, or "pick up", an edge, would look something like as at left.

Since the edge finder's moveable section kicks in the same direction as table movement, the following happens:

As the table is moved toward the column, so that the workpiece contacts the edge finder and kicks the moveable section of the edge finder, the operator's view of the edge finder looks as at "A" below from where the operator is shown in the sketch at left. Obviously, if you look from this direction, you better have good depth perception to detect the movement that occurs.

However, if the operator views the edge finder *from a position along the edge to be located* he is presented with the view as at "B" (far right) when the bottom of the edge finder kicks out. Here, no depth perception is required.

This viewpoint allows you to be more accurate than the normal view would provide.

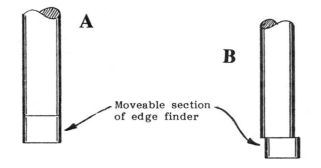

A

B

Moveable section of edge finder

This tip can be stated in two different ways:

1) View the edge finder along the edge to be located, not at 90° to it.

OR

2) View the edge finder from a position 90° from the axis of table movement, not along the axis of table movement.

Says Brian, "In tests conducted with a machinist friend acting as the judge, I was consistently, through several trials, able to locate and relocate an edge within less than 0.001" when I viewed the edgefinder from along the edge to be located. However, when I viewed the edgefinder from along the axis of table movement, I was only able to repeat within 0.006".

"Note also that prior to doing these tests, I'd never used an edge finder. I had only read about their use or saw them advertised in a catalog.

"The friend who served as the judge for these tests deserves the credit for this tip. He had noticed this phenomenon many years ago and - being extremely well read in the area of machining - wondered why he had never read of any reference to it in any book or magazine."

Lapping an Edge Finder

I'm told you can improve the performance of an edge finder (e/f) by lapping the mating end faces of its parts. I doubt a good one would need lapping, but on the cheaper ones it's probably a good idea. Carefully unscrew the tips from the internal spring, and chuck the e/f body in the mill. Put some oil on a small piece of 2000 grit wet/dry paper, back it up with a piece of flat metal about the size of a quarter, and present it to the spinning end of the e/f body on the tip of your finger. This will allow the abrasive paper to "float" in nice contact with the face of the part. A few seconds of this treatment on each element of the e/f should do it (you'll need to put a hole in the paper, and in the metal plate, to deal with the end pieces). Tim Smith tells me the best edge finders (and some other neat stuff) come from Herman Schmidt, 237 Burnham Street, East Hartford, CT 06108; phone 203-289-3347.   **See also page 139 and 251.**

Pickin' up an edge fast, and to "tenths".

Bill Lloyd buzzes me every so often from Chicago and we have a talk. One day he told me he has a way of picking up an edge that allows him to do it in a matter of seconds only, and to "tenths". I was all ears. (Bill ran jig borers for 20 years, so you can bet he knows a few tricks.)

Here's how he does it:

He puts an indicator in the spindle nose of the mill using a jointed mounting stem bent to suit - if using a plunger type indicator, the indicator would be set so that the plunger travels horizontally. (A finger type indicator, such as I have shown in the drwg at the top of the next page, would be a little more convenient.) He points the indicator at the face of the edge he wants to pick up, and moves the table to a position where the edge he's interested in looks like it's right under the spindle centerline. He then moves the indicator holder (its joints ain't rigid, just firm) so the indicator reads zero.

He then raises the quill a little, and spins it 180° by eye. He applies a gage block or similar piece of smooth flat steel to the face he just "read" with the indicator, thus "reversing" that surface and extending it above the top face of the workpiece. If the indicator does not now read zero as before, he'll move the machine table enough to split the difference, and repeat the test. This time he will expect to be dead on, or very close to it.

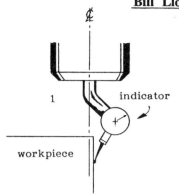

Bill Lloyd's way of picking up an edge

1

indicator

workpiece

For second reading, hold a parallel, gage block, or similar flat piece of metal against the face just "indicated," rotate the spindle 180°, and note the reading. Then move the milling machine table half the difference between the two readings to get the spindle centered over the edge of the workpiece. Repeat to check. It's fast.

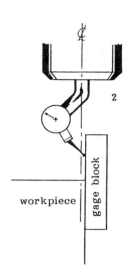

2

workpiece

gage block

SPUDS 'n Gravy
from **Marsh Collins, Crestline, CA**

I still don't have a "front-and-center finder" worked out for my Mill-Drill, to allow me to raise or lower the head of the mill in the middle of a job, and then be sure it's still looking right down the axis of whatever I was doing before I moved it, but I have worked out a few things that ease the strain. Some of the other guys who use mill/drill units might find these ideas useful...

Whenever I have to locate a precise point under the spindle nose, I make a **"spud"**, much as I've used for years on a drill press. In the lathe, chuck a piece of drill rod (or CRS, if that's what you have handy) the size of your collet, or any size that's handy, if it's for use in a drill chuck. Indicate it to run dead true, and then turn a sharp point with about a 20° included angle, leaving a shoulder about an inch up.

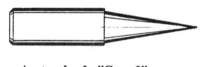

A typical "Spud"

I've made several spuds of different diameters and lengths to suit the various collets I have. When I need to line up over a center punch mark or small hole, I put a spud in the collet or drill chuck of the mill/drill, raise or lower the machine's head as needed, and then bring down the quill to center on the punch mark or hole, and lock the head. I then change over to a drill or milling cutter and I'm in business.

If the hole is larger than say 1/8", I take a bolt that fits snugly in the hole, or turn a button from a short piece of drill rod or scrap bolt to fit the hole snugly and center-drill the top, then use the spud to get lined up.

(In that case, I suspect a 60° point would be preferable. One can in some cases leave the job loose on the table, lower the conical point into the center hole - at which point the job will move into proper alignment with the spindle centerline - and then clamp the job down solidly._GBL_)

I also use a spud to check alignment of the workpiece with the table movement axes by lowering the spud to just a hair above the workpiece with a longitudinal line scribed on it, and wind the table from one end to the other, adjusting the workpiece until the line stays under the spud. Then clamp the job down solid, and you're in business.

Another use of the spud: when setting up to bore a hole or machine some other detail offset from a particular point, before I set up the workpiece I put a 1/2"ϕ spud slightly longer than the boring bar I will be using in the center hole of my boring head, with the latter's cross-feed at zero and with the boring head already in the machine. I place a small piece of flat steel with a center punch mark any place on it loose on the table and lower the point of the spud until it just enters the punch mark. I then rotate the head either by hand or at a slow speed and zero the spud point by adjusting the feed screw of the boring head. Once the point of the spud runs true, it (the spud) can be used to set up the workpiece, measuring from the spud point, or from a cylindrical replacement. This may strike some guys as pretty elementary, but I've encountered several machine-savvy people who'd never thought of doing it that way.

The "Sticky Pin"

There's another trick not maybe widely known. The Brits frequently refer to the use of a "sticky pin". Suppose you have an endmill in the milling machine, and you want to locate over a layout mark on a job, and then use that cutter on it. Why remove the cutter, or go to other time wasting steps, if not necessary? An ordinary sewing pin or similar can be stuck into a lump of plasticine (modeling clay), and the lump stuck on the end of the cutter. (Spindle off, thank you, for this!) Now switch on, and "milk" the pin so the tip is running true. Just as good as the pointy end of an edge-&-center finder, eh? Once you're located, pull the pin and the wad of clay off the end of the cutter, and start makin' chips.

●

MORE ON "OSBORNE'S MANOEUVRE"
from Bill Lowery, Wichita Falls, TX

Dear Guy,

I read with interest your notes on "Osborne's Manoeuvre" (**TMBR#2**, page 159). In the opening paragraph of that section, you give Mr. Herb Osborne credit for originating it. While I do not wish to steal any thunder from Mr. Osborne, I was taught this procedure by an old German high school machine shop teacher. I started my machine shop journey in 1944 in the machine shop at Central High School, Muskogee, OK, when I was 16. The old German's name was E.E. Klanke (pronounced Klanky). This old gent was something between a genius and a wizzard. Much to my regret I never got around to thanking him for his tremendous patience and the generous words of wisdom he shared with me.

With a couple of small exceptions explained below, Klanke's method was very similar to the way you presented it in **TMBR#2**.

To start with, we did not have an edge finder per se. And the machine was not turned on. In addition, he preferred I not use a drill chuck because it was not as true as a collet. My "edge finder" was simply a piece of drill rod gripped in the collet, plus a piece of cigarette paper.

The cigarette paper was placed between the drill rod and the job (say a Rotary Table) and the milling machine table moved until the cigarette paper was just tight enough between the two to be pulled through without tearing. Then, allowing for the thickness of the paper, the machine dial was zeroed and the procedure from there on was pretty much as you describe it.

Many years have passed and I don't recall any associated math such as you describe - all I remember is that I could readily see that his method worked. (Your explanation and drawing of same was great!) I still do it the same way today with one exception; I use an electronic edge finder - neatest little gadget you ever saw! Of course, in placing it in a collet, one must use some common sense, as the gripping pressure could crush the tubular device like a piece of macaroni.

This brings up the matter of "old" technology vs "new" technology (*), and who will replace the old timers when they go on to meet their Maker. They will take storehouses of hard earned knowledge with them. Now don't get me wrong - I'm all for the new technology. Some of the things I have seen at recent tool shows are absolutely mind-boggling. I recently attended one in Dallas, TX, and I felt like a kid in Toyland! It appears that gone are the days when a guy sat down to a drawing board and designed machines and tooling. It seems that all that is now done at a computer keyboard and I must say the results are truly amazing**.

> * (Here Bill drew my attention to HSM Sept/Oct 90, pages 6, 8, and 10 and the letters therein addressing the "old" technology vs "new" technology. GBL)

.... While serving my apprenticeship I was given a blueprint calling for a simple spacer of CRS, 3/4"φ x 3", with a 7/8" hole drilled through it's length. To even the most casual observer this was an obvious impossibility! However, our chief engineer was the guy who "signed off" all the blueprints before they were released to the machine shop. I realized that what he wanted was a 7/16" hole, not a 7/8" hole, as I already knew what the intended spacer was for. However, this guy was reluctant to let anybody tell him very much. And I must admit he was plenty sharp! A lowly apprentice boy would not dare to point out such a mistake to a man of such high esteem.

So I chucked a piece of 3/4" CRS and proceeded to drill the 7/8" hole through its length. I carefully collected all the chips, placed them in a Bull Durham tobacco sack, stapled it to the print, and forwarded same to QC (quality control). Of course, the QC boys couldn't wait to show our chief engineer my masterpiece.

The chief engineer decided he wanted my hide for breakfast. Next morning he came in the shop with the blueprint and product in hand. I pretended not to see him, and he marched to the office of the shop foreman. The two of them came at me looking mean and mad enough to stop a bulldog dead in its tracks. When he began to speak, I could feel my backside being ground to hamburger. But he couldn't pull it off. Instead, he broke up laughing, and simply said, "OK, kiddo, you got one on me." We shook hands and this was the start of a trusted friendship that provided a mutual exchange of ideas.

After that, I would on occasion get a call to his office. He would show me a sketch or preliminary print, and honestly and sincerely ask what I thought about this or that. And so it is - we learn from each other when we just stop and listen. Consequently I am still reading and learning and enjoying those machine magazines.

I am now retired, but I still enjoy my shop and machines. I am a skeet shooter and bird hunter and somewhat of an amateur gunsmith. I work with/for my shooting and fishing cronies, fabricating parts and/or modifying guns. We drink gallons of coffee and never lie about bird dogs, fishing or hunting -- well, not much anyway.

Bill

(This is not the first time I've heard about the 7/8 hole in the 3/4"φ bar, but this is the best told version. GBL)

> ** Do you want to hear about something that falls into that "truly amazing" category? How about making a "casting" from a polymer resin that hardens under UV light directed at it by a computer cogitating about a CAD drwg? As a rotating rod or something similar that the resin can adhere to is drawn up out of a bath of liquid resin, the UV light is directed at the liquid clinging to the rotating rod as it emerges from the bath. The UV light causes the resin to polymerize, and the liquid turns into a rigid material of the desired shape (from which an investment casting might be made, for example) as it is drawn up out of the bath. It may require more than one pass, but it would certainly fall into the mind boggling category, no? Apparently it's being done. Randy Orpe told me about this. GBL

MORE IDEAS FOR HAVING FUN WITH YOUR SINE BAR

If you frequently use a sine bar to set up a particular angle, or if you want to be able to set up certain angles without making up the necessary stack of gage blocks, why not machine, surface grind, and/or lap to size a single block of steel of the correct height to produce the desired angle(s)?

Better still, give the block different dimensions for length, width and height, so that the **one block can produce three different angles** useful to you.

Tim Flips his Sine bar

You will be well versed in the use of a sine bar for producing a desired angle (**TMBR#2**, page 76). But what if you have a job on the surface plate at some unknown angle, and you want to measure that angle - what to do then? **Tim Smith** pointed out to me that you can do it very easily, if the job configuration and size are such that you can lay your sine bar upside down on the angled face.

How to do it? Use your vernier height gage or similar to take a reading on the height of the top sine bar roll above the surface plate. Write down the reading. Let's say it's 7.493". Now move over and take a reading on the height of the lower roll of the sine bar - let's say it's 6.181".

Subtract the readings: 7.493 - 6.181" = 1.312.

Assuming you're using a 5" sine bar, the angle you want to know is now practically falling into your lap:

$$\text{Angle A} = \sin^{-1} \text{ of } y/R$$
$$= \sin^{-1} (1.312/5.000)$$
$$= \sin^{-1} 0.262,400$$
$$= \mathbf{15° \ 12' \ 45''}$$

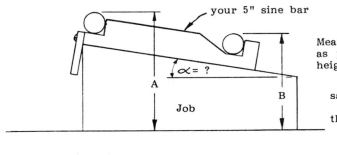

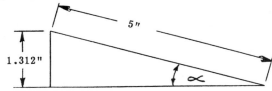

your 5" sine bar

$\propto$ = ?

A

B

Job

Measure A & B as accurately as possible, using a vernier height gage or similar.

say A - B = 1.312"

then 5" x sin$\propto$ = 1.312"

sin $\propto$ = 1.312"/5" = 0.2624

$\propto$ = sin^{-1} 0.2624 = 15.21252° = 15° 12' 45"

5"

1.312"

$\propto$

A LITTLE-KNOWN USE FOR A COMBINATION SQUARE

I have a fine old Starrett combination square, c/w center head and protractor head; this was in an old machinist's toolbox which I bought several years ago (**TMBR#1**, page 122). I have used this tool on many occasions, and while it is nicely made, and nice to have when needed, I never realized one of its possible uses until I encountered the following idea in a book called "Modern Toolmaking Methods" by Franklin D. Jones. This book is absolutely excellent, as is the companion work, "Accurate Tool Work" by Goodrich and Stanley. I have an original hardbound copy of the latter, circa 1908, bought from Bill Fenton several years ago. Thus, when I saw the book "Modern Toolmaking Methods" listed in the Campbell Tool Co. catalog (about $10) I knew immediately what sort of book it was likely to be, and therefore promptly ordered it.

Nor was I disappointed: it is a paperback reprint of the original book which appeared in 1915, and worth twice the price (as is "Accurate Tool Work", also available). And lest thee think to turn up thy nose at these old books, let me assure you that the fellas who wrote them, and the men they got some of their ideas from, did not, for a moment, need any lessons from us today. Those old boys knew what they were doing, and if you could emulate their work, well, I'd *almost* say you wouldn't need to be reading them funny books put out by that guy Lautard!

There is an interesting and unusual way of using a combination square to lay out an angle. I will use, because it is the easiest example I can bring to mind, the job of laying out an angle to be cut on the end of a piece of flat bar (it might be the end of a roof rafter, or it might be the end of a bar of metal to be cut thus, and then welded to another) as at right.

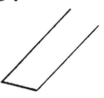

Let's say the angle required is 26° 19' (welded work is rarely so fussy, but let's just pick an oddball angle like that instead of something like 15°, OK?

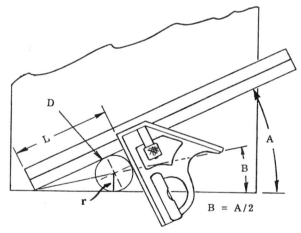

We need a disk "D" of some known radius, said disk to be placed as shown in the drwg at left. Then, by setting the square head a suitable (calculated) distance L from the end of the blade, we can have the desired angle "A" in the bag in short order.

$$L = (r/\tan A/2) + r$$

The desired angle "A" is 26° 19' = 26.3167°. Then A/2 = 13.1583° = B.

Now if we substitute "B" for "A/2", we can tidy up the formula somewhat, thus:

$$L = (r/\tan B) + r$$

and let's say r = 3/4"

Then $L = (0.75/\tan 13.1583°) + 0.75 = 0.75/0.2338 + 0.75 = 3.2081 + 0.75 = \underline{\textbf{3.9581}}"$.

If we wanted to increase L to say 5.6", we'd have to use a larger disk, which, if you calculate it, turns out to want to be one of radius r = 1.0611", = **2.1222"ϕ**.

Now what becomes immediately obvious is that we would much prefer to have **a few convenient sizes of disk** cut from say CRS bar, **at say 1/2", 3/4", 1" and maybe 2"ϕ**, and then set our blade length "L" to whatever our calculations indicate is required for a particular angle, rather than selecting an arbitrary blade setting that is probably little or no easier to make accurately, and then making a disk of some special ϕ that would produce the desired angle.

What would we do to lay out a 22-1/2° angle? Which disk, and what blade length L?

Let's run the calculation with a 1"ϕ disk (i.e. r = 1/2")

L = (r/tan 22.5°/2) + r =(0.5/tan 11-1/4°) + 1/2 = 0.5/0.203452 + 0.5 = 3.0137".

Having 9" of a 12" blade stickin' out the back end of the head of the square would probably not be very handy, so we might reconsider our choice of disk and go to the 2"ϕ disk. Then what would L be?

L = (1"/0.203452) + 1 = **6.0273"**.

Ok, that should be fine. Now let's consider the effect of an error (inaccuracy) in setting the blade to stick out the calculated amount.

Say we set it at 6" (i.e. off by 0.0273"). What angle would we get using the 2"ϕ disk we've decided to use?

Turns out the angle would be 2 x 11.0309932° = 11° 18' 36" x 2 = **22.6199°**.

(This would mean an error of about 1 part in 188.)

"Hey, never mind all that error stuff for a minute. How come you multiplied by 2 there, Lautard?"

Remember, we divided the desired angle by 2 way back when we started, so now we have to double our result to find out the angle we will get.

"Oh. Okay. Carry on."

Thank you.

What if we set the blade out to 6-1/32"? (This is "to the nearest 32nd," and obviously closer than straight 6" to the 6.0273" we ought to be at.)

Then, the angle we would get would be 22.482958° = **22° 28' 59"**.

(This time we have an error that amounts to about 1 part in 1320.)

Thus, by careful setting to the nearest 1/32nd inch graduation on the blade, we have come within about 1 minute of a degree of the desired 22-1/2° angle. A standard vernier protractor can be set to 5' of a degree, so we have likely beat that all hollow. Not bad!

Now, probably you've been thinking I'm being ridiculous, showin' all those numbers to the right of the decimal place. Well, yes, they do look ridiculous, but I just put them there to get the proper flavor of the calculation as the typical calculator sees it. If you start rounding off this type of calculation too early, or try to do all the calculations by rounding everything to 3 decimal places as you go, your answers will be booggered.

One of my guys down in the San Francisco area phoned me one evening about something or other, and we talked for a few minutes. He told me he'd never really understood trig in high school, but that as a result of the stuff on same in **TMBR#2**, he'd got quite interested in it. Motivated by the realization that trig would be useful to his shop activities, he'd got a book on it, and had learned how to do trig problems. He said he now uses it all the time. *Maybe somebody's readin' this stuff after all!*

Lubricating Milling Machine Spindles

I had a letter one day from Jim Callas, of San Rafael, CA. He'd ordered my booklet on oiling machine tools, but then wrote back with a question about how to lubricate the spindle of his mill. As I recall, he was concerned that his mill offered no means of oiling the spindle, and/or the outside of the quill. I wrote to him, and got back a short note as follows:

"Dear Guy.... Thanks for your splendid reply of July 10. I hate to think of you spending so much effort to answer my questions, but you could and should incorporate much of this info into a future publication. ...."

Jim went on to say that machine tool sellers in his area seemed interested only in selling machines, and not at all in providing useful answers to customers' questions after the sale. He said the most notable advice in the instruction manual that came with his mill was, "Not put (the machine) in the sunshine place."

I therefore decided to edit my reply very slightly, and put it in the sunshine place. Here it is:

Dear Jim:

Thanks for your note of July 3/91 re lubing milling machine spindles.

On my milling machine, which is a Taiwanese B'port copy under the name "FIRST", (sold as an "Alliant" in the USA, I believe), the drive pulley surrounding the spindle runs in well-greased* ball bearings, but there are also 3 other points on the head where one is to apply oil; 2 of them are fitted with Gitts-type oil cups. As for the type of oil to use, I would be inclined to use the same oil one would use on a lathe spindle. The FIRST owner's manual calls for the following:

1) clutch and bearing sleeve (on top of head, around top of drawbolt):- 5-10 drops twice weekly, of KUO-KUANG R68 or ESSO FEBIS K53.

2) Headstock matching Quill Holes: - fill oil cup twice daily with Gulf Way 52.

3) Countershaft Gear & Worm Gear Cradle:- Fill oil cup twice daily with VACTRA #2 or SHELL TONNA 33.

> * I know they're "well greased" because I helped grease 'em. When I first began to use the new mill, the head would frequently emit a disturbing whine. The dealer's service man came to see about it. He'd seen the problem before, and we fixed it: we pulled the spindle drive housing apart, dug the bearings out, and re-greased them. It still emits a high pitched hum once in a while, but that goes away if I switch it off briefly and then turn it on again.

My previous vertical mill was a "Kao Fong" Model KF-VMC. Its spindle ran in SKF ball bearings, but there was no provision for introducing oil or grease into the head anywhere. This always bothered me, although possibly without just cause. There was also no way to get oil into the headstock casting to lube the outside of the quill, which slid so elegantly in and out of same. Bob Haralson told me to lower the quill, wipe it spotlessly clean, and smear it with white petroleum jelly (i.e. vaseline), and then raise and lower the quill several times before raising the quill and wiping off any excess. This, he said, would be all it would ever need. How often? I suppose that doing this once a year or so would be fine. Now for the shocking part: I owned that mill from 1979 - 1989, and never did that, and the quill, although it looked dry (as opposed to oily) never looked like it needed or wanted any lube, and it moved as nice and slick when I sold it as it had when new.

I've seen other Taiwanese-built mills, more or less comparable to my Kao Fong, under other nameplates, which DID have oil cups on the headstock, presumably to feed oil to the spindle bearings, or the quill, or both. Certainly, all else being equal, one would prefer that the machine be so fitted, but I think my KF-VMC was a better machine overall, compared to the ones I saw like it that did have oiling provisions on the head.

Guy

Snow, Fire and Speedometer Cables
from **Joe Katz, De Land, FL**

Dear Guy,

Re the Kerosene Blowtorch you describe in **TMBR#2**: I've used one of these for a number of years. I inherited it from my father, and it was old when I was a little kid, before WW2. Hotter than H., it was always a bear to light and keep pressure up. When I started to use it regularly at the motorcycle shop (BMW of Daytona) to heat alloy heads for removal and installation of valve guides, it became obvious that I needed a better way to light it off.

What I did was to first throw away the hand pump, and replace the pump cap with a brass cap with a tire valve. Now I could pressurize it with shop air at 70-90 psi. ** Second, for preheat, I played a propane torch on the jet holder and the heating coils, and was soon in business.

*** you'd want to make very certain that you have a good sound tank before doing this!* GBL*

Replacing valve guides in alloy BMW heads requires the entire head to be raised to 400°F. The Kerosene torch would do this easily - in fact the first time I used it, I nearly melted the fins off. Kerosene, or diesel oil, has far more energy, in BTU's/gallon, than does gasoline. It just isn't as volatile, and thus is harder to light.

Cutting threads in the lathe: I've always wondered at the 'spring' tools recommended by the older books. It seems obvious that any spring downward would engage the flank of the thread - thus ruining the job. On my present lathe, a M300 Harrison 13", I use Aloris quick change tooling. The Aloris threading tool is as solid as the proverbial outhouse, and threading with it is a pleasure.

I frequently cut a thread in preference to using a die because it is so much fun. I keep my compound set at 90°, and feed directly in with the crossfeed. I start with a couple of fairly heavy cuts, 10 thou on the dial (= 5 thou actual), then taper off to two, then one, for finishing. I've cut threads from M8 x 1.0 to 5"-12 tpi, internal and external, with no trouble. I use liberal applications of "Hubers", a heavy, special purpose cutting oil.

Chain drilling: at page 125 in **TMBR#2**, you mention this topic, Guy. The fellas might like to know about Starrett's #118 Spacing Center Punch, which is helpful in this task. The #118 is a typically super Starrett center punch with a spring leg on the side. The spring leg has a spring loaded point that is set in the first punch mark to locate the next. The spring point permits the actual punch to be held vertical for best control. Spacing can be from 5/64" to 3/8". (As a result of Joe's letter I bought a #118. I haven't used it yet, but one thing it will be used for is laying out uniformly spaced rivet holes on a project that I hope to include in **TMBR#4**.GBL)

Back in the early '50's I worked for about a year in a small shop in upstate New York - Rogers Machine Works - in Alfred, NY. Rogers' sole product was the Rogers "Perfect 36", a vertical boring mill with a 30" chuck and a 36" swing. A wonderful place, old fashioned integrity, every man his own inspector - if the piece was right you put your own "mark" on it; if it was wrong - you scrapped it. I have many stories from my time there. Let me see if I can tell one or two.

There were a number of youngsters working at Rogers, all about my age (early 20's), and most of them were practical jokers. One morning I noticed that two of the most dedicated, and original, jokers were missing. I suspected that they were up to something and, since I didn't want to be the goat, I went off to look for them. They weren't up in the overhead (most of our machines were belt driven from overhead line-shafts), and they weren't on the shop floor. They weren't in the office, or the grinding room, the tool crib, or the lunchroom. I searched for them every time I started a long cut and could safely be away from my machine. About 2 hours after I had first missed them I found out where they had been. As an office worker - the purchasing agent as I recall - came walking through the shop, at the intersection of the two main aisles a giant snowball fell from the heavens and engulfed the poor guy. It drove him to the ground, and though it inflicted no permanent damage, he nearly froze before he could fight his way free of all the snow.

Our shop had a gently sloping roof with a clerestory section running above the main aisle. The jokers had been up on the roof for several hours rolling this snowball to a size that would just fit through one of the 4'-square windows in the clerestory. They then waited above the main intersection for a suitable pigeon to walk by down below. When the victim appeared, they swung the window open and rolled the snowball in. As I recall, the only punishment they earned was the job of shovelling up and removing the snow.

One of our old-timers was a fastidious fellow who always wore a tie with his chambray work shirt. On the job it was always inside his shop apron and presented no danger, but it was a challenge to us youngsters. We made several attempts to snip it off, but the old fellow was just quick enough that, short of overpowering him, we could not dock his tie. Our final strategy involved about ten of us. Just as lunch break was ending, we got into a pushing, shoving, laughing, tripping, pre-arranged free-for-all. As this revolving mass engulfed the tie-wearer, his main thought was to keep from being trampled, and he forgot to guard his tie. One of the boys whipped out a pair of shears and snipped at the tie - but the shears jammed halfway through the cloth. For all these months, the old fella had been wearing a speedometer cable threaded through his tie to protect against just such an attack! We gave him best for that, and he wore his tie with honor from then on.

Joe

--~f~---

A METHOD OF BLACKENING OF STEEL
from Claes G. Halfvarsson, Lillhardal, Sweden

When I want a nice deep black coat on my homemade tools, or on gun parts I make for my friends, I do as follows:

In a tank made of black iron (not stainless), for each liter of water I put in as follows:

1.1 kg NaOH	(Sodium hydroxide)	
0.55 kg NaNO$_3$	(" Nitrate)	
0.55 kg NaNO$_2$	(" Nitrite)	

Added to the 3rd printing: When using this solution, one must add water from time to time to replace water which boils away due to the high heat involved. *One must be exceedingly careful* when adding water to the blueing solution, because the caustic solution wants to boil the added water, and make it spit back out of the tank. GBL

Heat the above solution up to 140° Celsius (284°F).

Put the parts to be blackened in a wire basket, or otherwise provide for a secure means of holding them. After 5 to 10 minutes, take the parts out and have a close look at them. This amount of time is usually enough.

The temperature is quite critical - if it is not right, I find I get a red coat over the whole piece.

From the "Don't Overlook" file:
Galvanized sheet metal is available as scrap in endless quantities at any plumbing-and-heating shop, and can be made into various useful items for around the shop, and the rest of the house. If you buff it up with superfine steel wool, it looks pretty nice, or you can paint it.

Added to the 5th printing:
I've been told that if you want to paint gavanized sheet metal, the thing to do is to etch it with vinegar first. I haven't had occasion to try it yet.

GOODIES AT ROCK BOTTOM PRICES
from **Charles Thornton, Readyville, TN**

Dear Guy,

A ready supply of drilling, cutting, threading tools, etc. can often be had from tool and cutter grinding shops. They almost always have used or abandoned tooling at rock bottom prices. I have equipped my entire shop this way.

Shop equipment - lathes, mills, drill presses, etc. - can be found at machine tool rebuilders, and can be bought used, abandoned or reconditioned. Also, the fellows in this business can do your shopping for you if you tell them your needs.

Department of Defense surplus sales (via sealed bids) can be very lucrative, with equipment often selling at scrap prices!

If you have a friend who is a diesel engine mechanic, obtain several fuel injector plungers from a Cummins or Detroit diesel. These plungers can be ground into some of the best center punches and prick punches you have ever used.

Concrete reinforcing steel (re-bar) makes excellent punches, and turns well (after chucking and smoothing outside). It is usually about 40-45 Rockwell-C as is.

Last, I have obtained a number of measuring devices from a local retired instrument maker. Seems that he and his associates have accumulated lots of used/broken instruments to repair, to ward off boredom in their retirement. My new friend "Sarge" was happy to equip me, and give the old tools another start in life.

Charles

Working Up a Good Center

What's the difference between a prick punch and a center punch? Bill and I got talking about this one day last Fall. Everett Arnes had said something to me about it, and Bill reminded me of it all clearly enough to permit me to elucidate the matter here.

A prick punch has a much finer point on it than a center punch (c/p). The latter has to be blunter, for strength, and to form the sort of indentation that will guide a drill point.

Suppose you want to place a c/p mark at some particular point on a fussy job. Just close won't do. The first thing to do is to scribe in layout lines. Where these intersect is where the c/p mark is to be. If you had one, you could then use an optical c/p (see below re same), but if you didn't, then the next thing to do is to take a sharp prick punch (or your scriber), and lightly drag the point of same along one layout line until it drops into the intersecting layout line. You will feel this occur, and perhaps hear it, also.

Now lightly press the prick punch down onto the material where you think the c/p mark should end up. Hold the prick punch vertically while you do so. Now check the result with a magnifying glass.

> (Starrett makes a very nice little layout hammer, complete with a little magnifying glass built into the head. Very cute, but practical, too. Phil Lebow sent me one for Christmas '90. I just about flipped when I opened the package. Then I turned it over and found he'd had my name engraved on it! At that point, you coulda knocked me over with a layout hammer.)

If this preliminary mark is well located, you can proceed to deepen the mark with a light tap on the prick punch. If it still looks ok, go after it with a regular c/p, and deepen it to suit.

The next thing to do, since this is a fussy hole, is to draw out one or more concentric circles using the newly located center and a pair of well-sharpened dividers, and then put a light c/p mark at each of the 4 points where the original layout lines and the max. scribed OD of the hole intersect.

This is done so that when you begin to drill you can see if the conical hole the drill is making is concentric with the desired center location. If it is not, you want to know while only the conical tip of the drill is still in the surface of the job - once it is in to the full ϕ of the drill, the hole location is pretty well a *fait accompli* (which, being loosely translated, means that by that time there's nuthin' you kin do about it any more.)

Any eccentricity of the drilled hole with respect to the scribed circles will stand out like a sore thumb. If not concentric, you can use the c/p or a cold chisel to make a mark down the side of the conical depression being started by the drill. The cutting lip of the drill will bite deeper into the material at this mark, and thus draw the drill over in that direction. There's no need to try to draw it all the way over at one go, either.

If you proceed carefully, as outlined above, you should be able to work up the starting of a hole on a desired location very accurately by what are essentially hand methods.

Of course, all this is much easier if you are setting out to drill a 1/2" hole than if it is say 3/16".

A Small Center Punch

0.300"∅
5/16" sq.
0.312"∅
5°
0.106"∅
1/8"
5/16"
1-5/16"
1-1/4"
2-7/8"

Make from 5/16"
square drill rod

Bill has a whole mess of center punches in a block the size of a coffee can - all for different jobs. I don't have so many, but I have made a couple of small ones, for fine work, along the lines shown in the drwg at left. You can make the point to suit your wishes, from a fine prick punch point, to a conventional 60° conical point. I've made them from some 5/16" square drill rod. (Square drill rod, or what is effectively the same thing, is available from Brownells and other outfits.)

The first time I was down to visit Stewart Marshall, he showed me a very nice little punch for aligning a center punch mark in one workpiece with a hole in another, or transferring a hole center from an existing hole (in say a clock frame plate) into the piece below it, which you'd presumably put there and got all nicely lined up with the top one the way you wanted it. I could see that this would be useful to me, so when I got home, I ordered one from a large mail order machine shop supply house. When it came, I was less than happy with it -- it was certainly not in the same class as the one Stewart had. Back it went, and I ordered one from MSC (Cat. # 06488506; see bottom right corner, p. 962, in the 91/92 MSC cat.) This one is the same as Stewart's. Cost was about $2 more than the one that got sent back.

I mentioned using an **optical center punch** above. This is a real aid to accurate layout work where hole locations are to be center punched. Mine is the On-Mark brand, and I like it. In use, the cork-faced base piece is set over intersecting layout lines on the work where a center punch mark is wanted. The exact point of intersection is centered under the dot visible in the magnifying optical insert. The optical insert is then removed, and the hardened steel center punch dropped into the vacated hole. When this punch is thwocked with ye layout hammer, the resulting mark is about as close to where you wanted it as can be achieved by any hand means. Mine resides in the top of my toolbox in its little black and white storage bottle, along with other frequently used tools.

CHOOSING THE RIGHT GOODIES

Have you ever wondered why Starrett has 3 different patterns of spring type calipers in their line? The 3 are the Toolmakers' (with round legs), the "Fay" (square legs), and the "Yankee" pattern. The latter are made from flat stock, are offered in 6 sizes from 3" to 12", and are the least expensive. The Toolmaker style is for work (or users) where a fine, delicate feel is particularly wanted; 4 sizes from 2 to 6" are made. The Fay pattern are a little heavier duty, but are only offered in 3" and 6" sizes; I have a 1938 Starrett cat. in which the Fay pattern are shown in 6 sizes from 2-1/2 to 8".

A comment in that old Starrett catalog bears noting here: never caliper work while it is rotating. Quite aside from any possible safety considerations, such a measurement will not be accurate, and in most cases will be very misleading. The reason is that it takes only slight force to spring the legs of a caliper; if the tip of one leg is brought into contact with a rotating workpiece, and then the other, the friction with the work will likely draw the second tip of the caliper over the work, giving an entirely false reading. It seems elementary, but I have seen men do it the wrong way. A very delicate touch can tell you much with a caliper of this type, but handled wrongly, they will tell you nothing.

BB bought several types/sizes of Starrett calipers, after digging through my collection of firm joint and spring bow calipers. One he got was a 6" #42 Lock Joint hermaphrodite caliper. I never noticed the fine adjust feature mentioned in the catalog description for this item (and I see it applies to several other similar in the line), but I like it - it sure beats whackin' a regular firm joint caliper against everything in sight to bring it to a desired setting.

We were speaking of using a combination square for laying out angles elsewhere herein. Have you ever noticed that big #8 combination square that Starrett makes? No, it's not just a 24" blade in a regular size head - it's about twice the usual size *all over*. Take a look in the cat. and see it.

Something else that bears mention in connection with this same topic is the matter of which type of heads to choose when buying a combination square - cast, or forged and hardened? The cast heads are fine if the tool is to be used for woodwork, but for metalworking applications, always go for the forged and hardened heads. If you were to buy the cast type, and give it 6 months steady use in a sheet metal shop, you'd wear a groove in the working face of the square head, effectively ruining it.

Granite Surface Plates - Black or Pink?

According to Starrett, black granite is the least porous type of granite. It is also stiffer than pink granite, hence is the better choice if you want a plate to stage heavy items on for inspection. Pink granite contains quartz (black doesn't), hence is harder, and thus more resistant to wear. So pink is the better choice if you're going to use it as a layout plate, sliding tools around on it all the time. Interestingly, a Starrett Crystal Pink plate of a given size costs less than a Starrett black granite ditto, and Starrett pays the shipping anywhere in the Continental US (this I just found out from J&L Industrial Supply). Starrett Master Pink plates are not made in Toolroom grade, which grade is ok for virtually anything for which any hsm is ever likely to want a surface plate.

Long *vs.* Short Level Bubbles

Walt picked up a very fine used Starrett machinist's level at a garage sale. The bubble seemed kinda short to him, and he asked me about it when I was there one day, since I have one that is similar, although is by Moore & Wright, not Starrett. I asked the local Starrett rep, and he told me the vials in these levels are filled with alcohol. As the level is warmed, the liquid expands and the bubble gets smaller. They are checked at 70°F at the factory. Many's the carpenter, etc. who has left his tools overnight in the truck and found the bubble real long in the morning.

Shop-made File Handles that FIT Your Hand

Bob Patrick is a industrial arts teacher here in Vancouver. He wrote an interesting article in *Projects in Metal* (Dec.'89, page 8) on how to make file handles that **fit *your* hand.** The basics of his presentation are shown here, with permission of the author and Joe Rice, Editor of *Projects in Metal.*

Now obviously, you can buy file handles that are not bad, and they're not expensive, either. But you can make handles that really *fit* your hand.

Not worth the bother, you say? Maybe. But if you ever try a handle dimensioned just exactly right for your hand, I doubt you will ever be entirely satisfied with store-bought handles again.

And since whatever handle you elect to put on a file is the one that will probably stay on it for the rest of your life, why not make handles that suit your own hand?

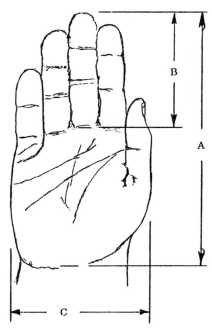

$$L^A = C + 1.88"$$

$$L^B = L^A - 2.75"$$

$$D = (A - B) \div \pi$$

For GBL's hand, **A = 7.63", B = 3.25", and C = 4.5"**

Therefore

$$L^A = 4.5" + 1.88" = 6.38"$$

$$L^B = 6.38" - 2.75" = 3.63". \text{ and}$$

$$D = (7.63" - 3.25") \div \pi = 4.38" \div \pi = 1.39"\phi$$

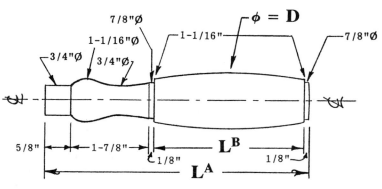

The drwgs and formulae above show the process Bob developed to dope out a hand-pleasing shape for file handles. Measure *your* hand, plug the numbers into the formulae, and then make your handles to the dimensions indicated by your calculations.

Now before you go turn up a bagful of them, stop and think for a minute what you're going to do for **a good ferrule**...

Stewart brought along several file handles he'd made just before he came up to visit one time, and they were nice. He turned them from Lignum Vitae (Latin for "living stone", so called because it is so hard and tough. It also turns like cold butter, but don't breathe the dust from it!) Stewart had made these handles on his great grandfathers's old Barnes pattern maker's lathe, which - although about 110 years old - is still capable of working to half a thou.

Stew uses **brass compression nuts as ferrules.** These can be got at any good hardware store, automotive parts store, etc. Turn the ferrule end of the handle blank down to just about the ID of the compression nut's thread, smear some 5 minute epoxy into the nut, and screw it onto the file handle blank. Turn the OD of the nut to suit your own tastes, and then turn the rest of the handle to fit your hand, per dimensions arrived at from the procedure above. Put the first one on your favorite file, and try it - feels like the handshake of an old friend when you pick it up, don't it?

File Cleaning

I had a call one evening from Bob Loveless, the well known knifemaker in Riverside, CA. He told me he's started keeping his files wet with oil lately, having read same in TMBR#1 (pages 7 and 8), and has found that about the best way to clean files is on a fine soft wire wheel on a bench grinder. He says it cleans 'em up like nuthin' else - he's chucked out all his file cards.

An Unusual type of File
you might find useful to know about

Perma-Grit files are made by braze welding Tungsten carbide grit to a steel backing. They can be cleaned with any solvent, and - where the steel backing is not bonded to a piece of wood, as in some cases they are - you can clean them with a propane torch if need be (e.g. suppose you got one loaded up with paint, or resinous wood, or ??). Exactly what place they might have in your shop I'm not sure, but you can't use them if you don't know about them. The maker is D.G. Products Co., 1302 Berwin Street, Dayton, OH 45429. One source is Micro-Mark, 340 Snyder Ave., Berkeley Heights, NJ 07922-1595.

———— o ————

Not Everybody Has the Same Interests

I am acutely aware that many of the people who read my books possess metalworking skill levels far beyond anything I will ever attain, and I well know they don't need my instructions for making a drill sharpening jig, or a knurling tool. Nevertheless, (even) such guys buy my books, and then phone or write to say they enjoyed them or found some useful idea(s) therein. This I take as a rich compliment, because I never set out to write a book a master tool and die maker would find of interest.

I'm also aware that not everybody who reads my books shares my interests. The following letter, from a reader who asked that his name not be used, illustrates the diversity of activities and occupational orientations of "my guys". Hopefully, you'll also find an idea or two buried herein that *you* can use.

Dear Guy:

I'm a farmer. I don't build clocks, model engines, or do foundry work. I use my shop equipment mostly for rebuilding farm machinery. My first lathe was a 12" x 36" Atlas (Craftsman), bought in 1966. I now also have a Taiwanese Chin Chung VO-AIS vertical milling machine, which I bought a few years ago. I'm still a beginner on that. Once in a while I make a tool or puller for a special need. The following ideas may be useful to some other readers of your books.

If you need a long 60° center for a lathe headstock, chuck a piece of bar and turn a 60° point on the end. Leave it undisturbed in the chuck, and use it as you would any other center.

If you ever have to pull a pin from a leaf spring in a truck or trailer, cut a strong wood (or steel) bar to fit between the spring's end rolls. Then remove the ignition key, jack the vehicle up, and block it up safely. Then put the wood or steel bar into place between the spring's end rolls as shown (see drwg below).

You can pull the pin out easily when all the tension is off it, using a slide hammer adapter that will screw into the grease zerk hole. This idea is a labor saver if the vehicle provides no room behind the spring end so you can get a punch in there behind the pin to drive it out.

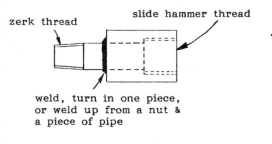

zerk thread   slide hammer thread

weld, turn in one piece,
or weld up from a nut &
a piece of pipe

To make sure you don't strip the zerk thread, which is a tapered pipe thread, screw this piece tightly into the zerk thread.

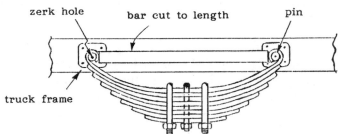

zerk hole   bar cut to length   pin

truck frame

Something that would be very useful to many mechanics would be info on how to build an instrumented "test bench," with provision to substitute parts on a grid board wired in to a working alternator, so you could test other alternators by substitution of parts. The drive could be from any machine with sufficient power and speed, plus it would require a 12 volt battery. It should have meters to check AC voltage, DC voltage, AC amps, etc.

Dust is the enemy of the Delco-Remy alternators (12 volt/60 amp) used on many farm tractors. They're also often used on automobiles. (Delco-Remy alternators usually have a hex recess in the end of the shaft.) I buy used alternators at machinery auctions. I take them as they come, and rebuild some from others by cannibalizing.

Speaking of auctions, I like to take my camera along to farm auctions. If I see an idea I like, I take photos, because I can't afford everything I like, even at auction prices. From a few snapshots I can get a good idea of sizes, ratios, etc., in case I later decide to build such an item.

I don't have any patents or special projects I have made. I guess you aren't into farming at all*, but here's the address of a paper you'll like very much. *(FARM SHOW, P.O. Box 1029, Lakeville, MN 55044; phone (612) 469-5572)* No advertising, but they do show a lot of farmers' shop-made stuff - genius at work! Some new farm equipment just on the market, some miniature tractors occasionally.

Finally, here's a tip you won't find in any lathe manual:

When drilling from the tailstock with taper shank drills in the lathe, always put a taper shank plug in the lathe's (empty) spindle nose taper. If you don't do this, and if the bit should be pulled out of the tailstock, it can jump forward into the headstock spindle socket, with ruinous consequences.

"GDP"

* You might be surprised.... People's lives take strange turns. I have a degree in Agricultural Mechanics, and I spent many summers working on an uncle's ranch. I would not trade my memories of those happy summers for a barn full of the finest machine tools in the world. GBL

As for the idea of a test bench for alternators, my friend Jake Wiebe built one which he uses on a regular basis at work. An alternator can be clamped to it, and driven by an electric motor and a V-belt, for testing. If enough guys were interested, details could be included in **TMBR#4**. Let me know....

ON MAKING & FLUTING CUTTERS, AND ON PARTING OFF

One of my guys in Australia wrote to me asking for more info on making and fluting cutters, and on parting off, which he says always gives him the vapors.

The following, from my reply, which he seemed to like, might be of some value to others.

Dear John,

You asked about making cutters, more specifically how to mill in the flutes and then how to relieve them.

Make a "cutter block" - that's what I call it - as at pages 86 & 101 in TMBR#1. Use it to hold your cutter blank. The "cutter block" allows you to readily "index" a cutter blank held in it for milling 1, 2, 3, 4, or 6 equally spaced flutes.

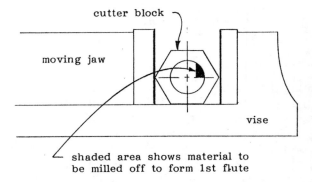

cutter block

moving jaw

vise

shaded area shows material to be milled off to form 1st flute

Make up your cutter blank. Stick it in a cutter block. Stick the cutter block in your milling machine vise, horizontal, with the business end of the cutter blank (i.e. the part you want to flute) sticking out past the side of the vise. I like to have it stick out to the right of the vise.

Let's say you want a 3 fluted cutter.

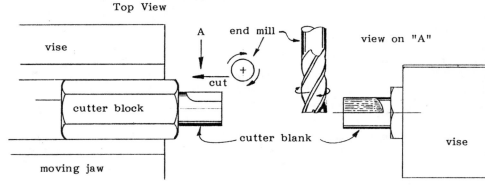

Top View

vise

cutter block

moving jaw

A

end mill

cut

cutter blank

view on "A"

cutter blank

vise

Use an end mill to mill in one flute. See drawing at left. The length and depth of this cut are determined by calculation or from a large scale drawing of the cutter - anywhere from 2 to maybe 10 times full size.

Having milled the first flute, turn the cutter block say 120° (= 2 flats on your hex cutter block), and repeat, going in to the same depth, etc. Rotate the cutter block 2 more flats, mill the 3rd flute, and there's your fluting all done.

(1) file secondary relief

(2) file primary relief

milled away in previous stage

leave a land about 0.020" wide not touched by your filing operations

(3) stone by hand after heat treating

You may wish to mill in the secondary relief, as well. If so, rotate the cutter blank in the cutter block as necessary, re-tighten the set screw, and mill some more.

Quite often - more often, I would say, particularly for small cutters - you will hand file the necessary reliefs. If you're going to do it this way, leave the fluted cutter blank in the cutter block when you finish milling the flutes. Blue the fluted end of the cutter blank with marking out blue, and put it at some convenient orientation in your bench vise (see **TMBR#1**, page 39) and hand file the reliefs so the cutter will cut.

The layout blue lets you see where you are filing, and can be renewed at any time during the filing operation if need be.

The above advice should be helpful. If you take a whack at it, you will very soon get the hang of it.

As for the essentials of parting off, they are these:

Grinding a Parting Tool Blade

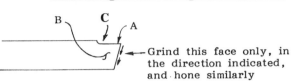

1) Cutter sharp, and properly relieved. If using an Eclipse type parting blade, grind it on the front only. Never grind it on top (a) (as a rule) or on the sides (b). The only time to grind the top is when the front has been ground back to (c), or if you need to because the job diameter is too big for it the way it is when it comes to hand.

Grind this face only, in the direction indicated, and hone similarly

Don't grind at A or B (as a rule - but there are exceptions)

2) Cutting edge exactly at center height or a hair below.

3) Minimum overhang (chuck/job, and toolpost/parting tool) sufficient to do the job.

4) Tool axis parallel to tool travel. (I set the tool by eye, and then widen the cut by withdrawing the cutter, moving it sideways, and advancing it into the cut again. Fussier tool setting is sometimes required.)

5) Speed moderate. I would use about 200 rpm for parting say 3/4"ϕ steel, and I'd probably jump the speed when the work diameter drops below 3/8". As a rough guide, try 1/2 of whatever speed you would use to turn a job of the same diameter.

6) Oil the cut. I use an acid brush (10 cents each when bought a couple of dozen at a time, and each one lasts for a long time) or an old toothbrush, plus heavy brown cutting oil.

7) Steady feed pressure. If it wants to chatter, drop the spindle speed and/or *increase* the infeed pressure a little. It takes a bit of experience to know whether or not to up the infeed pressure.

If the above are all copacetic, you should have no trouble. I use my parting blade in a Quick Change Toolpost I made which is a copy of a Myford Dickson QCTP. (I mention this by way of saying I don't use a rear toolpost; which is not to say that I do not think the latter are a good idea. All I'm saying is that I get along fine without one.) I think I've snapped a piece off the end of a parting blade maybe 3 times in 17 years.

Snapping a parting tool usually results from chips jamming in the cut, or feeding the tool too hard. Stop and withdraw the tool, and brush the chips off with the acid brush if they don't clear out of the cut on their own, which they probably won't. The chips (in steel), when they are coming off nicely, are tightly rolled, or partly broken curls.

Guy

Tim's Getting Sharper

Tim Smith bought a Glendo Accu-Finish unit, which you'll have seen advertised in HSM, and he says its slow turning diamond wheels allow him to put an edge like a straight razor on lathe tools, with the result that they produce a far better finish on the work. He's right - I got one too. I like its ability to grind known angles on tools, and to deal indiscriminately with carbon steel, HSS, carbide, etc.

HOW TO CUT THREADS WITHOUT A THREADING DIAL

A lot of machinists think that if a lathe lacks a thread cutting dial, you have to keep the halfnuts engaged, and reverse the spindle to move the carriage back for the next pass, throughout the entire task of cutting a particular thread, so as not to lose the relationship of the saddle to the leadscrew, which the threading dial otherwise lets you re-establish at will.

I've heard that lathes with metric leadscrews do not lend themselves to threading dials at all, or they do so in only a limited and recalcitrant manner. Whether this is true or not I don't know.

My Super 7 is not metric; it has a threading dial, and it works like a charm. Having to keep the halfnuts closed from start to finish of a thread cutting operation would be a nuisance, so far as I'm concerned, so any method that obviates the need to do so is worth knowing about. Here's something to try...

Before you start the first pass of a thread cutting op, put a stop on the lathe bed to the right of the carriage. Or use a mark on the bed, either one made with say a piece of chalk, or made by the footprint of the carriage in the oil on the bed. Or use the tailstock as a saddle stop.

Next, with the saddle up against this stop, turn the chuck over by hand until the halfnuts drop into engagement with the leadscrew comfortably.

Having done this, put a chalk mark on the chuck, and another on the leadscrew handwheel dial, and for each of these marks make a reference mark on an adjacent part of the lathe.

Start the lathe, and make your first pass down the area to be threaded. Pop the halfnuts at the far end, and move the carriage back to your stop. At this particular position along the lathe bed, the halfnuts will only drop into place on the leadscrew in one relationship with the thread on the leadscrew, or will be out of phase with it by one full turn of the l/s. Either way you're in business again, so far as I can figure.

Here are a couple of other things that you might not know about screwcutting...

If you put the topslide over to 30° and put the cut on from the topslide when screwcutting, here's how to establish, in advance, how far in to run the topslide to reach the necessary depth to get a complete thread:

1. Advance the cross slide to a convenient point that brings the tip of the tool close to the surface to be screw cut, and zero the cross slide feedscrew dial. I like to have the handle on the cross slide dial at 3 o'clock when the c/s dial reads zero, because I can return the c/s feedscrew handle to this position most quickly and easily.

2. Advance the top slide until the point of the tool touches the work, and zero the topslide dial.

3. Back up the c/s about 3/4 of a turn, or more if need be, and then advance it until you are as far out from the OD of the job as the depth of the thread relief groove (i.e. 1/tpi x sin 60°). For a 14 tpi thread, this would be 1/14 x 0.866 = 0.062", so (on my lathe) I'd advance the c/s until the dial read "38", which would be 0.062" out from the workpiece OD.

4. Advance the topslide until the tip of the tool again touches the job. Note the reading. It should be about 0.071" for a 14 tpi thread, but it will not necessarily be exactly this amount - it may be, but it will not necessarily be.

5. Back the topslide out again past zero, and then bring it back up to zero. It will now not be touching the work.

6. Advance the c/s feedscrew until its dial reads zero (at which point the tool tip should just touch the work), and begin the screwcutting operation. When the topslide dial reads 0.071"

again, you should be done.

Ease Up on the Last Cuts
Bill Fenton once told me to take about 5 thou at a pass in the early cuts of a threading operation, with one or two "free" cuts (with zero tool advance) every so often to work the spring out of the tool. As you get down to the end of the job, reduce the cut to 4 thou, then 3, then 2 and 1 and 1/2 a thou.

On a heavier lathe, another approach is to rough out the thread as quickly and in as few passes as possible, thus leaving yourself plenty of time for the fussier work of getting it right on the final size, and with a nice finish. If you squander your time on roughing out of the thread, you'll have little time for the fussy part, or you'll take too long to do the job, which will make whoever is footing the bill quite unhappy. This of course concerns the guy who's doing it for money more than the guy who's doing it for himself. But there is a principle here that bears thinking about.

(There's another principle that bears keeping in mind at the same time: If you bust the toolbit tryin' to rip the stuff off too fast, you'll lose more time regrinding the tool and picking up the thread than you might ever gain by rushing the job. Knowing how heavy a cut the tool can take requires experience.)

Getting Re-Established
You can take the toolbit out of the lathe, regrind it, pop the halfnuts, disconnect the gear train from the spindle to the leadscrew, and otherwise do everything you might think you ought not to do in the middle of a screwcutting operation, and still come out with a whole skin. The trick is in knowing how to re-establish the tool tip in the cut with all the dials set where you had them before you had to pull out.

Say you got all knocked into a cocked hat as above, *but before you backed off either feedscrew, you noted the readings on the dials.* Put the work back in the chuck. Put the tool back in the toolpost. Get it back square to the job. Get the halfnuts back onto the leadscrew. Roll the spindle forward far enough to get the tool tip somewhere along the thread you were cutting, and get rid of all the backlash. Now juggle the c/s and topslide feedscrews until the tip of the tool is back in the V of the thread. Never mind what the dials say while doing this. What you want is to get the tool where it'd take off a 'tenth' or less if you breathed on it, but nothing if you didn't. Now, when this happy state is achieved, re-set the dials to their previous readings, and carry on.

I learned this when I was cutting a thread for something for a chap in our church who is what I would regard as overly optimistic about what the Good Lord will do for anyone who asks, no matter how stupid the request. I wasn't sure the thread was deep enough, and I had no way to test it, because the mating part could not be got near it while it remained in the lathe. I explained this, and said that if we took it out and it didn't fit, we were in trouble.

"Well, pray for wisdom, and take it out."

"In this particular case, Bert, anybody who would do that would have to be about the most naive suckah since Alice went down the rabbit hole."

But I took it out of the lathe... and it didn't fit. I stood there wondering what to do next, and all of a sudden it came to me that I could pick up the thread as above. *There's probably a lesson in there somewhere...*

I had some other tips on screwcutting in **TMBR#2**, page 111/112. On the next page, you'll find some more thoughts on the matter.

Cutting Metric or Special Threads on an English Lathe
from **Dennis Danich, St. Paul, MN**

The problem with using transposition gearing (i.e. with a 127 tooth gear in the spindle/leadscrew geartrain) when cutting metric threads is that it is difficult to engage the halfnuts at the exact spot you first started the thread. It is possible to close the halfnuts 32 times on a standard eight pitch leadscrew, with an 8-division threading dial. Therein lies the problem.

I suggest proceeding as follows:

1. Set the lathe to cut the desired thread.

2. Position the tool ahead of the starting point of the thread. (Back the cross slide out a bit first.) Now jog the motor till number 1 comes up on the threading dial. Close the halfnuts, wiggling the saddle a bit left or right to help, and then position the micrometer stop on the right side of the saddle and secure it to the lathe bed.

3. Open the halfnuts, adjust the tool to the work, and zero the cross slide dial. Return the saddle to the stop. Start the lathe, and when the threading dial comes to 1, close the halfnuts. When you come to the end of the portion to be threaded, open the halfnuts, back off the cross slide, return the saddle to the stop, re-zero the cross slide, infeed the compound, close the halfnuts when the dial is on number 1, and take another cut.

4. Continue thus until the thread is finished.

Using the Saddle Stop for Odd or Difficult Threads
I also use this method for cutting an internal thread to a shoulder. I cut these threads on the back side of the work, with the lathe running in reverse, feeding out. To cut a left hand thread, flip the tool over, cut on the back side, and feed out with the machine running forwards.

Using this method, a thread can be cut right up to a shoulder, without the danger of a smash-up, because when you return the saddle to a stop and cut away from a shoulder you have an easier time with the halfnuts.

I also do all my boring on the back side with the tool upside down.

Dennis

------- -------

IN CASE YOU EVER HAVE TO MACHINE COPPER....

I had a call one day from one of my guys in upstate New York. He told me he has a little machine shop in his back yard, wherein he makes his living. Somebody had brought him a job which involved turning down a slug of copper which had been swaged onto an aluminum shank. After a while he'd tried about every trick he knew of to get a nice finish on the copper, in turning it, but no luck. (This fella is about 70, and has been in the trade since before I was born, so he has a certain amount of experience to draw upon!) Finally, he told me, he remembered that in TMBR#1 (page 18) I'd said milk was the stuff to use for tapping copper, so he zipped into the house, got some milk, and zipped back out to the shop, where he daubed it onto the job with an acid brush, and..... "Lautard, you would not *believe* the finish I got!"

Bill Fenton says that the brown sulfurized oil (Oster oil) used in pipe threading machines is also good for turning copper (commutators for example). Use a round nosed tool and don't go too fast, and you can get a nice finish. The sulfur in the oil turns the copper black, but this can be cleaned off afterwards.

MULTIPLE IDENTICAL CASTINGS IN EPOXY

I was down to see BB one day, and he showed me something that could be useful to know about. He needed multiple patterns from which to have castings made for his "B/Mount", a very slick model aircraft engine mount he developed and now makes. He had made the original pattern from wood, and had had several aluminum castings made from it at a local foundry. He now wanted a pattern that would allow the foundry to make a sand mold to produce the mount castings 10 at a time.

Rather than waste time trying to make several new wooden patterns, all identical, he milled a cavity in a stack of 3/4" thick Medite about 5 layers deep. The cavity was just a rough shape big enough to leave about 1/4" of air space all around his wooden pattern. He attached the latter to another piece of Medite, waxed* it, and clamped everything together so that his original wooden pattern was centered in the oversized cavity he'd milled. Let's call this the top end of the mold.

> * Apparently, if you don't use some kind of release agent, you will have a **terrible** time trying to get the castings free of each other; "Aero" floor wax has been found to be about the best stuff to use as a release agent.

He then poured some kind of black epoxy in through a couple of holes drilled in to meet the cavity from the other (bottom) end of the stack. After letting the epoxy set for some hours, he removed the wooden pattern, and now had, in the stack of Medite, a cavity that was an absolutely perfect female copy of his pattern.

The next step was to wax the interior of the cavity, and pour in more epoxy from the top end. When that set, out came the first of 10 male copies of his wooden pattern, all identical, without any effort to make them so. These were attached to a "board" to create a new, multiple pattern.

Now you may not want to make a multi-off pattern for a foundry casting, but... and this is just the first item that's come to my mind to make - you could make your own machine handle knobs, either in hard black epoxy, or Brownells' AccraGlas, or even some more resilient castible rubber/latex material.

Now your response may well be: "Why make something you can buy for a buck?"

I agree, but if you want a machine handle of a shape you can't buy, this would be a way to get it, and have a very professional looking result. How about a knob with a turk's-head worked onto it, in solid black, or even cast in stages, using 2 colors of epoxy? Other things one could make would include knife handles, handles for storage boxes, a handle for a rifle cleaning rod, a duplicate of a factory rifle buttplate, etc.

Footnotes:

The "black epoxy" BB used is a polyurethane casting resin called "Ultra-Cast FR1177". It is produced by Fiber Resin Corp., P.O. Box 4187, Burbank CA 91503; phone 1-800-624-9487. It contains silicon, and is therefore quite hard on cutting tools you may have occasion to use on it - e.g. drills, etc.

MAKING USE OF SURFACE TENSION

One of my guys wrote to say he's into making ship models, and mentioned the need for jigs and other methods of making lots of similar parts. The following excerpt from my reply contains ideas that others may find useful.

".....One method you should look into is lost wax casting. This technique allows reproduction of very fine detail, and the making of many identical parts. My friend Bob Eaton, of Blaine, WA does a lot of lost wax casting in his basement, in brass, gold, etc. He made a magnificent 1/8 scale model of a Civil War "Napoleon" cannon in brass and walnut. Many of the brass parts were investment cast, and would have been difficult to make by other methods. At least one bolt was cast with a ready-to-use thread, about 3/16"ϕ - and this was a square thread (not an Acme thread, but similar).

"Bob made most of the bolts for this model by silver soldering a cast head of the correct shape to cast/machined shanks of various lengths as required. Of course the head has to be nicely centered on the shank, and he told me how to do it:

Set the bolt head casting upside down on a charcoal soldering block. Hold the bolt shank vertical above the inverted head casting, and silversolder the shank to the head. While the solder is still molten, lift the shank up just a hair - if the temperature is right, the molten solder will lift the head up off the soldering block with the shank, and the surface tension of the molten solder will automatically center the head on the shank. Don't expect it to work for you the very first time you try it, although it may. Bob is a pretty fair hand at fine soldering.

"You could use your lathe and/or mill to machine "machinable wax" into patterns for some of the parts you need."

NOTE: Bob still has all the molds to make the metal parts for this model, and is willing to make sets of castings available to other hsm's who might want to build a similar model. (Full drwgs are also available.) If you are interested in obtaining a set, contact me, or Bob directly.

HOW TO KEEP YOUR DRILL CHUCKS CLUCKING

Drill chucks are mostly taken for granted, until they give some trouble. Then you either chuck 'em, fight with 'em, or fix 'em.

It's worth knowing how to fix them, because for about half the cost of buying a new one, you can return a used chuck to good-as-new condition, right there in the comfort and privacy of your basement shop. All you need is the know-how, some simple disassembly & re-assembly tools you can make for yourself, and whatever parts need replacement.

The first kind of trouble you can run into is in trying to get a chuck off its arbor. If it was put on there right, it's going to stay there until you WANT it off. Which means it ain't gonna drop off into the palm of your hand just because you snap your fingers at it.

Maybe we should back up a step or two, and consider first the matter of....

How to put a drill chuck on its arbor

A taper mount drill chuck must be seated very solidly on the end of the drill chuck arbor if it is to stay there, and not come "unglued" during use. The latter, if it happened, could - and likely *would* - be very bad news.

The proper procedure for putting a drill chuck on an arbor is as follows:

Check the taper hole in the chuck, and the taper plug on the arbor, for burrs, warts, pimples, etc. If any are found, stone them off.

The inside of the taper hole in the back of the chuck, and the outside of the short taper on the drill chuck arbor, want to be clean and DRY - no dust, no oil, no sweat from your hands, no nuthin'. I wipe the mating surfaces with rubbing alcohol on a piece of toilet paper.

Then put the arbor into the back of the chuck, and give the other end of the arbor a good, sharp, direct clout with a rawhide mallet, brass hammer, or, if you must use a steel hammer, put a piece of hard wood between.

How do you get it back off again?

Some chucks have a tapped hole inside, such that a screw can be inserted into the nose opening (where drills etc. usually go) and screwed into said hole, thereby forcing the chuck off it's arbor. But it isn't common to find such a tapped hole in a chuck. (You can put a tapped hole in a chuck yourself --- providing you do it before you put the chuck on an arbor!)

In some cases there will be an untapped through-hole in the back of the chuck. A pin can be put into this hole and seated against the end of the arbor, and the latter can sometimes be driven out thusly.

BB and I got into all of this one day - he had a chuck he wanted off its arbor. What to do? "Chuck removal wedges," I told him. This piqued his curiosity no end - BB just had to know what they were.

I explained that they were merely a pair of slotted wedges which you fork over the back of the arbor right behind the drill chuck, and by tapping or squeezing the wedges, the chuck can be popped off with no sweat.

BB soon learned some more - that there was not one single machine shop supply house in all of Vancouver that had even heard of this esoteric item, let alone had a set in stock. BB couldn't wait for a pair to come by mail, so he made them. I was ordering something else from MSC, so I ordered a pair. The next day BB brought his around to show me, commenting with some embarrassment about the tool marks on them.

"Never mind - they're probably the only pair of chuck removal wedges in the whole city. Did they work?"

BB admitted that they had done the necessary with distinction.

"Well then, hang 'em up on your wall and be proud of them."

Mine came a while later, and before they would fit - i.e. fork around the base of the Jacobs taper on my #2MT arbor, they'll need to be filed out a little wider in the fork opening (they're soft enough to file).

In heavy use, chuck removal wedges will get chewed and beat up pretty bad; they work harden, and spread, and eventually crack, typically when you are trying to squeeze the forky part back together a little. They are not made to last forever, but they will probably never wear out in the average hsm's shop. Since I still haven't filed mine to size, we figure BB still has the only functional pair of chuck removal wedges in Vancouver.

> (When I ran the above past BB, he said he never asked me about them - he'd seen them in the MSC catalog. I dunno... I remember it all differently.)

Disassembly of keyed chucks

Next, the whole chuck may need to be dismantled for cleaning, or replacement of worn or broken parts.

Jacobs chucks are disassembled for servicing by pressing the geared sleeve off the chuck body. This can be done in a bench vise, or in an arbor press if you have one. You will need to make a disassembly fixture to use in the pressing operation. The fixture is simply a piece of steel - could be a short slug of CRS faced on both ends, or a piece of 3/4" CRS plate - with a hole bored in it about 5 thou over the diameter of the chuck body where it sticks out the back (arbor) end of the geared sleeve. I have seen an illustration of a chuck being pressed apart with the chuck's sleeve sitting on two parallels, but a ring would be much more stable.

Some people say you need another ring to set on the nose of the chuck body so that the pressing force is applied to the nose of the chuck, not the ends of the chuck jaws. Other people say you can press right on the jaw tips. Making the second ring would be no sweat, and maybe smart.

Once the chuck's geared sleeve is off the chuck, the internal parts are readily removed, and the necessary work carried out. Start with a good clean-up of everything. Don't drop nuthin' on the floor. If none of the parts are worn or broken, put each piece back in place in the chuck body in the proper order, and in its proper place, and press the geared sleeve back on.

NOTE
Not all keyed chucks are the same. At page 41 in the July/August 1985 issue of Fine Woodworking (FWW) there is an excellent article by Richard B. Walker on how to service keyed chucks. Among much else the author has to say, he explains that while most Jacobs chucks have the ring gear in one piece with the tightening sleeve, in some chucks (presumably mostly by other makers) the sleeve and ring gear are separate pieces. If you have ideas about dismantling a chuck of this latter type, you need to know that the sleeve presses off in the opposite direction from that of other (normal) keyed chucks.

A footnote to the above article:

In FWW, Jan/Feb '86, page 6, Tom Tilson, of Lima, OH points out something that is not commonly known: if you are repairing several chucks at once, don't dump all the replacement jaw parts (i.e. 3 jaws plus the split nuts) into a pile, as the parts are **not** interchangeable from set to set. Keep each set of replacement jaw parts *separate unto itself as a set*, or you will spend your next vacation trying to sort them back out into the groupings they were in when you got them. *(Do we hunt up obscure but useful info? You know it.)*

How to give your Keyless Chucks a Deluxe Overhaul

Keyless Albrecht chucks are another matter. They come apart quite easily if you know the trick of it, but if you don't, you could spend all day diddlin' around with one and still be no further ahead.

When I was looking into this, I ended up talking to Ed Schmidt at Albrecht, Inc., in Hauppauge, NY (that's pronounced Hop-og, or Haw-pog, by the way) and he told me a few things. He'd had a call from a shop foreman 2 or 3 weeks previously. A brand new Albrecht chuck had been issued to a millwright. The millwright somehow spun a tool shank in the chuck, damaging the tips of the jaws somewhat. New parts were ordered, and when they came they were given to the millwright to install.

This was not an entirely good idea, because the millwright didn't know how to get the chuck apart. He ground a flat on the "hood". No result. He put a pipe wrench on it. No luck. He put a pin up the nose of the chuck, thinking that the hole in the back of the chuck was a through hole clear to the end of the arbor. And he pushed on the pin quite hard. This also was not a good idea - the hole leads to a shouldered ring in the back of the chuck, and if you push on it too hard it usually breaks into a great many pieces. This one did. By the time the foreman decided to call Ed Schmidt, the chuck was a write-off.

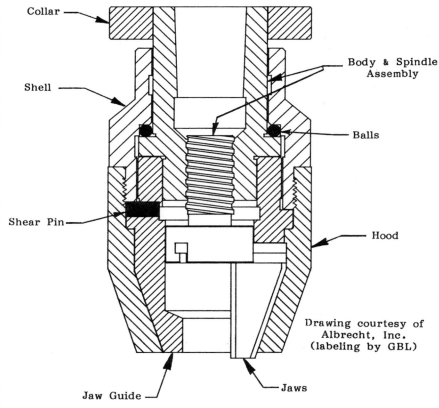

Drawing courtesy of Albrecht, Inc. (labeling by GBL)

Albrecht has a sheet you can get that gives the bare bones version of how to service keyless chucks, but I got the real inside dope on how to do a deluxe overhaul job the service sheet don't say nuthin' about. (Not that they're trying to hide anything from you - it's just that nobody digs up and dishes out the good stuff for you like I do....)

Have a look at the cross section drwg above, and I'll lay it on you like a French bathing suit, first - which is to say, just covering the essential points - and then we'll get into the little extras...

How to service an Albrecht keyless drill chuck:

Remove the collar.

Grip the shell of the chuck in soft/protected vise jaws.

Use a strap wrench to remove the hood. The thread is a right hand thread. As shown on the drawing, the split line is where the smooth and knurled areas of the chuck meet.

Once the hood is off, slide the jaws out of the jaw guide and then remove the jaw guide.

Bring the body and spindle assembly out through the "jaw" end of the shell to expose the balls. Obviously, don't lose any of the balls.

Clean all parts with varsol or similar solvent. Replace any worn or damaged parts, and reassemble in reverse order.

> *Note: Do not lubricate the spindle/body when reassembling the chuck. I'll tell you why below.*

Now for the extra goodies....

Where to grip the shell? Grip it by the small diameter part, because in that area it's supported against crushing by the body. Obviously you're not going to try to squash it, but it's better to grip it here than on the larger diameter, because if you grip it on the larger diameter you could distort the shell, which will only make it that much harder to unscrew from the hood.

If your vise has hardened serrated jaws, protect the shell with some slips of sheet aluminum or sheet copper. (For same, chop up a piece of copper water pipe; I've mentioned that before.) If you want to be even fancier, get yourself a piece of aluminum or steel flat stock maybe 1/2" thick, and in it bore a hole that is a close fit on the small diameter portion of the shell. Deburr the edges of the hole, and split it so that when you stick the shell in it, and stick everything in the vise, it'll clamp up snug on the whole circumference of the shell.

As noted above, use a strap wrench on the hood. If you don't have one, an old V-belt wrapped around the knurled part may suffice. Again, what you don't want is pressure at 2 opposite points on the hood, which would distort it. Here too, if you wanted to, you could make a wrench from a piece of aluminum say 1/2 or 5/8" thick by maybe 9" long. Bore a hole near one end that'll be a nice fit over the knurled part of the hood. Drill it for a clamping bolt, and then split it. Stick this on the hood, tighten it up, and give it a careful shot with a hammer, and you're in like Schmidt.

(You don't really need these special "grabbers" unless you're doing a lot of chucks, but you might want to make them up for yourself, just for fun, or to do chucks for other guys as well as yourself. Ed said the real tough ones to unscrew are those where the chuck is jammed up tight with a drill shank still in the chuck. That's when you really need the helpers.)

Once you get the hood/shell joint loose, it should unscrew by hand the rest of the way.

Now, *why not lubricate the body/spindle* during re-assembly?

Ed told me these parts are lapped together, and a certain degree of friction between them is required for proper lock up and function of the chuck. Lubrication will lower this friction and the chuck will not operate properly.

If the chuck opens up in use, it is, in the first place, a sign that the chuck needs to be serviced, and if, after you have serviced it, the chuck still wants to open up in use, check the condition of the body and spindle.

If the threaded bearing surfaces of the spindle are not good, the parts should be replaced. If the bearing surfaces are good, but worn very shiny, the body and spindle can be re-lapped to each other. This is an easy process that should put everything in good order. **This is part of the deluxe job the service sheet doesn't tell you about.**

Use medium-coarse valve grinding compound to re-lap the body and spindle, working them together a few times with moderate axial pressure in both directions (push on it and pull on it). When you think it's right, clean out ALL the grinding compound, and then clean it out again.

To complete the deluxe overhaul, next inspect the tapered portion of the inside of the hood. If you find 3 mirror-smooth wear lines on this surface, from years of use, take a small piece of about 300-grit emery paper, poke it into the hood on the end of your finger, and twist the hood on it. The object is to break up these over-polished areas so there is room for some lube (see below), much like the cross hatched pattern desired when you hone an engine cylinder.

Use a little light grease to lube the track where the balls ride, also put some grease inside the shell on the undercut (in the small diameter area), and inside the jaw guide where it fits over the body. Finally, apply a very thin layer of grease in the tapered part of the hood. When you're done, there should be no grease on the spindle or back surface of the jaw guide. If you get any grease on the latter area, wipe it off - what you want there is metal to metal contact, or the chuck will have a spongy feel.

And **that's** how to do a deluxe overhaul.

For a parts list for Albrecht chucks, phone 1-516-273-1010 and ask Ed or one of his guys for their sheet on how to dismantle Albrecht chucks. They'll mail or fax it to you. And they can supply whatever new parts you need through a dealer near you.

Replacement parts for Jacobs plain and ball bearing keyed chucks, as well as disassembly, servicing and re-assembly instructions, can be obtained from your local industrial distributor or mail order machine shop supply houses.

Salvaging a Drill Chuck
from **Ed Claypoole, Butler, PA**

Dear Guy,

My wife Tracy ran a 1/2" drill (very sharp) into a too-large pilot hole the other day (odd, because she's usually better than me at paying attention). Naturally, it stuck in the thick flat material and really ground the bit and the Jacobs chuck into a mess. I couldn't get the chuck apart, and didn't have an ID grinder to get inside to grind the working faces of the jaws. A local guy with considerable chuck rebuilding experience said it was trashed.

After looking at it for 2 weeks on the bench, I got an idea. I chucked a 1/4"ϕ ceramic rod in the lathe and closed the drill chuck gently over it. Some high rpm's combined with a gentle lengthwise sliding while occasionally hand tightening the chuck, and I soon had those hardened jaws reground and honed in line. An indicator later showed almost zero run out, in fact the chuck was now fully as good as my others.

Hope this helps someone out with a terrible mess.

Ed

Dateline: Foster City, CA

RETIRED WOODWORKER PRODUCES NOTEWORTHY FLAT ON STEEL BAR

I have engaged in an internal debate of considerable duration regarding whether or not to re-print herein my article from HSM on making a steel beam trammel. After much thought, I decided not to, because it is available not only in the January/February, and March/April 1985 issues of HSM, but also in the HSM book, "Projects *Two*", page 152.*

Published by HSM, "Projects *Two*" is one of a series of hardcover books in which are reprinted many past HSM articles. Unless (and maybe even if) you have a complete collection of HSM, these books are well worth their price, because among the multitude of articles in each volume you are almost certain to find something that alone is worth the price of the whole book. For example, in Projects *Three*, there is an article on making a very low height 40-ton shop press that I want to talk to my friend Jake about... (also in that same book is an article on making a desk lamp in the form of a scaled-up toolmaker's surface gage... and you *know* the guy who wrote it...)

* (If you want but can't readily get at the article from either of these two sources, feel free to contact me.)

One day I received a very nice letter from Bob Wolfe, of Foster City, CA. In it was a photo - see below - and an explanation of how he had produced, essentially by freehand grinding methods, the flat on three beams of differing lengths for the steel beam trammel he'd made based on my article. His method not only worked, but produced a flat that varied less than a thou from end to end of a 16" beam. I think you will find his method interesting, and possibly useful to yourself in this or some other project, so here it is.

"....I discovered that by making light passes with a reasonable amount of even pressure, a uniform flat could be hand ground to within about a 0.001" tolerance on a steel rod.... "

From **Robert W. Wolfe, Foster City, CA**

Dear Guy:

Since I feel I have come to know you personally through reading your various books, I thought you might be interested in the following.

In 1980 a good friend got me interested in metalwork - I had worked exclusively with wood for nearly 30 years. By late 1984 I was hooked. I subscribed to HSM, and about that same time bought a metal lathe, although I didn't really know much about it.

Recently, while browsing through my HSM collection, I came across the first issue I have (Jan/Feb '85), and took in, for the first time, the fact that you were the author of the article on making a steel beam trammel. I can still remember the arrival of that first issue in the morning mail - I immediately sat down and began leafing through it. When I came to page 46, my eyes lit on the photo of your steel beam trammel. What a neat first project for my then-new lathe, and what a great addition to my tool box!

My elation was soon tinged with frustration, and then dismay, as I read through the article. Knurling? I didn't have a knurling tool. And producing the flat on the beam? I had neither milling machine nor surface grinder, nor did I know anyone who did. Somehow it didn't seem right, after plunking down all that money for the lathe, to have to pay to have one of my first parts made for me by someone else.

122

Further, the author suggested that if farming out the milling and grinding were necessary, to first rough out the flat "by whatever means you can devise...", without the slightest clue on how that might be done! (I musta said that in a weak moment. GBL)

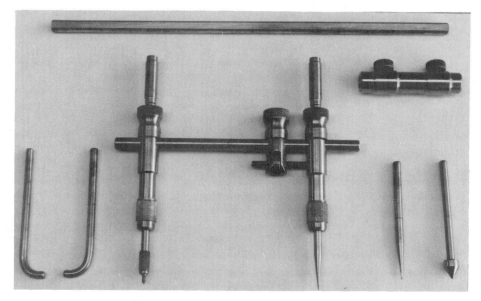

My friend showed me how to set up a pointed toolbit on its side in the lathe toolpost, and rack the carriage back and forth to cut straight flutes in place of knurling, but he was of no help on how to produce the flat on the beam, given the tools at hand.

I mulled over the situation for about a week, and eventually began to realize that it was not as critical that the flat be even in width (depth of cut) as it was to have it all in one plane. Why not try to produce the flat by running the rod past a grinding wheel? The question was, how to hold the workpiece for this operation?

Entering my workshop early on a Saturday morning, I noticed a length of aluminum angle stock and a length of 1/8" x 3/4" steel flat stock in my scrap pile. It struck me that if I were to solder the beam material (I planned to use 5/16"ϕ CRS) to the edge of the steel strip in such a manner that the beam would ride on the toolrest in front of the grinding wheel, the solder might be strong enough to hold the workpiece.

I picked up the aluminum angle and propped it up so it looked like a "V". I cut a 6" length of 5/16" CRS rod, and faced the ends. After cleaning it thoroughly, I laid the rod in the bottom of the aluminum "V", cut a piece of the 1/8" x 3/4" HRMS flat stock, brightened up one edge and one flat side on the belt sander, and slid the flat stock down one side

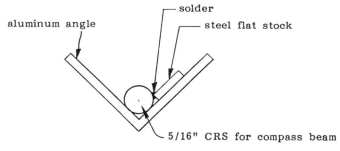

of the "V" against the rod. Being careful not to overheat the pieces, I found I was able to lay in a nicely flared solder bead the full length of the junction between the two pieces.

While waiting for the pieces to cool, I wondered if grinding the flat on my CRS beam might cause it to bow, as you mentioned a beam made from drill rod might do. And if it did, would it break the soldered joint?

However, after making a careful setup to get the tangent point where the grinding would begin as near 90° as possible to the soldered support strip, I tried a few light passes.

The result looked good, so I made a few more passes, keeping the workpiece cool so as not to melt the soldered joint. After a few more passes, measurement showed that I was getting there. I was pleased to see that I was also achieving a surprisingly even grind. I was, however, not entirely happy at the cupped rather than flat surface I was producing, due to the curvature of my grinding wheel.

Determined that any machine shop I took it to for finish grinding was going to see that I had done my part in "roughing out" the beam, I worked the piece down to within a few thou of the specified 0.2815" dimension. I then took it over to the workbench to scrutinize it in detail and check it over with my dial caliper.

By placing a short length of wire in the cupped section just ground, I could measure the variation along the length of the beam. I was amazed to find that these variations were on the order of 0.001" over the whole length! Needless to say, I was more than pleased with my super steadiness of hand.

Studying the piece some more, I began to consider how to flatten out the cupped surface.

Many years before, I had made a "surface grinder" of sorts by mounting a 6"ϕ x 3/4" stone on a short 1/2"ϕ arbor for use in my drill press. For many years I had used this with various jigs to apply a fine finish grind to the bevelled area of plane irons, chisels, planer blades, etc. I recalled that when I mounted the wheel, I had spent much time to get it to run true axially, using large cardboard washers sanded to a very slight wedge form, so that the wheel could be adjusted to remove any wobble, or face runout. I had also taken the time to dress it round. Both operations had been done on the drill press, and index marks had been placed on the arbor and the chuck so that it could be put back to run true again at will. It was effective for what it was designed to do, but the maximum workpiece length that could be dealt with was about 4-3/4" - not enough for a 6" beam.

Studying the matter a little longer, I decided to mount the stone in the drill press, and clamp a fence to a board gripped in the drill press vise, the latter being bolted down just off the end of the face of the wheel as in Fig. 2. I set the fence against the back side of the flat stock to which I had soldered the beam, tilted the board slightly to get the "cup" square with the face of the stone, and set the drill press spindle speed to about 450 rpm. After an initial pass, I loosened the clamp holding one end of the fence, gave the fence a slight nudge inward, tightened the clamp, and made another pass. To my great delight, it seemed to be working!

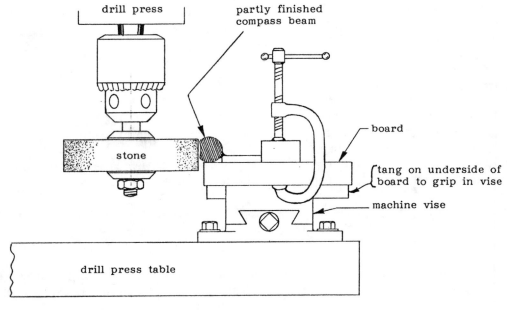

Cupping of primary grind
on workpiece is exaggerated

As I repeated this process several times, the material being removed became greater and greater, and it became more and more difficult to move the beam past the stone. As the grinding approached the bottom of the "cup" it took 5 or 6 passes at each setting of the fence to get the stone to spark out completely.

(At one point I stepped up the spindle speed to 700 rpm, but that immediately began to produce vibration marks in the workpiece, so I backed off.)

It was at this point that the real mechanics of what was happening became clear: it was not my super steady hand that deserved the credit after all. When grinding a piece of rectangular stock, the width of material being ground remains constant, hence the depth of cut is directly proportional to the amount of pressure being applied. When you try to grind a flat on a piece of round stock, however, to remove material at a given rate on each pass requires significantly increasing amounts of pressure as depth of cut increases, because the width of cut (hence amount of material being removed) is also increasing. Thus, on any given pass, a high spot on the beam has less cross section area, and for a given pressure, will be removed at a greater rate than at the nominal depth, while a low spot will have a greater cross section area/width of cut, and the wheel will cut it at a lesser rate than at the nominal depth.

By making light passes with a reasonable amount of even pressure, a uniform flat can be hand ground to within reasonable tolerances - a far cry from those suggested by the 4-decimal place dimension on your drawing, perhaps, but reasonable nonetheless, and it was all my own work.

I was still a couple of thou high when all indication of cupping disappeared. At this stage I decided that a little high was better than a little low. So I unsoldered the beam from the piece of flat stock, and after cleaning off all vestiges of solder, I took the calipers to it again: it measured 0.2825" at one end, gradually tapering up to 0.2835 at the other end. There appeared to be a slight, very short dip about 1/3 of the way in from the high end. The other interesting thing was that when I placed it on my table saw top, there was no sign of bowing.

I pulled out an 8" stone and began to work off some of the scratches and work down the high end. After several passes over the stone it became clear that the flat was "flat" for its entire length except for the tiny low spot 1/3 of the way in from the end. I couldn't measure the depth of the low spot - it just showed up in the stoning process as an area where the grinding marks would not work out.

(I wondered if the original CRS rod might have had a miniscule kink there, which - if it did - would explain why it measured the same there as elsewhere along the rod. I asked Bob about this, and his reply was that if there was a kink, he was unable to detect it. He said he figured the blemish developed while the cupped surface was being ground down to a nice flat, and was due to some flaw in the setup used, possibly the fence, which was rather light. When he later made a longer beam from a piece of drill rod (see below), he used a longer base board, and a heavier piece of mat'l for the fence, and found no such blemish on that one. GBL)

I was quite excited over what I had produced, and because it was now approaching noon, I called my friend to come over for lunch so I could show him my handiwork.

He showed up in a few minutes, mike in hand, and was soon engrossed in inspecting the beam, measuring and staring, holding it up to the light, and measuring some more. The only thing he said during all of this was to call attention to the small defect near one end, and to note that it was a tad high at one end. I asked him a couple of times what kind of measurement he could place on the defect, but he never responded. When my wife slipped the grilled cheese sandwiches in front of us, I asked him if he would like a beer. He said, "Yes, after I watch you do it again."

"Heck," I said, feeling a little cocky, "nothing to it. It's all in the touch!"

So we picked up our sandwiches, thanked my wife, and headed for the shop. All the setups were in place, so it didn't take long to go through the same routine, this time with a piece of CRS rod about 11" long. I had the job done in something under an hour, including maybe 5/10 minutes with the stone to try to get rid of the tiny low spot, again about 1/3 of the way in from one end. The variation in depth of cut over the length of the beam was just a bit less this time. And again, no sign of bowing along the length of the beam. Possibly heating the bar to solder it to the holder was also stress relieving the bar.

I finally handed the second beam to my friend for his inspection. After a few minutes and several comments about the small blemish, and no response to my repeated request for his measurement of same, he said, "Well, I'll be damned. I'll have that beer now."

About six months later, I needed a longer beam for use in connection with one of my woodworking projects. I decided to use drill rod this time, primarily to test my assumption on the bowing problem. I chose a length of 16" because the distance between inside faces of the grinding wheels on my bench grinder is 17".**

I proceeded in the same manner as before, except that to give me more stability during the initial grinding I clamped a piece of 1-1/2" aluminum angle stock about 36" long to the workrests on the grinder. The results this time came out the same as before, including a "no bow" condition. On this beam I took the additional time and effort to get the overall measurement closer to the 0.2815 your drawing indicated. I wound up with 0.282 to something less than 0.283 over the 16" length of the beam.

> ** *(I asked Bob if by removing the unused wheel, he could have made a beam of greater length. He wrote back, saying* "Yes, but I didn't want to remove the wheel and have to re-balance and re-true it afterwards. Also, after doing the initial write-up I sent you, a job came up where I needed a longer reach than my 6", 11" and 16" beams would give me even if I coupled them all together. I spent some time looking at that unused wheel, and thinking (which should always precede action if one wants to spare oneself a lot of trouble!), and it finally occurred to me that what I really needed was a beam of sufficient length for the job at hand, with a relatively short flat at each end, rather than a flat the full length of the beam. If you use a coupling *(Detail "D" in my article.*GBL*)* to lengthen the beam, the end that goes in the coupling doesn't require a particularly precise flat. By grinding a flat say 3 to 4" long on one end for the tram, with the beam oriented in such a way that the tail of the beam is away from the unused wheel, there is no problem in producing the proper size flat. Then just make a short flat, maybe 1" long, with the beam running in front of the unused wheel.
>
> "I clamped the heel of a toolmaker's clamp on the worktable next to the unused wheel, to keep the beam from running against the wheel while grinding the short flat on the beam. There was only one other change and that was in the surface grinding operation on the drill press to remove the cup grind. Because the flat does not run the full length of the beam, the 1" length would have to be fed in the "hungry" way, and the longer flat at the other end would have to be backed out in the "hungry" direction. I decided that I would not use a fence, but would just freehand both ends. It turned out quite easy to do both, with good results." *)*

In re-reading your article recently, and studying the drawings, it crossed my mind that you just might have had your tongue in your cheek when you called out that four decimal place measurement for the beam! However, I occasionally find myself fondling my original 6" beam and recalling the satisfaction of a highly accurate job carried out entirely by hand methods.

...about that 4 decimal place dimension on the beam flat:

The beam starts out as 5/16"ϕ material. The purpose of the flat on it is to make both tram heads look in the same direction no matter where we slide them along the beam and tighten them up.

How big a flat? Well, say about what we'd get if we peeled about 1/32" off the side of the bar. But you can't say, "Mill or grind off 1/32nd of an inch." - it just wouldn't have the right ring to it. So I put it in decimal inches, and wound up with 0.2813, which I rounded to 0.2815. Either that or when I measured the beam of the Starrett one, I measured it as 0.2815".

But it ain't all that critical, if you think about it. What I should've said, either to myself or in the text of the article, was "....call it 0.282" ± a couple of thou."

However, I still like Bob's approach to grinding the beam by hand methods, and that he was able to come out so close to a desired dimension is only part of what makes his approach so interesting. I suspect that if one cared to do so, one could lap the flat down to an even depth over the whole length of the bar, but when he'd already got it within the thickness of a cigarette paper from one end to the other, and nice and smooth, why bother? GBL

.... and a Pencil Lead Holder

I also made a pencil lead holder for my Steel Beam Trammel. This required drilling a two-diameter hole the full length of a 2-5/8" piece of 3/16"φ drill rod, drilling #48 (which is the diameter of draftsman's pencil lead) as far as possible from one end, then drilling #31 from the other end to meet. (I botched one, but the second one came out ok.)

The other part of this job was to make a Chuck scaled down from the Chucks made for the main Trammel Heads. I used a 10-32 thread in this smaller Chuck. The finished Lead Holder works well and didn't take very long.

Below is a sketch of my Lead Holder, and measurements taken from it. The following is a description of the sequence in which the various operations should be performed, with explanations of special considerations.

Pencil Lead Holder Body...

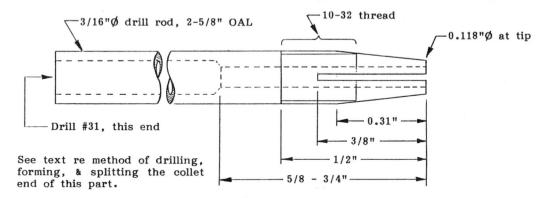

Cut a 2-5/8" long piece of 3/16" drill rod, for the Body/collet of the Lead Holder. Face both ends and use a center drill to mark the centers on each end of the rod with a small dimple. After that, the first job is to drill a hole the full length of the 3/16"φ drill rod.

As I said, the Lead Holder is intended to take standard 0.076"φ drawing leads, so begin with a #48 (0.076") twist drill. (I'd be overpowered by the urge to start with say a #50 (0.070") drill, and then finish with a #48 drill.GBL) If you're lucky, your #48 drill will give you an absolutely straight, and slightly oversize hole - not likely, but worth a try.

With the lathe set to its highest speed, drill #48 down the axis of the drill rod for about 3/4". Use a light touch on the tailstock h'wheel, and liberal amounts of cutting oil. Then reverse the workpiece in the chuck, and drill in #31 from the other end to meet your #48 hole. The depth involved (about 2") will exceed the fluted portion of a #31 drill, so frequent withdrawal is required to clear chips.

Once the two holes meet, and all burrs and chips are cleaned out, test the collet end of the holder with a pencil lead for clearance. If there is any bind at all it will be necessary to "ream" the hole with a #47 drill.

> NOTE: If the lead still binds, the hole can be opened up to #46 or #45 drill size. Beyond this, the closure limits of the collet will be exceeded, and a new Lead Holder body should be made.

When the above is done, screwcut the "collet" end of the Body 32 tpi, and then set this piece aside for now. We'll do the taper on the end to form the collet nose in a few minutes.

Begin making the Chuck by putting a fine knurl on about 5/8" of the end of a piece of 1/4"⌀ CRS. Leaving the Chuck on the parent bar for now, use a parting tool or similar to delineate the finished OAL of the Chuck, and face the outboard end. Clean off the knurl back from each end in proportion to, and to match the appearance of, the Chucks on the main Trammel Heads.

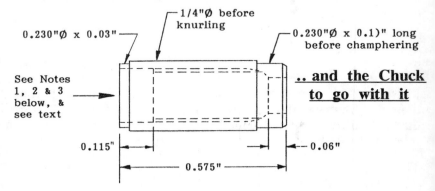

.. and the Chuck to go with it

1. Drill #30 (0.1285"⌀) right through
2. Drill #21 approx. 0.51" deep, tap 10-32
3. Drill #11 (0.191"⌀) approx 0.115" deep

Now poke a #30 drill in about 3/4" deep, follow with a #21 drill to a depth of 9/16" (i.e. to within about 1/16" of what will become the bottom end of the Chuck), and finally put a #11 drill in to the depth indicated on the drwg for clearance around the full diameter portion above the threads on the Lead Holder Body. Put a #10-32 taper tap in to the point where the tip of the tap just begins to touch the #30 hole.

Part the Chuck off at specified OAL, and put the Body of the Lead Holder back in the lathe chuck with the threaded end out. With a fine cut file, carefully work the threaded tip of the collet down to the same general contour as the tap used to thread the Chuck, bringing the very tip down to 0.118"⌀. As the tip is being worked down, test the fit of the Chuck frequently and continue filing until the tip of the collet just begins to protrude slightly from the end of the Chuck.

Next, split the end of the collet portion of the Body to form collet jaws. I used an X-Acto razor saw to make the required cuts, by hand. This saw can be bought for a couple of dollars at any well stocked hobby or model shop, and gives a kerf of about 0.015". To clean out the chips and burrs caused by the saw cuts I made several alternate passes of a drill and the saw, and then used dental floss for the finishing touches - surprisingly, it works quite well in tight spots like this.

Now, when you assemble the Chuck and the Lead Holder Body, the jaws of the collet should be seen to be working through their full limit as the Chuck is tightened and loosened.

The last operation in making the Holder is to give the inside surface of the collet jaws a form that will maximize the area of contact between the jaws and the pencil lead. With the Body still chucked in the lathe, run the Chuck on until the jaws of the collet just begin to close. Make an erasable mark on the Chuck and a corresponding mark on the Body. Now tighten the Chuck up until the jaws are closed to their limit. Make another mark on the Body opposite the mark on the Chuck. This should be something like 1/2 to 3/4 of a full turn of the Chuck past the first mark.

Back the Chuck off to a point 1/2 way between the two marks. Check to make sure that the diameter of the hole in the collet is less than that of the pencil lead. If not, tighten the Chuck slightly until it is. Now, with a #48 drill in your tailstock drill chuck, ream out the point of the collet. Clean up the chips and the job is done.

I wrote to Bob about some of the above, and said, just by-the-bye, that, "...the idea of using an X-Acto razor saw for slitting is (good) news to me. I recently made 2 identical items, and had to make 5 in all, as the first 3 got trashed in trying to slit them with a slitting saw in the B'port. I got along fine on the first cut, but when I rotated the job 90° for the 2nd cut, the job had not sufficient rigidity to resist the cutter, with the results as shown at right: the outboard half of the part being split would bend away from the cutter. I eventually made the 2nd cut in each one by hand with a fine blade in a jeweler's saw frame."

Bob came back to me with a very good idea:

"Maybe I can spread a little happiness: Next time you need to carry out a slitting operation similar to the one you described, try placing small wooden wedges (say toothpicks) at several locations in the first slot formed. Then wrap a series of loops of light copper wire (say 20 gauge) around the workpiece along the section to be cross split, and twist each loop tight. The saw will cut each loop in turn as it progresses up the shank being split, and as it does so, the next loop will take the load imposed by the cutter. The small wooden wedges with their high friction coefficient with steel have the effect of resealing the first slot so that you are making the second cut with much the same rigidity in the workpiece as when you made the first cut. It works.... most of the time."

Casting about for something different to do?

Randy Orpe, of North Fork, CA, visited us last October. Randy has several used military vehicles, which he uses for everyday transport, haulage, etc. He needed a Tulsa winch bell casting for his GI Dodge pick-up truck. This casting is about the size of a roasting pan big enough to make a 25 lb. turkey nervous. The manufacturer's price on a new one would've run about $300 or so. Randy was able to borrow one from a local guy, and used it as a pattern from which to make his own casting, in aluminum (same as the original). The project was entirely successful. He bored out the hole in his casting, pressed in a bronze bush, bored that to size, drilled one other hole, and the job was done.

Randy got his foundry stuff from an outfit called "Pyramid Foundry" or something like that. The sand he used (Petro-bond) even picked up the chipped paint on the borrowed housing, and reproduced it on his casting!

Why is this in here? To point out that it is perhaps more realistic than you might think to make your own castings for projects that might otherwise be largely out of reach.

Footnote to the above: there is a sculptor here in Vancouver who does his originals in a material that is *very* easy to work with, and then has them cast in bronze. The possibilities of adapting his techniques for hsm types are intriguing. Stay tuned....

LOW-COST SHOP-MADE SAND BLAST EQUIPMENT FOR YOUR SHOP
based largely on info supplied by
Marshall Collins, Crestline, CA
with some further info from **Vic Baldwin**, **Phil Lebow**, and
Gunsmith Kinks, Vol. II, courtesy of F. Brownell & Son, Inc.

Ever wished you had your own bench-top or stand-alone sandblast outfit? A store-bought cabinet sandblast outfit will set you back anywhere from $300 to $1000 or more. But you can make an equally effective outfit for yourself for very modest cost, and you can probably even make some money with it, if you want to.

When I speak of this equipment as a "sandblast outfit", you should of course realize that it can be used with just about any kind of dry gritty stuff - sand, glass beads, ground up corn cobs, ground walnut shells, etc.

> You didn't know about walnut shells? An aircraft maintenance mechanic told me that when they want to clean a jet engine, they fire it up, and when it's goin' good they throw a bucket of ground walnut shells into the intake. Cleans it out like a dose of salts, apparently.

Some years ago, Vic Baldwin, a machinist friend here in West Vancouver, made himself a bench-top sandblast cabinet so nice that if you saw it, you could be excused for thinking it came from a major manufacturer of such equipment. He showed me parts he had finished in it, some using fine glass beads, others using ground walnut shells, and the results it gave were all anyone could desire.

As I have indicated before **(TMBR#1**, p. 172), I like a soft satiny glass bead blast finish, and I've long felt I'd like to have a sandblast cabinet. Two factors deterred me: the lack of a place to put one, and a decent air compressor to run it. Obviously the need for an air compressor is easy to fix (just add $*), and some changes we recently made around here have produced some spare space that could accommodate one, so when Marsh Collins offered to send me some info on his sandblast outfit, it got me to thinkin'…

> * Jake Wiebe & I are working on a new air compressor for my shop. I'll likely have a little chunk in TMBR#4 on how to put together a compressor outfit, where to buy the various bits and pieces, how to do the plumbing, etc.

I buzzed Vic and asked if I could come over to have another look at his sandblast cabinet. He told me he'd sold it, but said to come on over and see his new one, which he said was better, in that it took very little work to make.

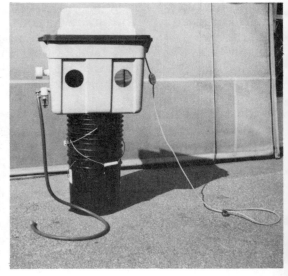

Turned out Vic had used a Rubbermaid container - which he said he got on sale for under $10 - and modified as in the photo below, by cutting 2 hand holes in the side (about 4-5"ϕ), and a 8" x 12" opening cut out of the lid, which serves as the loading and viewing port. Although the photo shows a 2nd, and smaller Rubbermaid container, which is fairly transparent, inverted over top of the hole in the larger container's lid, Vic says this has not proven to be a good idea - the best way to close the hole in the lid is to simply lay a piece of Plexiglass or ordinary window glass over it; through this you get a lot better idea of the results you're achieving on the job.

I was surprised Vic didn't have long-sleeved rubber gloves attached to his sandblast box; he just puts his hands through the hand holes. He says no sand escapes through these holes (which are obviously partially blocked by his arms when he's using it.) This puzzled me at first, but after a while I realized why the sand doesn't escape: when he is about to use the outfit, he hooks a Shop Vac to it via an adaptor attached to the left wall of the box. The Shop Vac puts a negative pressure on the interior of the box, so air is coming *in* through the unblocked portion of the hand holes. Vic also has a light inside, and of course the air supply and sand pickup tubes (plus more suction air for the Shop Vac) entering the box through smaller holes.

And, as Vic says, it don't matter if you kick it or knock it over, 'cause there's nuthin' to break except the light bulb.

This would be a good simple setup for someone with limited shop space and/or who would only use a sandblast rig infrequently. It is quick to make, easily stored out of the way, and set up in minutes when wanted.

Marsh Collins' sandblasting outfit is based on a portable sandblast kit he bought 20/25 years ago. (Maybe from Sears, he says, but not sure on that.) His cabinet looks more like what comes to mind (my mind, anyway) when the term "sandblast cabinet" is used. Marsh made it from plywood, solid wood, metal, pop rivets, etc., and put the sand hopper from the portable sandblast kit under the cabinet. Marsh says he mostly uses playground sand* for ammunition. Spent sand drops into the hopper-style bottom of the cabinet until he pulls out the sliding door to run it back into the sand hopper. A home-made shop vacuum cleaner is connected to a sand trap on the upper right end of the cabinet, and - as with Vic's rig - takes off whatever sand/dust may stay airborne when he's using it.

> * If you use playground sand, it would be a good idea to wash it to remove the fine dust fraction. You can do this pretty much as I described the washing of bone meal in the section on casehardening herein, as the principle involved is the same: the finer fractions are carried off by the wash water; what you want is the part which remains behind. When the wash water in the pail clears in a few seconds after agitation, fill the pail, give it another good stir, let it settle for a few seconds, pour off the water and suspended dust, and then spread the sand out to dry in the sun.

A cabinet like Marsh's would be easy to build. Glass beads can be bought from Brownells, Inc., as can long sleeved gloves, although such are not a necessity - you can use ordinary grocery-store scrubbing gloves.

The more I looked at Marsh's drwg, which just about filled one 8-1/2 x 11 page, the better I liked it. I have managed to get the same info onto 2 pages!

Save even more money.....
A store-bought portable sandblast outfit such as Marsh's is based on, will typically set you back $50/75. But you can make your own portable sandblast kit and save 90% of that. Making the actual sandblast gun is a simple job - see below - and a 5 gallon plastic pail can be used for the sand hopper. One source of such a pail would be just about any house construction site. A variety of construction supplies comes in these pails, including drywaller's "mud", and if you can get the guys to fill one of these with water as soon as they empty it, you'll end up with a nice clean pail.

Two ways to make the "gun" part are shown in the drwgs below.

Get the necessary rubber hoses from an auto parts store. (Marsh says to use 3/8" common air hose. Whatever - it has to fit onto the appropriate attachment points on your blast gun.) You'll need 2 or 3 feet for the sand supply line (Marsh suggests more like 5-6'), and a similar or longer piece to connect to the air supply line coming from your air compressor.

There's no need to do any "plumbing" on the plastic pail - the sand supply line can just be dropped in the top of the bucket so it reaches down into the sand.

If you want to go with the Rubbermaid cabinet idea, you could similarly just throw the sand in the bottom of the cabinet and bury the sand hose in the pile - just let it lay on the bottom. You may have to stop from time to time to herd spent sand over to the pick-up point. Then, just store the sand, gun, sand hose, etc. in the plastic cabinet until needed again.

Marsh Collins' Sandblast Cabinet - General Arrangement Drwg

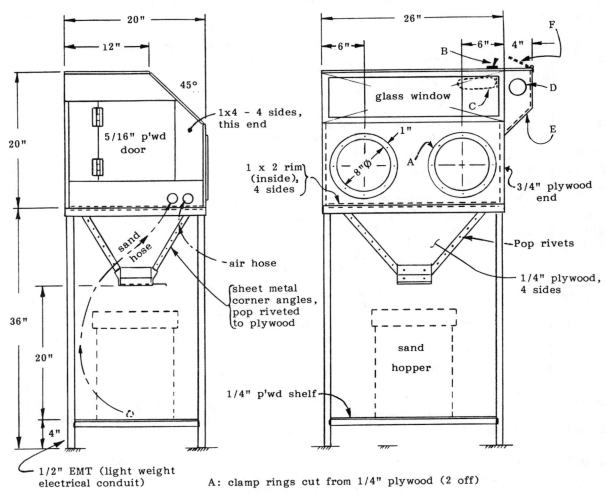

A: clamp rings cut from 1/4" plywood (2 off)
B: toggle switch for light and shop vac
C: interior light & socket
D: hole to suit shop-vac hose
E: 1/4" p'wd sand trap w/ baffle (8-9" long)
F: 1/4" hinged plywood lid

If you want to make your cabinet fancier, say by going to sheet metal construction, figure out your panel sizes, and get a sheet metal shop to cut the pieces to spec for you. You don't even need to bend the pieces of sheet stock - they can be screwed (or pop riveted) to an internal frame of angle iron or wood*. Gasketing all the joints with strips of sheet rubber or silicone sealant should give a pretty leak-proof cabinet. Make the window easily replaceable (probable cost: $1), because after some use it'll get frosted up. After everything's done, paint the parts (dark green or grey) and then assemble it. I'd be willing to bet that when it's done, you'll have as nice a sand blast cabinet as you could ever want, and probably as nice a one as you could buy. I think it'd be a fun project.

Detail 1 - Typical Section at Rim

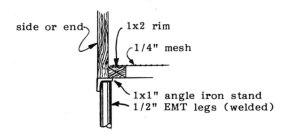

side or end
1x2 rim
1/4" mesh
1x1" angle iron stand
1/2" EMT legs (welded)

Detail 2 - Sand Hopper Gate

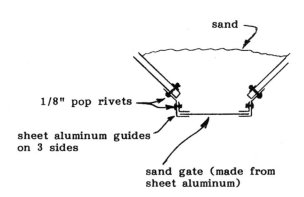

sand
1/8" pop rivets
sheet aluminum guides on 3 sides
sand gate (made from sheet aluminum)

Detail 3 - Section thru Window & Gloves

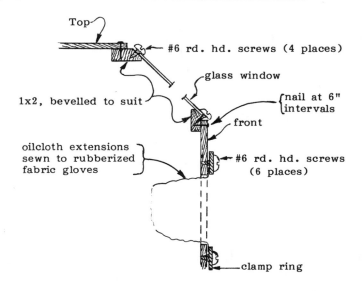

Top
#6 rd. hd. screws (4 places)
glass window
1x2, bevelled to suit
nail at 6" intervals
front
oilcloth extensions sewn to rubberized fabric gloves
#6 rd. hd. screws (6 places)
clamp ring

Detail 4 - Section thru Sand Trap

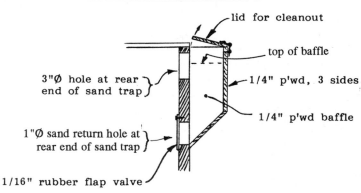

lid for cleanout
top of baffle
3"Ø hole at rear end of sand trap
1/4" p'wd, 3 sides
1/4" p'wd baffle
1"Ø sand return hole at rear end of sand trap
1/16" rubber flap valve

NOTE: holes between cabinet & sand trap are at rear of sand trap. Baffle is forward of these holes. Shop Vac hole (see D, in General Arrangement drwg) is at front of sand trap. Air is drawn into sand trap from cabinet; entrained sand will drop out for return to cabinet via sand return hole; air is sucked out thru hole D at front of sand trap.

* (Footnote to paragraph at bottom of previous page) Use say 1/4-20 button head or regular socket head cap screws, plus nuts and lock washers to fit. (A fella could tap 1/4-20 holes in the angle iron, but there'd be a lot of them, so buyin' a box of nuts would save muchos work.)

I was talking to Walt just a few days after I got Marsh Collins' info. Walt told me that a few months previously he'd gone to a local sheet metal fabricating shop to see if they would cut him out the parts for a sandblast cabinet. The guy offered to do it for a flat $65, or "time and materials", whichever Walt preferred. Walt opted for time + materials. When he came back to get his stuff, the total bill was ..... $25! This confirmed my own thinking that if you want to make yourself a real slick, professional looking sandblast cabinet, you could very easily do so, starting the job by having the main components knocked out for you for a surprisingly modest cost.

Additional comments digested from Marsh's letter will be found on the next page.

You should wear gloves if a sandblasting session is going to run very long, because after a while the sand starts to irritate the hand holding the work. But the gloves need not be fastened to the cabinet - as noted above, just ordinary grocery-store scrubbing gloves would do fine.

NOTE: the following quote from a Jan. 7/93 letter from Brownells to myself should be useful:

"... our sand blast cabinet gloves measure about 5-1/2"φ at the cuff, and 5"φ about 2 inches down from the cuff.

"... One of our tech support people used a couple of pieces of galvanized stove pipe, and attached it to (his) cabinet. He then slid the gloves over the pipe and put a big strap around each one. These gloves are not flanged, and I would not think they could be mounted directly to the cabinet."

"... Hope the above is helpful..."   Ken Raynor, Brownells, Inc.

Marsh uses a Hansen quick-coupler on his air input line. His sand hopper is about 9"φ, by about 14" high to the screen*. The bottom of the Sears sand hopper is conical, molded integral with the sides, and has a long-radius copper elbow molded right into the bottom of the cone, the skirt forming a base for the hopper, and the sand hose going out through a hole in the skirt. The top has a screen molded in, the screen being removable to dump trash. If you opt for the Rubbermaid cabinet, and/or the 5 gal. bucket idea, punch a piece of window screen into a rough cone and set it over the top of the bucket for the same purpose.

* (NOTE: always screen the sand going back to the hopper, as flecks of scale and goo that are blasted off the work can clog the sandblast gun nozzle).

The stand under Marsh's cabinet was made from thin wall conduit because he had lots of scraps of same salvaged from construction jobs. The stand could just as easily have been 2" x 2" wood, plywood (3 sided, in that case), or whatever.

Much the same idea is shown in Gunsmith Kinks II, at page 171 and over the page, where the fella who contributed the idea says that the bigger the air compressor you can afford, the better your sandblast outfit will operate.

You can even make your own sandblast gun, from a few short pieces of black iron pipe and pipe fittings, at very low cost. If you prefer to be a little more elegant, you can make a fancier version. Drwgs for both types appear two pages hence.

The way these guns work is this: high pressure air coming through the air supply pipe is speeded up greatly as it goes through the reduced diameter of the orifice in the gun. As the air speed goes up, the pressure drops. This creates a partial vacuum in the gun, which causes air and entrained sand to be sucked up the sand supply line. When the sand jumps out of the sand supply hose and into the gun, it enters the jet of high speed air coming through the orifice, and gets shot out of the gun's nozzle at quite a speed - I haven't tried to figure out how fast, but probably 300/400 mph - and there's yer basic "sandblast". Mighty effective!

(NOTE: *The drwgs for a sandblast cabinet and two types of sandblast gun on the following pages are adapted from Gunsmith Kinks, Vol. II page 424 and 425, with the generous permission of the Copyright holder, F. Brownell and Son, Copyright 1983).*

Some Ideas on Making Money with a Sandblaster

If you want to make a buck with your sandblast rig, get in touch with your local gunsmith. Typical gunsmith shop charges for glass bead blasting a barreled action plus the other metal parts of a rifle (trigger guard, floor plate, bolt handle, etc.) is $20/$50, so if the local gunsmith doesn't have his own bead blast facilities, you could offer to do it for him (or to let him do it himself using your rig) for something under those prices, thus leaving him room to make

A Sandblast Cabinet
that looks like it came from a factory

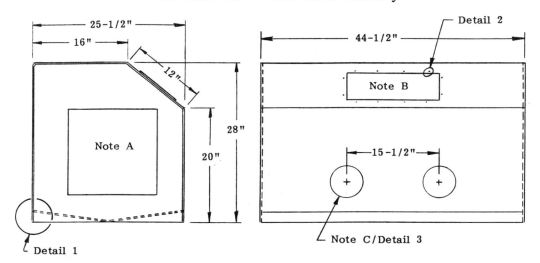

Make from 11 gauge sheet steel. Brake 44-1/2 x 72" piece of material as indicated to form back, top and front in one piece. Shear sides as shown, 2" taller than front and back. Shear bottom of cabinet to 25-5/8" x 44-1/4" and crease on long centerline, to collect sand, and to prevent warping while being welded in place.

NOTE A
Make Door about 15" x 15". Cut out piece, and re-attach with hinge and latch. Door required at one end only.

NOTE B
Cut a 7-1/2" x 14-1/2" opening for a window. Cut glass to 8 x 15". Cut four 1/4" wide strips of inner tube rubber, and secure glass to cabinet as at Detail 2.

NOTE C
Cut 2 holes about 5"ϕ for hands. Fit Brownell's Neoprene Gauntlet Gloves or similar if desired, per Detail 3. Check hole size against glove cuffs.

Detail 2

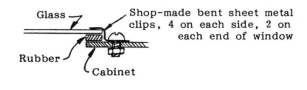

Detail 1

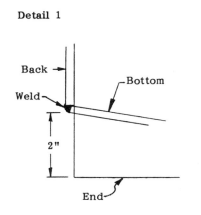

Detail 3

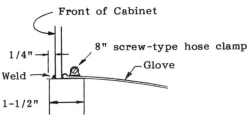

Sheet metal cuff formed to fit glove
& hole in cabinet. Weld into cabinet.

135

A Sandblast Gun made from Black Iron Pipe Fittings

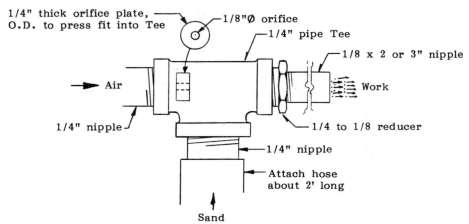

1/4" thick orifice plate, O.D. to press fit into Tee

1/8"Ø orifice

1/4" pipe Tee

1/8 x 2 or 3" nipple

Air

Work

1/4" nipple

1/4 to 1/8 reducer

1/4" nipple

Attach hose about 2' long

Sand

A Sandblast Gun Machined from a Block of Steel

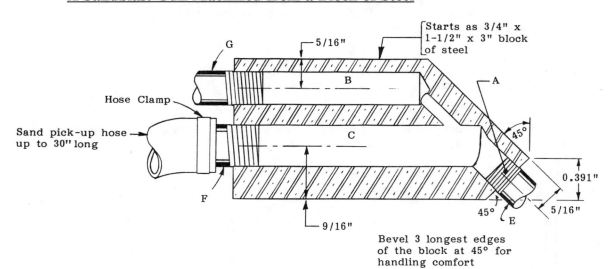

G

5/16"

Starts as 3/4" x 1-1/2" x 3" block of steel

B

A

Hose Clamp

45°

Sand pick-up hose up to 30" long

C

0.391"

F

45°

E

5/16"

9/16"

Bevel 3 longest edges of the block at 45° for handling comfort

A   Drill 1/8"Ø x 1-3/8" deep, drill 11/32"Ø x 5/8" deep, and tap 1/8 NPT.

B   Drill 11/32"Ø x 1-7/8" deep, and tap 1/8 NPT.

C   Drill 7/16"Ø x 2-1/2" deep, and tap 1/4 NPT.

D   This dimension will give a 5/8" wide flat for the nozzle hole.

E   1/8" x 2-1/2" pipe nipple

F   1/4" x 1-1/2" pipe nipple

G   as E; feed air @ 50/60 psi thru a 1/4" stopcock, put 1/4 to 1/8" reducer on nozzle side of stopcock, & mate with G.

136

something from it also. Plus there are the motorcycle and car buffs, probably most easily reached through dealerships and car clubs respectively.

There is also the possibility of doing custom sandblasting for artistic and architectural purposes - in glass, metal, and wood. One guy I know hereabouts cleans smoke/soot off fireplace fronts and interiors. He goes to the customer's home, tapes up a tent-like plastic sheet enclosure for himself, his machine and the job, and goes to work. When he's done, he vacuums up any mess that his vacuum cleaner didn't get while he was doing the job. Then he comes out of the tent, removes it and his equipment, collects his money, and leaves. He also uses this no-mess-when-I'm-done approach to do signs and art work on wood or glass doors, and windows in banks, offices, restaurants and private homes. Back at his shop, he'll etch fancy stuff to your liking on drinking mugs, mirrors, etc.

This fella uses an adhesive-backed sheet vinyl material about 1/16" thick, called "Buttercut". This stuff is sold by artists' supply houses for use as a "resist" for sandblasting designs into wood, stone and other materials.

If you want to sandblast some lettering into a piece of wood to make a sign, for example, find an alphabet in a type face you like, photocopy it several times at whatever size you want, and cut out the letters required for the words you want.

Stick a piece of Buttercut down on your sign blank, and then stick the lettering down on top of the Buttercut. (Letter spacing makes or breaks such jobs, as I have indicated elsewhere herein.) Carefully cut around each letter with an Exacto knife or similar. You can then pull away the letters, or the background, depending upon which you wish to sandblast.

Stewart Marshall told me that at about the time the wreck of the Titanic was located, he made up some very weathered looking 'miniature' "Titanic" name boards. He used 2 x 12's about 12' long (the real ones would've been several times that size), and sandblasted the lettering in, slopped on a bunch of paint, sandblasted them some more, and so on until, when they were done, they looked like they'd just come up from the wreck. He sold them to several seafood restaurants, and had fun doing it.

None of the above is metalwork, but if you did find you could get some paying work of this sort, and dropped all the money into a mason jar under the shop floor boards, one day you might pull it out and find some extra shop toys in it.

A Black Velvet Finish for Gun Parts
from a letter from **Marsh Collins**

I put a smooth "black-velvet" finish on the rifles I build, instead of bluing them in a hot salt bath. Looks pretty nice - others might like to know about it.

(Sunlight reflected from highly polished metal parts can be seen for long distances. This can be a sure game-spooker. If you are building a hunting rifle, a matte finish therefore merits serious consideration. GBL)

I smooth the parts down with emery, a fine flap wheel, etc. until all tool marks, etc. are gone. I plug the barrel at both ends to protect the bore and chamber, and then give the barreled action and other parts a nice smooth sandblast, inspecting to make sure it is EVEN everywhere. Don't touch the metal with bare hands or anything that could have grease or oil on it - I use paper towels to handle all the parts. (This is not a job for walnut shells - there would probably be bits of (oily) walnut meat in among the shells. GBL)

After the parts are sandblasted, I clamp the barreled action in my vise (with aluminum jaws) by the magazine well, to hold it steady. I heat the entire surface with a propane torch until a drop of water will sizzle on the metal. I let it cool for a minute or so to even out the heat, and then use Gun Parts Inc.'s (formerly Numrich's) 44-40 cold blue (or Brownell's Oxpho-Blue), applied liberally with swabs made from white paper towel cut in 4ths and folded to a pad about 1" wide and doubled. I pour the 44-40 into a clean plastic margarine tub or similar, and put it close to the vise so it's handy. That way, I can dip the swabs and keep rubbing the liquid on until the metal cools and is no longer vaporizing it.

(When swabbing on the 44-40, keep your face out of the vapors; do it at arm's length or wear close-fitting plastic goggles. The fumes are mildly corrosive, so if they rise up into the rafters and there find your prize varmint rifle, or a favorite faceplate, you will regret it later. Better not to do it in an unventilated shop. Better yet, do it outside. And wear rubber gloves, or wash your hands as soon as you finish.)

I let the parts dry for a couple of hours, then go over them thoroughly with fine steel wool - stainless is best, but ordinary steel wool* is also ok. I want to get all the loose black powdery coating off that will come off. After that, I finish up by rubbing well with cloth or a paper towel, still keeping any grease or oil away from it.

* (NOTE: see **TMBR#1**, page 173, re washing regular steel wool in liquid dishwasher detergent to remove the preservative oil therein. GBL)

Next, I let it sit for a day or two before going over it once or twice with cold 44-40, steel wooling thoroughly between each application. When the deep black finish no longer thins out with the steel wooling, I wash everything thoroughly in clean water to neutralize the chemical, and then heat with the torch just enough to evaporate any water, and wipe down again.

The last step is to heat it again until water sizzles (test on a spot that won't show) and then, while hot, spray with Triflon, Tef-Lube or any of the Teflon-bearing oils. When the part is cold, wipe off the oil to a dry surface, and the job is done.

I have rifles that I finished this way when I built them 20 years ago and have used regularly ever since, and they still look good.

"....it looked like a military Luger in mint condition...."

I had a call from Phil Lebow one day, and we got talking about the above. Phil said he'd once been asked to re-do a military Luger that someone had unwisely polished and re-blued. Phil said he boiled the gun's various parts in TSP to degrease them, plugged the barrel to protect the bore, then popped the parts into a plastic garbage can with some muriatic acid (hydrochloric acid, or HCl; get from any swimming pool supply house). He pulled the parts out frequently to check on the progress of the acid's etching action - what he wanted was to etch the surface to get rid of the polish, but to avoid letting things go too far and pit the surface of the steel.

In due course - with the aid of a certain amount of rubbing of some areas with degreased steel wool - he decided things had gone far enough, as everything seemed to have a nice even etch. He washed the parts and sent them out to a gunsmith for a trip through the hot bluing tanks. And when they came back? Phil said the reassembled gun, with a new pair of appropriate grip panels installed, looked like a military Luger in mint condition!

If you were going to try this on something, Phil suggested trying it first with the acid diluted 3:1 with water (3 parts water to 1 part acid - and always add the acid to the water, never the other way around). The reason for diluting the acid is that not all steels will etch at the same rate, and as indicated above, you just want to etch the surface, not pit it. Degreasing the steel wool is important, too. GBL

There is a further note re the same topic (etched/blued finish) at page 282 of **Gunsmith Kinks, Vol. I**. Etching in a diluted solution of Butyl Cellosolve and phosphoric acid, before bluing, results in a handsome and extremely durable black "parkerized" finish, according to Charles Degenhart, of Marysville, CA. See the book for the full dope.

Elsewhere in this same book there is a great amount of info on polishing and bluing, and in **Kinks II** there is a bunch more, plus many pages on electroless nickel plating. Apparently, with the latter process, if you want a satin finish, you can beadblast before the plating stage, and get a nice result, but..... you don't want to bead blast without first cleaning and degreasing the parts, because you can pound oil, bluing, silicone, etc. into the steel, which will play havoc with the plating process. Might be worth noting for the black velvet finish described above, too.

"You're still of the opinion that **Kinks I & II** are worth having in shops like ours, Guy?" Yes, I was just about to mention that.

THAT LAST HALF THOU

I've had 3 or 4 interesting calls from George Leverton, of Steamboat Rock, Iowa. As a rule he has some shop trick or idea to pass along for the brethren. One time, he told me he'd recently been fitting a breech block to a Winchester Hi-Wall receiver, and he eventually got the fit just about the way he wanted it, but not quite - it was still a hair too tight.

An old tool & die maker was watching him work. The old fella told him to put some layout blue on the job and let it dry. George did so, and then the old fella told him to file off the blue. It worked. By way of explanation, what happens here is that in filing off the blue, you will also take off a very small amount of metal - maybe half a thou. The blue is only to tell you when "enough is enough". And incidentally, it tells you not only when, but where, you can quit filing.

According to George's old pal, this was an old WWII-era trick often used by the toolmakers at Rock Island Arsenal, where he had worked. However, I doubt it was invented either then or there. I mentioned this to Stewart Marshall one day, and his response was, "Well shucks, La-Tord, any machinist knows that."

The funny part of the story, George said, was that almost as soon as the old guy had shared this tip with him, he was mad at himself for having done so.

This brings to mind something Bob Haralson once told me. "When I find a man in a shop I'm runnin' who won't show an apprentice boy anything he wants to know, I fire him."

Knowing Bob as I do, I probably needn't have asked him if he meant that he soon found some pretext for firing the guy.

"No. I **fire** him!" said Bob emphatically. I've chuckled over that many a time since.

More re Lapping an Edge Finder (see page 94)

Added to the 2nd printing: Dismantling an edge finder (e/f) is a little tricky. I wrecked mine finding out how to get it apart, but was able to re-compress the spring to the point where it was again operational. On mine, what works is to pull one end of the e/f far enough out that you can get something - e.g. a piece of 1/16"ϕ drill rod - in against the *end* of the last coil of the spring. If you thus provide some resistance to the spring turning when you try to unscrew the end component, success is yours. (Because I had stretched the spring out too far, I had lots of room, so used my left thumbnail. You don't want to do it that way.) Reassembly is easy (so long as you didn't over-stretch the spring!).

Please read, and understand, the following warning:

WARNING

If the info which follows inspires you to think about tuning up the factory trigger in a rifle, or making a new trigger to put on a rifle for yourself or someone else, you had better realize full well from right here to the end of the job that you are making an extremely critical part of the weapon. If it fails, and someone gets hurt as a result, you are going to bear full responsibility, so don't do it unless you are absolutely certain that you **thoroughly understand** what follows. And if you proceed, it is **YOUR responsibility to MAKE SURE, by CAREFUL, THOROUGH TESTING,** that the parts are made right, that they work properly together, that they are properly heat treated, and that when installed in the rifle the trigger **DOES** what it is supposed to do, namely hold back the firing pin until you wish to fire the rife - with no ifs, ands or buts. The trigger *must* do this or you might as well be carrying around a hand grenade with no pin, ready to go off at any moment. **The cardinal rule of weapons handling is this**: *never point a gun at anything you do not wish to destroy, and never fire unless you know what the bullet will hit.* If it hits your foot, or your son's forehead, you are not going to be very happy.

Since neither the author nor the publisher has any control over individuals using this information, **all responsibility for the safety and reliability of any trigger made to this general design rests solely with the user of the information.**

There is a lot of info here. Study **ALL OF IT** thoroughly.

Please re-read, and understand, the above warning.

TRIGGER TUNING and SHOP-MADE TRIGGERS
for
Mauser-type Bolt Action Rifles
from **Marsh Collins, Crestline, CA**

Dear Guy,

Some of the fellas might be interested in some info on tuning up factory triggers in bolt action rifles, and on how to make replacement triggers for Mauser-type bolt action rifles. The drawings below are typical of the triggers I've built - quite a number of them, actually - for myself and friends. There's nothing magic about them - they operate much the same as Timney, Dayton-Traister, etc., but I'm one of those guys who likes to "do it myself".

The info which follows is just what I'd tell anyone who wanted to improve his trigger's (hence rifle's) performance, was reasonably mechanically inclined, had some skill in using basic hand tools, and would enjoy having a decent trigger in his rifle without spending $100 to $300 for a gunsmith to do it for him.

I once read a gun magazine article in which the author said to trace the Ruger #1's original trigger (which I think is cast aluminum) onto a piece of steel in order to make a better looking replacement for the factory item. Tracing existing parts is not good enough!

However, let's look first at tuning up factory triggers in bolt action rifles, before going into how to make replacement triggers for military Mauser-type bolt actions.

Please re-read the warning that introduces this material.

How many times have you encountered a rifle whose trigger pull was such that you almost felt it required a chain hoist? And what kind of accuracy can you obtain if you're trying to hold the sights on a target while valiantly yanking at a balky trigger, or one that feels as scratchy as pulling a nail across a coarse file?

Many store-bought rifles are shipped from the factory with stiff trigger pulls in response to the many product-liability lawsuits we have seen in this country. With care and common sense the average shooter can learn to adjust that stiff factory trigger to get the maximum accuracy out of his weapon, with safety.

I believe that even among gun buffs there is a great misunderstanding of triggers - their function, their limitations and how they work. As the trigger is often the limiting factor as to the accuracy a shooter can obtain with any particular firearm, it might be well to first discuss certain "trigger basics".

A trigger system is simply a catch that restrains the firing mechanism until the shooter chooses to fire the weapon. How the trigger operates when you are trying to fire the weapon is where the rub comes in.

> NOTE: My comments about adjusting triggers are limited to bolt action rifles only. I would recommend against any tampering with pistol triggers. Except on target pistols, manufacturers generally provide no means of adjustment, and some even discourage disassembly for cleaning. Shotgun triggers are also generally not adjustable, so, other than bolt action shotguns, I'd suggest leaving these alone too.

Looking first at factory sporting rifles, there seems to be a growing and justifiable dissatisfaction with many off-the-shelf guns. Many times at the local range I have had shooters complain to me that they just can't get decent groups with an expensive new rifle. The first thing I try is the trigger pull; I usually find it to be heavy, and often scratchy as well. The trigger parts simply are not "finished" at the factory the way they were some years ago; the working surfaces have never been de-burred or stoned to smoothness. I have never seen this problem with Remington 700 series rifles, nor the Winchester Model 70, but I have encountered (and cured) it on four Ruger Model 77's, as well as on rifles by several other makers.

> Mentioning the Ruger 77 in this way is not meant to cast any aspersion on this particular design or on the Ruger line in general. If I were in the market for a "store-bought rifle," my choice would be the Ruger 77. I like 'em, and the trigger is a minor item to correct, particularly in view of the 77's excellent design, appearance, and overall quality.

It is a simple matter to remove the trigger group, disassemble it, and improve things with a hard Arkansas stone. BUT..... **be very careful only to smooth the metal and to preserve the original angles!**

First off, study your action and how all the parts go together, so that you'll be able to re-assemble everything correctly later. Be careful of springs under tension. Relieve the tension, if you can, before removing a spring. (And figure out how you'll get it back in place if you can't.)

> TIP: To keep from losing a spring you are about to remove, first tie a piece of string to it, with a nut or something heavier on the other end. Then, if it gets away from you, it won't go far, and will be much easier to find.

Keep all the small parts in a box or small plastic tub so they don't get lost. Work on a clean bench or table. Lay a white paper towel or cloth where you're going to work, so you can clearly see all those small parts.

Before trying to correct any other problems, stone the working surfaces smooth, as the smoothness of operation of all parts of the trigger will affect all other adjustments. Rough working faces directly affect the weight of the trigger pull, as well as uniformity of let-off from one shot to the next.

The next thing to address is the trigger engagement with the sear. If this engagement is insufficient, the mechanism will jar off easily with any bump or jolt. This is obviously an **EXTREMELY** dangerous condition which **MUST** be corrected, **or the weapon condemned.**

If the trigger/sear engagement is too generous, the trigger creep will be excessive and will destroy accuracy. There is usually a screw for this adjustment. Study the trigger mechanism to determine where this sear engagement adjustment screw is, and then adjust it until the amount of engagement is at a minimum level which thorough testing proves does not allow the trigger to be jarred off.

The next adjustment to make is weight of pull - which is done via the trigger spring. There's usually a screw for this, too, so study your action until you locate it. Using a small fishing scale, adjust it to your liking, again checking thoroughly that the trigger/sear engagement will not jar off.

> **NOTE**: Trigger pull weights of 2-1/2 to 3-1/2 pounds are fine for target work, but on a hunting rifle, I'd suggest something more like 4 to 4-1/2 pounds. You may *like* a light trigger, but you don't *want* one on a gun that is carried afield in circumstances where such a light pull could lead to the unintended discharge of a round that could cost the hunt or cause the death or life-long disablement of a shooting partner - wife, son, daughter, etc. **Use common sense** in setting the weight of pull.

> *As for trigger pulls down in the 2 to 5 ounce range, that is strictly for benchrest guns, and possibly - MAYBE - a chuck or crow rifle...... A pull of this sort is so light that a rifle so equipped should be considered UNSAFE as a field rifle, except in the hands of someone with enough sense never to close the bolt until he is about to shoot. Carrying such a rifle cocked and loaded under any normal circumstance is to beg disaster.*

After adjustment, bump and jar the fully reassembled weapon in every possible manner and direction to make sure it won't jar off in the field; that could be, at the least, embarrassing, and at worst, fatal. (If I were afield with someone whose rifle discharged accidentally for such a reason, even if it did no harm, it would be the end of the hunt right there, so far as I was concerned.) [Amen to that, brother. GBL]

> In the book "Gunsmithing Tips & Projects" **, there is, at page 214, an excellent article (from *RIFLE* Magazine, #14, March/April 1971) by Roy Dunlap, called "'Shrink' Groups with Some Home Gunsmithing". In this article, Dunlap has some other and equally cogent ideas on trigger tune-up work and other things to do to make a rifle perform better. I highly recommend you also see that item - it's well worth the price of the whole book if you want to know more tricks on how to do the job right. Even better: get every gunsmithing book you can get your hands on - Baker, MacFarland, Howe, etc. - and study - repeat, STUDY - what every one of them has to say about trigger work. You will no doubt learn something new and different from every one of them.

> Added some while later: Customer John Bringe tells me there is some very interesting technical/engineering info on triggers, calculation of lock times and more, in Stuart Otteson's book "The Bolt Action Rifle, Volume 1" (available from Wolfe). Wolfe also reprinted several past articles from *RIFLE* Magazine as a booklet, under the title "Benchrest Actions & Triggers". This document went into trigger design even more deeply, but it is out of print at date of writing (June '93). However, I mention it in case you might wish to borrow it from a library, had an opportunity to pick up a used copy, or wish to prod the publisher to reprint it. GBL

** mentioned later in this book

Military triggers are a different matter. Military rifles are designed for ruggedness and simplicity, not minute of angle (MOA) accuracy. Consider: a military bolt action rifle must have some play between the bolt and receiver walls to permit it to function even with sand or mud in the receiver. This also means that there is a certain amount of up-and-down play at the rear of the bolt where the sear-notch is usually located. Military triggers generally pull the sear down and out of engagement with the sear-notch to fire the weapon. With tension on the striker spring, this means that as the trigger is pulled, the rear of the bolt is pulled downward until stopped by the receiver walls, then the sear is dragged off the sear-notch. Any sand or other material that gets between the bolt and receiver wall can change this pull from shot to shot.

The better grades of sporting rifles avoid this by using either a "push-down" or a "roll-over" sear, where the sear either slides down, or rotates down and away from the sear-notch, releasing the striker. There is a subtle difference here that the reader should understand:

> On a military rifle, pulling the trigger pulls the sear down until it disengages from the sear notch, and the firing pin moves forward, firing the rifle.

> On a sporting bolt action rifle with a good "roll-over" or "push-down" (non-military) trigger, when the trigger is pulled, the trigger's tip ceases to support the sear; the sear then moves downward (with either rotating or straight line motion) under the force exerted on it by the firing pin spring via the sear notch...... and the firing pin moves forward, firing the rifle.

This latter type of trigger eliminates the uneven movement of the rear of the bolt, improving accuracy in two ways:

> 1st, by creating a smoother let-off, and

> 2nd, by allowing the bolt to support the rear of the cartridge more squarely throughout the firing action.

> Note: the roll-over type is also often referred to as a fall-down trigger. It is primarily the making of this type of trigger that we will look at below, and from here on we will use the term "roll-over" to describe this type of trigger. The Winchester M70 trigger system is referred to as an "over-ride" type, as is Timney's (and possibly some others) for the P-14 and P-17 Enfield.

This "roll-over" system is used in many commercially available trigger systems like the Timney, Dayton-Traister and Shilen. However, there is no great mystery about them, and anyone with the inclination and a modicum of mechanical sense can make an equally fine trigger at home, with a minimum of tooling.

I have built many such triggers, for many different types of military rifles, including a wild conversion of an 1888 German Commission Rifle to an MOA-shooting .223. They take some work to make, but the only machine tool I consider essential is a small bench drill-press, to keep the holes square with the work. For the rest of it, only a vise, a hack-saw, some files and ordinary twist drills are required.

**Please re-read, and understand, the warning
given at the beginning of this section.**

DIMENSIONS

When I make a trigger, I measure out all the existing dimensions necessary to fit THAT rifle, and go to it. There are three governing features that need to be watched:
1. the sear engagement,
2. the distance from the sear face to the pin which fastens the trigger assembly to the receiver, and
3. the measurement above this pin center to the floor of the sear channel in the boltway.

Fig. 1 General Arrangement Drawing
for a
SHOP-MADE TRIGGER for Mauser-type Bolt Action Rifles

[drawn with left hand Sideplate (nearest viewer) removed to show interior details]

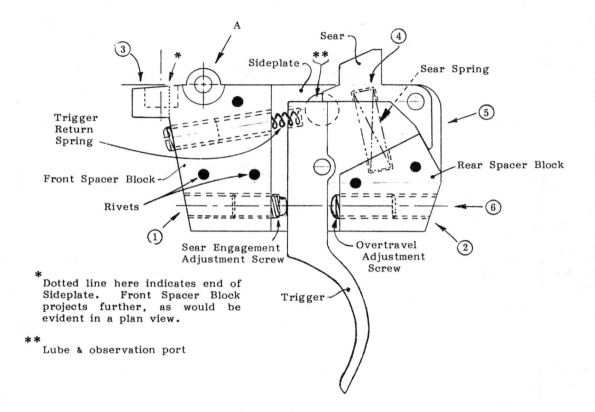

* Dotted line here indicates end of Sideplate. Front Spacer Block projects further, as would be evident in a plan view.

** Lube & observation port

Notes re General Arrangement Drawing

1. Slope at front may be required for clearance when trigger assembly swings down at rear when bolt is worked. Note that trigger assembly pivots at "A" - bottom left corner of trigger frame will move forward (i.e. to the left in the drawing).

2. Slope at rear is to provide clearance, if required, for "whatever" in trigger well in stock.

3. Hand file the slope forward of Pivot A by eye to provide sufficient clearance so the trigger assembly can tilt down far enough when the bolt is pulled back.

4. When the Trigger is pulled, the Sear rotates downward, forced down by the sear notch of the striker (firing pin), which moves forward under the pressure of the firing pin spring. Note also that when the Sear swings down, the angle of the Sear Spring centerline changes. Spring pockets entering the "spring" faces of the Sear and Rear Spacer Block at 90° to these faces allow the Spring and spring pockets to line up when the Sear Spring is compressed. Just make sure the pockets in Sear and Rear Spacer Block line up reasonably well, so as not to kink the Sear Spring.

5. This face of the Rear Spacer Block serves only as a stop for the tail on the Sear, so that the Sear will not flop over backwards.

6. An alternative arrangement is to put the Trigger Return Spring in this hole, and put the overtravel adjustment screw in the Front Spacer Block. Either way will work.

144

These all affect the bolt and safety functioning. For example, if 2. above is not correct, the safety probably won't work. **So, these measurements are considered critical.** Others are not. For example, the angles at the front and rear ends of the trigger housing are there to provide needed clearances so the assembly can swing down and slightly forward, and to clear the contours within the trigger guard assembly that might bind or obstruct the free movement of the assembly. Since these angles are for clearance purposes only, they are not at all critical. (Which means they don't have to be calculated or laid out in degrees, minutes and seconds. ±1° is quite fine enough.)

Also, the trigger finger bow (curved portion of the trigger lever) needs to be positioned so it looks right in the trigger guard. I usually also find it necessary to open out the aperture for the sear in the floor of the receiver and the aperture for the trigger in the trigger guard frame.

For these reasons, I invariably start by making a drwg of the military trigger assembly from the particular rifle I'm working on, and then duplicate the critical dimensions from that in my drawing of the new trigger assembly.

I initially dimension the sear engaging surface to get the maximum engagement - it can be filed or ground down as required more easily than added to. It must finish up high enough to reliably engage the sear notch on the striker without being so high that it cannot drop down fully out of the way when the trigger is pulled.

Virtually all the other dimensions and shapes are to suit available springs, screws and other hardware.

The main relationships I've seen are those between the sear and the assembly mounting pin, the trigger bar pivot location and the shape of the trigger to fit the existing trigger guard. All else is relatively adjustable, so long as the assembly comes out small enough and contoured to fit and function properly in the space between the receiver and the trigger guard.

Original Mauser Trigger and Sear
(Drwg is of parts from a particular rifle. Yours may be different. See text)

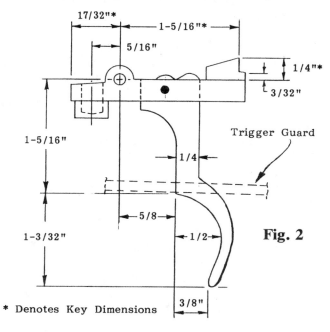

Fig. 2

* Denotes Key Dimensions

The angles at the front and rear of the Sideplates (see Figs. 1 & 3) are simply clearances, to suit the individual rifle being worked on. In the case of the one drawn here, the front slope was required to avoid interference with the magazine floorplate catch, and the one at the rear was to clear a humped-up shape on the trigger guard on the rifle that I was making it for. I've made some that were just squared off; on others I sloped it more sharply to clear a shorter well in the stock for the trigger.

On the GA drwg (Fig. 1) the pivot hole is shown 1/8" down from the top of the Sideplates. This was essentially an arbitrary dimension to leave 1/16" of metal between the top edge of the 1/8" hole and the edge of the Sideplates. This location doesn't take any amount of thrust or shock, so I don't expect any amount of wear in a lifetime, and I wanted to keep the pivot as close to the level of the sear as I could, to increase the downward component to the swing of the sear when released, reducing friction at the sear engagement. The number of degrees of rotation of the sear would be increased if the pivot were too close horizontally under the point of engagement, making a torsion spring almost mandatory, and I have an aversion to those; I've repaired too many firearms where a torsion spring was the offending part.

On the 1888 German Commission rifle I converted to .223, the sear location was forward of the trigger by about 1/2". I had to design a trigger that used a short push-rod to actuate the "secondary" sear. It sounds complicated, but in reality it wasn't. I simply made a short pivoted "secondary trigger" bar that the sear engaged with, and a push rod with pointed ends riding in a drilled hole in a middle spacer block. (See Fig. 7, following main drwgs.GBL) It's one of the smoothest triggers I have.

I have included only broad details of Marsh's "1888 trigger," and only to show how the general design can be adapted to suit a non-standard situation. There are in fact a couple of things about Fig. 7 that are not quite correct, but it will serve to convey the general idea. It is *not* intended to impart any understanding of how to make a trigger for the 1888 German Commission (or any other) rifle.GBL

The accompanying drawings detail a "roll-over" style trigger for a Mauser 98, but the basic idea, with appropriate changes in some of the dimensions, can be applied to any of the Mauser-based variants such as the M1903 Springfield, any of the Type 30, 38 or 99 Jap Arisakas, many Mannlichers, and some sporters, also Mauser based, where the maker has retained the military-style trigger system.

The dimensions shown apply to one particular rifle, and, because dimensions vary among rifles of generally the same make and model (as different contracts called for modifications, etc.), it is necessary to *start out by studying the action, and checking all the critical dimensions on the rifle for which the trigger is being made.*

Mausers and their clones are simple, but it seems that every manufacturer (Mauser, Loewe, FN, Brno, and the dozens of third-world countries that have either specified weapons made for them or have manufactured their own) had their own druthers re dimensions. Then there are the changes due to a change in caliber (magazine length, width, or depth) and any special changes to improve accuracy.

As I noted earlier, in Mauser-style actions, the bolt must have a reasonable amount of slop in the boltway to avoid jamming by small amounts of dirt. This allows the bolt to move up and down 1/32" to 1/16"; whenever a grain of sand or other crud gets in the boltway, the trigger pull (with a military-type trigger) will vary from shot to shot as the bolt moves down with the downward movement of the sear until stopped by the receiver floor, a grain of sand, etc. For military purposes this is OK - soldiers are seldom marksmen. For target shooting and hunting rifles, which

Fig. 3   <u>Sideplates</u> (2 off)
(1/16" CRS)

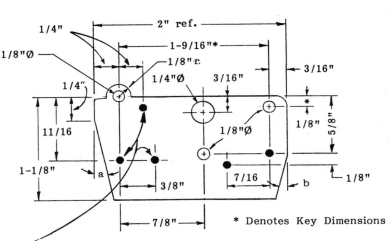

Rivet Holes (drill in one only initially and use that one as drilling template for Spacer Blocks and opposite Side Plate.

<u>Note re angles a & b:</u>

Not critical, and not required in all cases. Check what's required on the rifle to which the trigger is to be fitted, and then file or mill off sufficient material to permit trigger assembly to swing freely. Trigger guards of various Mauser-type rifles have different configurations in this area. This particular drwg depicts a trigger for a rifle that had a rise (bump) on the inside face of the trigger guard, hence trigger Side-plates (& Rear Spacer Block) were cut away to clear at the back end. The front angle was cut to clear the rifle's floor plate catch.

146

will rarely get the abuse a military arm is expected to get, we want better attainable accuracy, and for this, one big contributing factor is a good trigger.

HOW THEY WORK

The roll-over sear, like the Timney and Dayton-Traister - and mine - care nothing about the up-and-down position of the bolt, and consequently the sear notch. The sear and trigger are held in constant alignment within the trigger assembly, regardless of sear-notch height, and the actual let-off occurs inside the assembly rather than at the sear/sear-notch engagement. The "over-ride" trigger of the Win. M-70, and replacement triggers for the P-17 Enfield, require the sear notch to ride up the sear ramp and be "jammed" against the upper limit of the boltway until the trigger is pulled, releasing the sear, which is then forced downward by the slope of the ramp. Although these seem to work quite well, there is still a possibility for crud in the boltway to vary the let-off, as it is still somewhat dependent on the bolt's elevation. I guess that's mostly theoretical, but I prefer the roll-over sear principle for simplicity.

Installing a trigger of this type will require no modifications to the receiver other than possibly lengthening the openings for the sear (in the receiver) or trigger (in the trigger guard). These changes, if required, will be obvious, and can be determined by your own measurements.

Sear
(1/4" CRS)

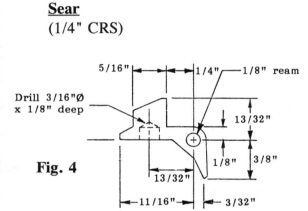

Fig. 4

The new Sear (Fig. 4 at left) is a right-angle lever. The firing pin spring puts a forward push on the new Sear's vertical arm, which engages the sear-notch on the rifle's firing pin, or striker. The horizontal arm of the new Sear engages the top end of the new Trigger by about 1/32", and the contact area must be stoned square and smooth.

The Sear Pivot Pin is a short piece of 1/8" drill rod, and is carried by the Sideplates. I have also used 1/8" welding rod for Pivot Pins, and found that it worked fine for years, but drill rod is better.

A small coil spring returns the Sear to the "up" or "cocked" position after the pressure of the sear-notch is gone and while the shooter's finger is still holding the trigger back. A tail, or stop lug, on the back of the Sear prevents the Sear from rotating up too far, easing removal and re-installation of the trigger assembly. (As you study the drwgs, you will see that once the Sear, sear return spring, and Sear Pivot Pin are put in place, this stop lug on the back of the Sear keeps the Sear from flopping around, and the spring cannot escape, thus making the assembler's life much simpler.)

The Trigger (Fig. 5) is simply a bar that props up the horizontal arm of the Sear. It is shaped at the bottom to accommodate the shooter's finger, while the top (sear contact area) must be square and smooth. The Trigger is carried on a Pivot Pin identical to the one carrying the Sear.

A spring returns the Trigger to its position under the sear nose when the Trigger is released by the shooter, and an adjustment screw allows the Trigger to return to the same position after each shot.

An overtravel screw behind the Trigger stops it after it has released the Sear. This prevents excessive movement of the trigger finger, which could affect the shooter's aim at the moment of firing.

Trigger
(1/4" CRS)

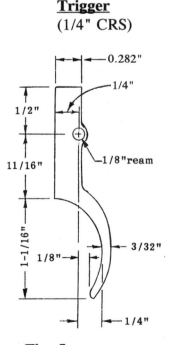

Fig. 5

The entire assembly is enclosed in a "frame" consisting of two Sideplates separated by two "spacer blocks" that carry the springs and adjustment screws. The Spacer Blocks and Sideplates are semi-permanently assembled with five small rivets. Rivets can be made from aluminum aircraft rivets, soft welding rod, coat hanger wire, or any other similar metal that can be mushroomed slightly by carefully directed hammer blows.

The Trigger and Sear are 1/4" thick, as are the Spacer Blocks. Therefore the Trigger and Sear sides will likely need to be thinned slightly by rubbing on a file or on a sheet of fine abrasive paper on a sheet of glass, so that they will move freely between the Sideplates once the trigger housing is riveted together. (See the **quickie tip, "Filing for 'Flat' and 'Finish'"** following this section.GBL

Fig. 6   <u>**Front & Rear Spacer Blocks**</u> (1 off, each)
                                     (1/4" CRS or Aluminum)

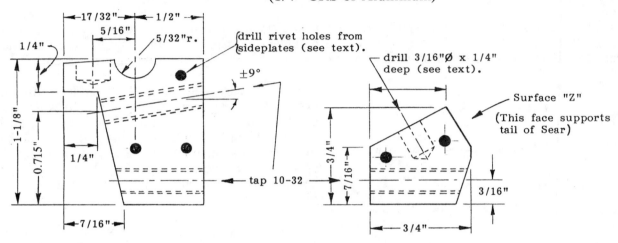

NOTE: Please see comments headed "Some final notes concerning the above," on the page after next, regarding procedures for bringing the Front and Rear Spacer Blocks to final profile.

The new trigger assembly is attached to the bottom of the receiver with a single pin (you may be able to re-use the original pin; if not, make a new one to suit) in the same lug and hole that carried the original trigger mechanism. A smaller spring than the original is placed in the hole ahead of this pin to force the assembly up against the bottom of the receiver except when the bolt is moved to the rear to reload and cock the rifle. This allows the sear-notch on the striker to pass over the Sear, which returns to its upright position immediately after firing; the entire assembly swings downward a fraction of an inch to allow the sear-notch to pass over the Sear as the bolt is drawn back, and catches it on the forward stroke of the bolt, which leaves the weapon ready for the next shot.

SPRINGS
As for details on the springs I use in my triggers, I take them from a spring assortment I bought about 30 years ago. Brownell's #69 and #71 spring kits would give the average hsm a lot of springs that will cover many jobs. (The #69 spring kit is most likely to provide suitable springs for triggers such as those described here.)

There's nothing really critical about spring specs - you just need an assortment at hand from which you can pick springs of a size that will go into the holes, and that are strong enough to do their job.

If you'd rather wind your own, the only one that is at all fussy is the trigger spring. It must be stiff enough to provide the adjustment range you prefer, and may take some "cut and try" before you get one that suits. (I'd be inclined to start with about 0.020"φ wire for the sear and assembly return springs, and 0.025"φ or possibly a little heavier for the trigger spring.GBL)

The trigger spring wants to be about 5/32"OD x 1" long. The sear spring's hole is not tapped, so the sear spring can be about 3/16"OD, and 7/8 or 1" long. Very little force is required to raise the Sear while the trigger is still back, so any strength that will do the job is fine.

The spring at the front of the trigger assembly should be 3/16"φ x about 7/16" long, and need be only stiff enough to hold the assembly firmly up against its own weight.

(There are some useful notes on spring winding in **TMBR#1**, page 154, and at pages 23-25 & 30, in **"Hey Tim, I gotta tell ya..."**. GBL)

MAKE YOUR OWN DRAWINGS
Dimensions, as previously noted, **are general, and must be altered as necessary to suit the individual weapon.** I have included a drawing of the original Mauser trigger from the rifle for which this particular trigger was built, showing on it dimensions which are critical and which must match on the replacement assembly. Working clearances must be provided, to allow the assembly to swing downward for operation of the bolt. I suggest that you **make an accurate drawing of both the existing trigger and the proposed new trigger, for two reasons**:

1st, to make sure everything will fit in the space available.

2nd, to gain a full grasp of how the trigger works and what you must do to make the parts.

HEAT TREATING THE PARTS
It is necessary to harden the Sear and the upper end of the Trigger after these parts are otherwise finished. If you choose to make them from cold rolled steel, you can harden them with an oxyacetylene torch adjusted to give a strongly carburizing flame: coat the parts with softened soap, heat to a cherry red, quench in salt water, and polish afterward with a fine wire wheel. The soap prevents oxidation. If you are not familiar with this method of casehardening, don't attempt it without first seeking some instruction from a good welder or welding instructor who knows his stuff. (Another way to go would be to use Hard-n-Tuff, or Kasenit, available from Brownells, Inc.)

NOTE from GBL: One could also caseharden the working areas of the Sear and Trigger using pack casehardening methods detailed in the story "The Bullseye Mixture", in **TMBR#2**, and expanded upon in the opening pages of this book, but that may seem like a lot of bother for two such small pieces.

In fact, I would hesitate to use this approach on the Trigger, which has a rather thin section around the pivot pin hole; the casehardening could penetrate right through in this area, with the attendant possibility of it ending up too brittle, and perhaps shattering at some later time - possibly with terrible consequences. To avoid this problem, you could mask off the area around the hole in the Trigger, either with localized copper plating or by covering this part of the Trigger with fire clay, as is discussed in more detail in "The Bullseye Mixture".

The other way to go is to make the Trigger and Sear from oil hardening ground flat stock or similar. An 18" length of Starrett 1/4" x 3/4" ground flat stock from MSC will cost about $10, and will make several triggers and sears.

Marsh's response to my thoughts as above was as follows: "....Use of high carbon tool steel would also be good to (consider) ... but I've never had any problem with breakage of either of these parts (trigger or sear) even using only a carburizing flame, as neither carries any great force or shock in the direction of the thin sections. I hold the sear at the hole with pliers while heating it, which acts as a heat sink, and only the two points get really red. The trigger I only heat at the top end."

The two Pivot Pins are called out at 1/8"ϕ, but can be anything near that size. Rivet sizes can be whatever is handy:- just drill to suit. The drwg calls for 1/16" CRS sheet for the Sideplates, but I've made several with aluminum Sideplates.

The 1/4" hole in the Sideplates is to view the sear/trigger engagement and for lubrication. I use a moly disulphide grease on the working surfaces, as it seems to stay put well.

> NOTE: On Arisaka rifles, the sear engaging surface must be flush with the shoulder in the receiver. Why? When the Arisaka safety is moved to the "safe" position, the striker (firing pin) is rotated so that the firing pin's sear-notch rests on the receiver shoulder (making this one of the safest safeties made). When you return the safety to the "Off" or "Fire" position, the sear notch returns to the sear engaging surface of the trigger mechanism. Hence, this surface of the sear must be in the right position fore and aft in the receiver to allow this to happen.

Now, if you make such a trigger for yourself, and your shooting friends admire the smooth and crisp trigger pull of your rifle, you can tell them, "I made it that way."

Marsh

> The above was about the end of Marsh's original material. The following notes were developed during correspondence between Marsh and myself as we knocked it into the suave presentation you see here!

Some final notes concerning the above:
In making the parts of one of these triggers, you will probably find it fastest, and easiest, to make the Sideplates first, and then, turning from them to the Front Spacer Block (see Fig. 6 above), machine only the top, bottom, and rear faces of this piece initially. Leave the front, or left hand faces of the Front Spacer Block hacksawn roughly oversize at this stage, and later file or mill* them to exactly match the profile of the Sideplate once the 3 pieces are riveted together. This way, the sloped portion at the front of the assembly (if a sloped area is even required to suit the rifle for which you are making the trigger - it may not be) will be flush right across, giving a more professional appearance. (Not that it matters much - the whole thing is hidden from view when installed in the rifle - but why not do a nice job of it?)

> * Marsh said that most of the "milling" on his triggers has been done with files, rather than a milling machine, which he only recently acquired. Filing requires elbow grease, but minimizes set-up time.

Similarly, you may find it easier to machine the bottom, left, and sloped left upper faces of the Rear Spacer Block and again just roughly hacksaw the rear, or right hand faces of this part to something near final profile initially. Then, clamp it in between the Sideplates (the Front Spacer Block is already riveted in between them) and drill the rivet holes in the Rear Spacer Block from the Sideplate holes. (Drill the holes in one Sideplate only, initially. Then at this stage, extend these holes right through both the Rear Spacer Block and the opposite Sideplate at one go, so the holes will be in exact alignment through the 3 pieces.)

Next, do NOT rivet the Rear Spacer Block in place. Put it in place with service pins - i.e. pins the same size as the rivet holes, but not riveted at all. Scribe the profile of the rear end of the Sideplate (sloped or otherwise, as required) onto the Rear Spacer Block, and file up to the line. This leaves Surface "Z" only (see note at far right hand side of Fig. 6) to be worked on, and here, because of the service pins, you can take the Rear Spacer Block out as often as necessary, and file away at surface "Z" until, when reassembled, the Sear is properly positioned by the shape to which surface "Z" of the Rear Spacer Block has been filed.

Rivet the Rear Spacer Block into place only when you are satisfied it is properly shaped to do its job. The rivet holes are shown as round black dots on all the drawings.

Note: Once the rivet holes are drilled through from one Sideplate, through the Spacer Blocks and through the opposite Sideplate, countersink the outside of each rivet hole in the Sideplates slightly with a center drill. Rivet the Spacer Blocks in place at the appropriate stage in the job, and carefully file the rivets flush with the surface of the Sideplates.

You will notice that I have shown flat bottomed holes for the springs in the Sear, Rear Spacer Block, Trigger, and Front Spacer Block. These holes don't have to be flat bottomed - they can be drilled with a regular twist drill, or cut with a slot drill (= 2-fluted end mill), or even bored with a suitable tool in a boring head in your vertical mill.

To properly locate the spring hole near the top of the Trigger, I would be inclined to drill it once the whole mechanism was assembled. I would use the tapping size drill to drill the trigger spring hole in the Front Spacer Block and extend the hole into the top end of the Trigger, before putting my 10-32 tap into the hole in the Front Spacer Block. I might then remove the Trigger and enlarge and/or machine a flat bottom in the spring hole, as noted in the previous paragraph.

The sear lifting spring is shown with a dash and double dot rectangle and diagonals, whereas I have made a more artistic (??) depiction of the trigger spring.

The foregoing suggestions arose from the extra insight I gained into the design while preparing the drwgs you see here, which are simply my re-renderings of Marsh Collin's very succinct (and complete) drwgs. What he got on one page, I could easily get on 3. Hope you've enjoyed it. GBL.

Please re-read the warning that introduced this material.

Filing for "Flat" and "Finish"
from **Jarvis Williams, West LA, CA**

Small parts look much better if they can be finished up with a file, with all filing marks going in one direction, but very often this results in a part that is no longer flat. It is hard to hold a piece in a vise and still file it flat. A lot of people say they can, but seldom when they are near a vise and file. However, if you will put the file in the vise and draw the part over it, a nice finish will result and the part will remain flat. The file must be sharp and clean and you can't expect to remove much metal, just improve the finish.

Jarvis again:
The best thing small taps - 0-80, 2-56, 4-40 - seem to do is break. If the tap is sharp and the tap hole is the correct diameter, they shouldn't twist off unless they bottom out. They will still snap off if they are bent at all. If the tap and wrench are held in one hand and the part in the other, instead of in a vise, the tendency to snap the little taps seems to be less, as you have a better feel for how things are going.

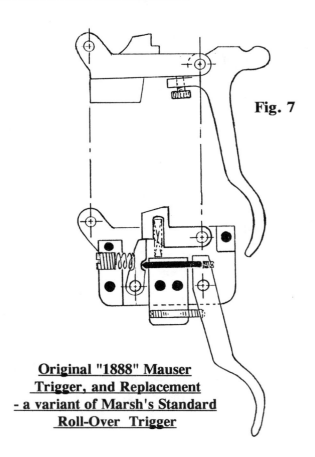

Fig. 7

Original "1888" Mauser Trigger, and Replacement - a variant of Marsh's Standard Roll-Over Trigger

AN EFFECTIVE SHOP-MADE MUZZLE BRAKE
more from **Marsh Collins, Crestline, CA**

Years ago I built a 300 Winchester Mag on a P-17 action. When rechambering or rebarreling Enfields and Springfields, I always set the barrel back one full thread and eliminate the funnel at the back end of the chamber. This is called "safety-chambering", and results in the head of the cartridge being completely enclosed and supported (against rupture) by steel*. Remington calls it "counterboring" and make like they invented it, but Arisaka was doing the same thing in 1905 - that's why Arisakas are so hard to blow up.

* 98 Mausers have a ring in the receiver that does the same thing.

Anyway, in rechambering this particular P-17, the chamber gage kept telling me I was done when I knew I still had some reaming to do. I finally discovered that the bolt was slightly non-concentric with respect to the receiver. The gage was forcing the bolt sideways into a pinch when it was in the action, but when the barrel was out of the action the mike said I still had several thousandths to go, and without the gage the bolt closed smoothly. Once I realized what the trouble was, the fix was simple: originally I had allowed about 10 thou clearance around the bolt head. Reluctantly I bored it out to 20 thou, which allowed the bolt to move laterally enough to cure the problem.

That rifle still shoots minute-of-angle groups, and the recoil is about like a .30/30, thanks to a muzzle brake I made for it. Muzzle brakes are duck-soup to make - see drwg below.

Most of the dimensions in a muzzle brake are not sacred, and one can be made from anything from hot-rolled mild steel to CRS to stainless. The one thing that I have found to be critical is that the expansion chamber must not be longer than the shortest bullet to be fired. Accuracy will be poor if it is longer - I figure that some of the escaping gas, as the bullet leaves the "true" muzzle, passes the bullet and has a tendency to randomly deflect it sideways.

The vent holes from the expansion chamber aren't critical. I generally drill #30 or 9/64". When I have tried using larger

A Shop-Made Muzzle Brake

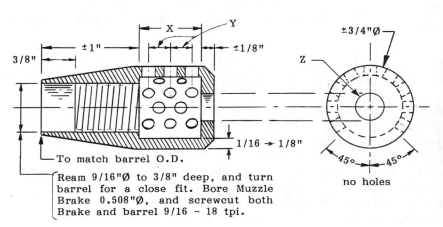

Ream 9/16"Ø to 3/8" deep, and turn barrel for a close fit. Bore Muzzle Brake 0.508"Ø, and screwcut both Brake and barrel 9/16 - 18 tpi.

X: Max X = bullet length.

Y: vent holes - anything from 1/8" to 3/16". Marsh says smaller holes - say #30 or 9/64"Ø - seem to work better. Make dimension "Y" such as to give an even spacing of the holes along the length of the chamber, for good appearance, and stagger the holes as indicated in the drwg above.

Z: Exit hole should be 15 to 30 thou over bullet diameter.

holes, say 3/16" or maybe 1/4", the recoil-damping effect seems decreased, perhaps because the large holes allow a too-free release of gas, reducing the effect on the flange at the front of the muzzle brake, where all the work is done. Having a flat face around the bullet exit hole is what does most of the work.

The vent holes are staggered and spaced to get a maximum of holes without getting them too close together - usually 3/16" center-to-center on a line is fine for something like a .30 caliber barrel. Note that there are no holes in the area about 45° each side of bottom dead center; this is to allow the holes on top to reduce muzzle-jump. The holes also allow use of a suitable nail or other rod to be used (carefully!) as a wrench when removing or replacing the brake.

I found that the size of the hole for the bullet to pass through is not critical; I use 5/16" for a .30 cal. bullet, for example. The key thing is to have an expansion chamber that extends forward from the "true" muzzle no longer than the bullet and to make sure the whole thing is in good axial alignment with the bore.

Although the brake is screwed onto the end of the barrel (as the drwg plainly shows), it is a good idea to incorporate a set-screw as well if the threads don't come out overly tight. On a couple of these brakes that I've made, where I have used a coarse thread instead of a fine one, they have had a tendency to start to unscrew after many rounds, and this could cause bullet-shaving if the brake loosens enough for the bullet to contact the "false" muzzle as it passes through, and.... bye-bye accuracy. I've had that happen, but outside of the total loss of accuracy -- I suddenly couldn't figure out where the bullets were going, after a session of near-MOA shooting -- there were no serious results. I removed the muzzle brake, cut the barrel back and re-crowned it in the lathe, rethreaded it with a finer thread, and made a new brake to suit. No more problem!

The 9/16"-18 thread called for at the muzzle/brake juncture is simply one that has been a convenient size for the barrels (30 caliber and smaller) I've put them on. Once when I put one on a heavy barrel, I used a 5/8-18 thread, as would be required for larger bullets - 375, 444, etc.

There's nothing magic about the length, but do cut a length of thread on the muzzle equal to about 1.5 times the diameter, plus a little more if you like, for good luck, as the brake takes a pretty hard wallop when it stops what would otherwise come back against your shoulder.

I put a pair of witness marks (punch marks) on top, so that when I remove the brake to clean the rifle, I can get it back on true. There's nothing sacred about the taper at the rear - all it is for is to give a smooth, eye-pleasing transition from the OD over the expansion chamber down to the OD of the barrel just back of the thread.

One last point: make sure the guys next to you on the firing line are wearing their hearing protectors - any muzzle brake tends to throw out to the side a lot of the noise that would normally be going out front.

A concluding note from GBL: see also Brownells' Catalog for various other muzzle brake designs.

A Further Brief Note on Rifle Triggers

Back to the matter of triggers again briefly: Reference was made near the beginning of the preceding section to a magazine article on making a better-looking trigger for the Ruger #1. If you want try making a new trigger for a #1, see the book **Gunsmithing Tips & Projects**, from Wolfe Publishing. I can't recommend the article in question any more highly than did Marsh Collins, although the book does have a lot of other ideas in it that you might find interesting.

If the rifle is intended for varmint hunting or other fine, deliberate shooting, there's another alternative that may interest you, and that is to fit it with a commercial **CanJar set trigger**. Installing the CanJar unit isn't necessarily a straight "plug-in" job, because the #1 has varied a little internally over the years, and you may have to do a little minor gunsmithing. The work required is well explained in the maker's instructions, and should be within the capacity of any hsm capable of building a trigger unit of the type described by March Collins. The CanJar trigger is available from M.H. CanJar Co., 500 E. 45th Avenue, Denver, CO 80216; phone (303) 295-2638. Talk to Mary White.

Rebuilt Rugers, Stuck Bullet Removal, and the *Ne Plus Ultra* in Reamer Stoning Information

The Harley's front wheel lead me down to Cliff La Bounty's place in Maple Falls, WA a couple of times last Summer. As I've mentioned elsewhere, Cliff makes a business of re-boring and re-rifling rifle barrels on a largely automatic rifling machine he designed and built.

Cliff showed me something that gave me an idea for something a guy could do for himself, and/or that might well result in a certain amount of quiet, down-in-your-basement business.

Cliff had a friend's gun in the shop to remove a cast bullet which had stuck in the barrel about 3/4" back from the muzzle. Cliff said that if he'd had the bullet out before I arrived, he would have let me shoot the gun, and I would've liked to do so, because it was somewhat out of the ordinary.

The gun had started out as a Ruger "Bisley" 6-shooter, but had been converted to a "5-shooter" with a .50 caliber bore. The cartridges were made from .348 WCF brass shortened to an overall length about right for a cylinder of typical Ruger length. Cliff makes the barrels in long sticks (30" or so), and Hamilton Bowen*, who builds these pistols, hacks off pieces so he gets 5 or 6 barrels out of one of what Cliff makes for him.

  * Bowen Custom Arms, PO Box 67, Louisville, TN 37777. Send $3 for current catalog.

The barrels are Mag-N-Port'd, and their outside profile includes an integral ring at the muzzle as seen in the photo at right. A new ramp front sight is screwed to the barrel, one of the screws being located in this raised ring. The distance from the front face of the Ruger frame to the rear edge of this ring on the barrel is precisely the same as the length of the gun's ejector rod housing. Thus, once installed, the ejector rod housing cannot move forward as a result of recoil. (I wondered whether such anchoring is needed or not, but Cliff assured me it *is* necessary. It looks good, too.)

Bowen makes the 5-shot cylinders to exactly suit each particular gun's frame in length, and - because the distance from base pin axis to barrel axis varies from one gun to another - he measures the gun, and bores the chambers on a pitch circle to match.

Next, he cross drills the head of the base pin, and drills on into the barrel, about 1/4" ahead of the front face of the frame, and then taps both holes simultaneously for a socket set screw of probably 4-40 size. (The hole in the barrel is a blind hole, obviously.) This screw prevents the base pin from jumping forward, which can happen when firing stiff loads in guns of the basic Colt Single Action Army design. Hamilton said in a letter to me that he didn't originate the idea of this screw, nor is he the only one to use it. He says it is simply a good idea in the face of heavy recoil and worn pin latches.

Finally, Hamilton adds a small plate in the frame recess beside the cylinder stop, (aka "cylinder locking bolt"), to take out all side-to-side play in the cylinder stop, so that when the gun is cocked and a chamber comes into line with the barrel and the cylinder lock pops up, the cylinder is virtually immobilized - i.e. zero rotational play is present when the gun is ready to fire.

Certainly some aspects of the above re-work are not child's play, especially the first time around. For example, modifying the hand and cutting the new cylinder's ratchet so that the cylinder will rotate 72° (for a 5-shot cylinder) rather than the normal 60° is no minor task that a guy would knock out before supper some evening.

However, the ballistic advantage of a shortened .348 WCF case belching out a .50 caliber bullet over a standard Ruger Blackhawk in .44 Mag or .45 Long Colt is not the only attraction of this sort of re-work. Some of the other modifications, or a variation on them, such as silversoldering a nicely contoured piece of steel to the barrel to help anchor the ejector rod housing, adding a screw to prevent the base pin from jumping forward under recoil, and certainly adding the little plate in the frame to take out side play in the cylinder stop - all these are within the capability of any competent basement hsm/gunsmith hobbyist, and could be performed on any Ruger Blackhawk. And as you may well know, a Blackhawk in .45 Colt can be loaded up to levels that would not be wise in a Colt Single Action Army of the same caliber, and would doubtless give that .50 x .348 cartridge a run for its money.

I think drilling two rows of six or seven 3/32"ϕ holes alongside the front sight would have about the same effect as Mag-N-Port'ing the barrel. Any burrs left in the bore will come out with the first few shots. Alternatively one could make a muzzle brake per drwgs elsewhere herein.

From a technical (rather than legal/regulatory) standpoint, I would not hesitate to do any of the above modifications to a Ruger belonging to myself or someone else.

I would do a little soul searching before I would tackle the job of making a new 5- or 6-shot cylinder precision-fitted to the individual gun, but I believe I could do it. I would space out the 5 or 6 holes for the chambers either on my 6" rotary table, which is very accurate, or I would do it by co-ordinates, using the DRO system on my mill. Lacking either of these aids, (and my boring head, which I would definitely use to ensure maximum accuracy of the chambers' alignment) I would resort to jigs, toolmaker's buttons, etc. Just remember, men probably less well equipped than you or me made the prototypes of the early black powder muzzle loading Colt revolvers, back about 1834. If they could do it, why should we not be able to try?

Bowen Custom Arms charges about $1,350 (on top of the customer's gun) for one of these full house re-works. Even if one had no inclination to build similar weapons, it would be relatively easy to get oneself in a position to offer the base pin lock screw, and the cylinder stop item.

You could probably charge $75 to $100 or more to do both these modifications, probably $20/25 for the base pin job alone. A lot of shooters like to have such small/minor "improvements" done to their guns, just for the pleasure of spending a buck or three to make their "toy" different from what the rest of the guys have.

For some people, the appeal of steel exceeds the allure of aluminum by a wide margin, which gives rise to another little job one can do on the Ruger Blackhawk, and this is to replace the original aluminum grip frame with a steel grip frame. Ruger sells unfinished steel grip frames with both round- and square-backed trigger guards. They used to sell a brass grip frame also, but I am told they no longer do, which is a pity. With suitable power polishing equipment, one could make short work of polishing up the brass one from the "as cast" state in which they were shipped; the actual work of fitting it to the individual gun took about 1/2 an hour. The same work on the steel version might take slightly longer, plus it would need to be blued.

And see also Gun Digest #32 (1978), page 225, for - *ahem* - an extremely good article on tuning and slicking up a Ruger Blackhawk to be as smooth as the Super Blackhawk is when it leaves the factory. (There are two minor errors in that article. In the illustration at page 227, there is a "1" missing from one of the arrows directed at the hammer strut. On the next page, under the middle right hand photo, the caption calls for "...careful hard stoning.." of the gun's hammer notches - this should read "... careful hand stoning....".)

Anyhow, this might give you some ideas, commercial or otherwise. If you do it for profit, make sure you're all nice and pink and clean in the government's eyes.

Footnote: Some months later I visited Cliff again, and had an opportunity to shoot the Hamilton Bowen gun. I don't know the details of the loads we tried, but they kicked like a mule. Cliff said I was the first guy he'd seen fire it one handed. (Don't be impressed - maybe I was just the dumbest one so far.) Three shots were plenty for me. 10 or 15 shots would probably leave your hand sore for hours. Cliff says the gun is phenomenally accurate. A gun like this, and a location where you could shoot against a safe backstop like a bald ridge or a rock bluff 5/600 yards away, would make for what I would regard as very serious fun!

Removal of Stuck Bullets from Barrels

I asked Cliff how he would go about removing the stuck bullet referred to at the beginning of this section. He said it isn't very difficult, but that it is a job he prefers to do when no one is around to distract him.

The first thing you have to know is whether the stuck bullet is a cast lead bullet or a jacketed bullet, because this makes a big difference in how you proceed. If you don't know - and you may not, if the gun belongs to someone else who has brought it to you for the removal job (the owner may not know, either) - then you **must** proceed as if it is known to be a jacketed bullet.

NOTE

It goes without saying that you would absolutely NEVER try to shoot a stuck bullet out, because to do so will virtually guarantee a bulged or burst barrel, possibly a wrecked gun, and - very likely - serious personal injury. Even if someone was so foolish as to try to shoot the obstruction out, and be so fortunate as to come out of it with the barrel only slightly bulged, and therefore with its accuracy probably intact, the resale value of the gun would be much reduced. Therefore, **NEVER** try to shoot ANY bore obstruction out. Stuck cleaning patches? Burn 'em out with a red hot wire, repeated as often as necessary.

Removing Cast Lead Bullets

If you are definitely dealing with a cast lead bullet, it can be driven out of the bore with little resistance. Put a short, burr-free CRS or brass rod into whichever end of the barrel is most convenient, and carefully tap it with a soft faced hammer, driving the bullet out the other end of the barrel. The rod wants to be just under bore diameter (or wrap it with masking tape to bring it up to near this size), and should have both ends nicely chamfered to preclude damage to the barrel.

Always use a one-piece rod, or a sequence of longer and longer one-piece rods. Do not put a series of short pieces down the bore one on top of the other.

Removing Jacketed Bullets

Like lead bullets, jacketed bullets care naught which direction they come out, but they must *come* out - they cannot be driven out. Jacketed lead bullets have much greater resistance to passage through the bore than cast lead bullets. If you try driving a stuck jacketed bullet out, you will swell it and get it stuck worse. **Here's what to do instead:**

Turn a piece of steel to a nice slip fit in the bore, making it about 1 to 2 bore diameters long, and then drill say a 1/8 or 3/16" hole up the middle. Deburr all over, and drop this bushing into the barrel. Now carefully drill a hole right through the bullet. (You may have to make an extension for your drill, depending on where the bullet is lodged in the bore.) Remove the drill and the bushing. Put a coarse threaded steel (not brass) wood screw into the hole in the bullet, and screw it in solid. Make up a simple small "slide hammer" tool, hook it onto the screw, and draw the bullet out. Like I said, never try to push a jacketed bullet out - *pull* it out.

This strikes me as a job any hsm should know how to do (and now you do). It is also a job you could do for others, either as a favor, or as a chargeable service.

NOTE: there is further info on this topic in the NRA Gunsmithing Guide. Not all of it agrees with what is here. For example, they say to use drill rod, not soft metal (i.e. not brass, and not CRS) to drive out bore obstructions. They also say to put some lightweight oil into the bore at an early stage in the proceedings, which strikes me as a good idea.

Some Basic Reamer Information

At one point during one of my visits, Cliff stopped to stone a rifling cutter, and then examined the results under a binocular microscope. He let me look, and explained to me what I was seeing. When he had stoned the cutter the way he wanted it, there was a wire edge showing. He got rid of this by passing the edge of a penny (coin) along the edge once, after which we looked again, and it was gone. Apparently you can do the same thing when hand sharpening reamers.

Cliff also told me the following, which I found quite interesting:

Reamers sometimes show an unequal amount of cuttings in their several flutes, when pulled out of a hole. Such reamers are of lesser quality. The uneven cutting arises because the maker has finish ground the reamer to simultaneously sharpen it and bring it to final size. On such a reamer, each flute is ground in turn, to produce the primary relief behind the cutting edge, and by the time the last flute or 2 comes up for grinding, the grinding wheel has worn enough that the last flute(s) is/are oversize, or in any case, not the same as the first flute or 2. The result is uneven chip removal, as noted above.

Better quality reamers are fluted and then cylindrically ground to size, leaving a cylindrical land about 0.002" wide behind the cutting edge. This reamer can then be used as is, but.....

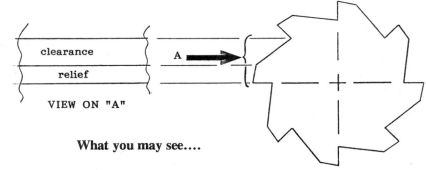

What you may see....

If such a reamer is then stoned to produce a 2-1/2 to 2-3/4° primary relief on this land, right up to the cutting edge, you have the ultimate reamer. (Cliff also said that this angle is so well established as being almost universally correct that there is no need to consider trying other angles for any normal work. 2-1/2 to 2-3/4° is THE angle to use.)

When you look at a properly stoned reamer, it will look approximately as drwg above.

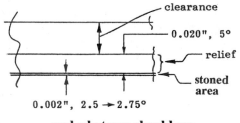

... and what you should see

But what is not visible - or at best only visible if your eyes are good and you know what you're looking for - is the stoned area noted above. Since this area is typically only maybe 2 thou wide, you could be forgiven for not even noticing it at all. If it's there, and if you know what to look for (as now you do), you should see it. Behind that is the secondary relief - typically about 5°, and much wider - and then there is the clearance behind that - maybe 10°.

So you may at first see only two flats, but there are actually three, if the reamer is properly stoned. And if you know how to light the job properly (as you will find out a few pages further on, in what Keith Francis has to say on the topic), that little stoned area will light up like a neon sign.

Cliff had a little pocket microscope about the size of a tire inflation gage, with a graduated reticule engraved on the lens - when you look in it, the grads are superimposed on the item being studied. The grads are 0.002" apart... Unfortunately these critters reverse the image you see, which can be a little confusing at first, but they are useful for seeing what has been done to a reamer or other cutter, either by you or someone else. Of course, if it's all messed up, it wasn't you.

Travers sells these under the name "Micro-mike" for about $18 at time of writing (mid '92). I would say the 20x one is much the better choice for our kind of work than the 40x one. Cliff's was a 20x. The MSC cat. shows one at about $58, which is probably better - they call it a "Zoom Pocket Microscope". Magnification is 16 to 32x, and it has a significantly greater field of view than the Travers unit.

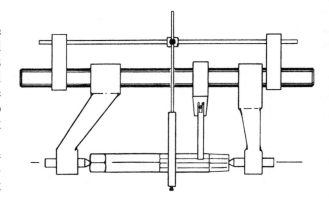

Interestingly, the basement machinist can, with care and a toolpost grinder, make excellent reamers of the better class described above, and can stone them to final sharpness in a fixture of the sort roughly approximated in the drwg* at right - an end view will be shown in a few pages, where it will serve to illustrate something else as well. Frank Mittermeier, Inc. sells a fixture of this type, but one can be made easily enough in the basement shop. You can also set a reamer up between centers in your lathe, and stone it there, as we will see on the next page.

* (Says the sharp eye'd reader: "'Approximated' is right! That guy Lautard musta done that drwg very late at night. If he wants to stone it that way, he'd better either be workin' on a left handed reamer, or he'd better take the reamer out and flip it end-for-end between the centers! The tooth rest finger won't find much to push on right now if it's a conventional right hand reamer." Sorry about that, fellas!)

Cliff said that such fixtures do not need to be made with enormous precision. What they do need is to be very precisely ADJUSTED in use, so that the reamer edge to be sharpened is parallel to the guide bar, so that one's stone rides at the same angle all along the flute. Cliff's reamer stoning fixture incorporates a height adjustable V-block, by which means he can adjust a reamer's cutting taper to be parallel with the guide bar so he can stone that area. In re-boring a rifle barrel, Cliff *pulls* his reamers through the barrel, so his extension shank on the reamer is silver soldered to the tapered end of the reamer, and he does not want to have to remove this extension tube every time he has to sharpen the reamer.

Re staggering reamer teeth: To aid in preventing reamers from chattering, the flutes are often unevenly spaced. See "Modern Toolmaking Methods*," and/or Machinery's H'book for information on how to produce this effect during manufacture. It is preferably achieved by means of the dividing head, rather than by moving the milling machine table off the reamer centerline. Regardless of which of these two methods have been used to stagger the flutes, the correct 2-1/2 to 2-3/4° angle can be stoned at the cutting edges as detailed below.

* There's some very interesting and informative material on reamers in "Modern Toolmaking Methods," by Franklin D. Jones. I have mentioned this book (a Lindsay reprint) elsewhere herein.

Setting up to stone the desired 2-1/2 to 2-3/4°
relief angle on a reamer

As I mentioned a few paragraphs back, you can set a reamer up in your lathe, and stone it there. Whether you put it in a dedicated reamer sharpening fixture, or in the lathe, setting things up so the stoning operation produces the desired 2-1/2 to 2-3/4° relief angle is carried out in much the same way. In some respects, the lathe is the easier "fixture" to do it in, because it is big and massive, and will sit nice and still while you make your set-up.

The procedure, when done in the lathe, is carried out as detailed in the notes under the drawing below. In the drwg, the reamer (a 4-flute job, for the benefit of our overworked draftsman!) is chucked in the lathe, with tailstock support, and..... well, look at the drwg, and then read the notes. The first 3 notes, 1., 2., and 3., correspond to the numbers 1, 2 and 3 on the drwg.

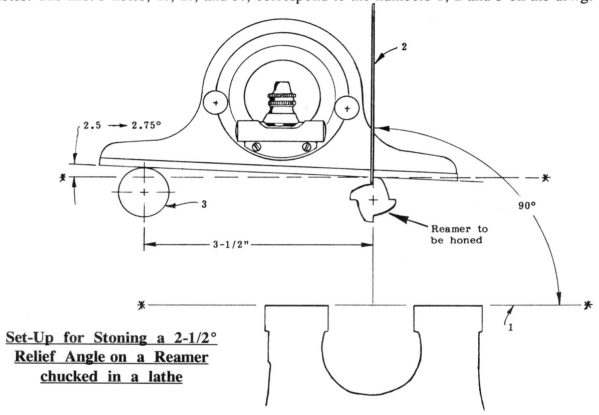

Set-Up for Stoning a 2-1/2°
Relief Angle on a Reamer
chucked in a lathe

1. The lathe should be set up so it is level from front to back.

2. Turn the reamer & the chuck so a 6" rule placed against the front face of a flute is vertical.

3. Put a straight piece of 1/2"φ drill rod or similar horizontally in a boring bar holder in the toolpost, so that it is parallel to the reamer and the lathe axis, and about 3-1/2" away from it.

With the front face of one flute of the reamer now vertical, as established at (2) above, adjust the bevel protractor head of your combination set to 2-1/2°, and lay it across the top of the reamer and the rod (item 3). Raise (3) until the level bubble in the protractor head shows "level". The angle from the top of (3) down to the cutting edge of the reamer flute is then 2-1/2° as desired.

Now substitute a triangular or square fine India stone or similar (see below) for the protractor head, and stone along the flute with it. To keep from stoning the top of the piece of drill rod, you could put a piece of paper under the stone at that end, and slide it along with the stone.

If you don't have a protractor head, there is another way to get the same result. Simply put a level across the rod (3) and the reamer, and adjust the height of the bar in the toolpost till the level reads "level". Then raise the toolpost 0.153" (or slip a nice fitting sleeve 0.306" larger in diameter onto the bar in the tool post) and you will have the same 2-1/2° slope.

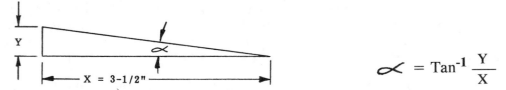

$$\alpha = \text{Tan}^{-1} \frac{Y}{X}$$

Tolerances affecting the above set-up:

As noted above, if the rod is offset 3.5" from the lathe axis, we must raise the rod 0.153" to give us 2-1/2° relief on the reamer. Assuming the rod offset is on spec, then it might be well to consider what happens if you raise the bar too much or too little.

if you raise the bar 0.125", the angle will be 2° 02' 43" - too little;
if you raise the bar 0.156", the angle will be 2° 33' 22" = ok;
if you raise the bar 0.165", the angle will be 2.699° = 2° 41' 57" = still ok;
if you raise the bar 0.175", the angle will be 2.862° = 2° 51' 45" - too much.

So call it 0.160" plus or minus about 8 thou to keep the angle in the desired range of 2.5° to 2.75°.

There's more.

MORE ABOUT REAMER STONING JIGS
from **Bill Webb, Kansas City, MO**

I've had several very nice letters from Bill Webb, a riflesmith of some 40 years experience, down in Kansas City, MO. You'll get a pretty good idea of why I enjoy Bill's letters from the following, which is boiled down from a couple of them....

Dear Guy,

Enclosed are some photos of a little reamer stoning jig I built about 20 years ago. It's an improved version of Red Elliott's old stoning jig. Great for stoning commercial reamers to SAAMI minimum dimensions; also for sharpening, or finishing up your own reamers after grinding. Use medium and fine India's for most of the work, and a hard Arkansas for finishing. Surprising how much stoning it takes to remove 0.001"!

My refinements to the basic Red Elliott stoning jig consist of the addition of an indicator and a micrometer head. The indicator reads the radial values of the flutes at the cutting edge, while the micrometer head controls the position of the indicator along the reamer axis.

The indicator is used to check radial progress - i.e. flute height. It is attached to a fixed vertical rod set into a ground rectangular steel block. The block, c/w indicator, is removed during actual stoning.

Rough stoning is done diagonally across the flutes, finish stoning is done along the flute. Diagonal stoning leaves a contrasting finish to better see what the stone is doing. When I want to see how my stoning is progressing, or to compare one flute against another, I set the block in place against the rail which you can see just back of the reamer in some of the photos, and the indicator tells me what I want to know at that point on the reamer. The rail is a strip of 1/8 x

1/2" ground flat stock attached to the base with its axis parallel with the axis of the reamer centers.

I can return the indicator block to the same position (i.e. to any desired position) by setting it on the ground fixture base, with its front edge up against the rail. The block, with the indicator attached, can be slid along the rail to a pre-selected position and stopped against the micrometer head. The latter is attached to a fixed holding block, which is clamped to the base and is moveable for different ranges.

The micrometer stop not only provides precise, repeatable linear positioning for the indicator block, but is also useful for detecting the amount of taper in the reamer body. This is done by selecting a point near the shoulder, and then setting the indicator and micrometer head to zero. Next, back the micrometer head out 1", and then slide the indicator block along till it again contacts the micrometer stop. The reamer is rotated and the high point on the flute(s) is read. Since these readings are being taken 1" away from the first set of readings, it is then very easy to figure out the taper, in inches per inch (per side).

The primary reason for these additions is to help maintain flute concentricity while stoning. Each flute should give nearly identical indicator readings at the cutting edge. An indicator check should be made at the pilot, or pilot mandrel* shank to be certain that the pilot, the center (i.e. the center hole at the end of the reamer), and the periphery of the flutes are all con-

centric. If the pilot shank and its center are not concentric, then a small V-block 1/4" thick is attached to the front center block, and the pilot is allowed to ride in this V-block. The V-block has a slot in it for a screw, and is vertically adjusted to cradle the pilot, using the regular front center to roughly pre-position the V-block. After positioning, the center is reversed end for end in its hole, and its flat butt end is pressed against the pilot end of the reamer to maintain pressure to the rear, while the pilot (or pilot shank) rides the "V" block.

* A minor point of terminological hair splitting. If a particular reamer is intended to be used with a pilot bushing, then the end of the reamer would be referred to as a mandrel (which goes inside the bushing). If the reamer has a full diameter (solid) pilot, then it would be referred to as "the pilot". Got that?GBL

Some reamer makers grind their reamers without benefit of a driving dog. They utilize a "driving center," which is in the form of a 60° pyramid rather than a conical center. If you encounter such a reamer, you should lap the burrs formed by the driving center from the rear center hole (which is obviously where they will be).

I should perhaps revise my statement about finishing with hard Arkansas stones - that may be a little "overkill". A fine India stone is all I presently use to finish off a reamer, and I can see no difference in the surface finish of the finished chamber, etc.

I'm not familiar with the ceramic stones you mention. (Brownells' 1/2" sq. x 6" in medium, fine & extra fine.GBL) I have a couple of hard white Arkansas stones, which I seldom use. The size I use in my jig is 3/8" square x 4".

Another application for such a jig would be in filing relief on home-made reamers, etc. before hardening. I haven't tried the jig for this, but I think you could epoxy a piece of narrow pillar file onto a 3/8" block, and use it like a stone.

Notice in this photo that the stone and the holder in which it is clamped have been rotated relative to the square shank at the back end of the stone clamp. (Compare the stone's orientation to that shown in the photo at bottom left of previous page.) This explains why there are 3 screws on the stone holder: two to clamp the stone, one to lock the holder to its shank at any desired orientation. It would appear from the reamer visible under the stone that Bill has rotated the stone to work on the "shoulder" portion of the reamer. Keith Francis has a somewhat different approach to stoning this part of a reamer, as you will soon see... GBL

In this photo, the stone and its holder/guide shank have been set aside out of the way while the indicator is being used.

I didn't originate the principle of this jig, nor am I certain who did. My dad used the same basic scheme back in the 40's to file relief on his homemade reamers, and then stone them after heat treating. I don't know if he thought it up, or copied it. Stoning jigs are mentioned at p. 366 in Landis' book "Twenty-two Caliber Varmint Rifles". He gives primary credit to Red Elliott and Sam Wilson. I've since seen a rudimentary version of such a jig in one of the gun magazines. You might want to give Keith Francis a call - he knows far more than I ever will about reamer stoning jigs...."

By the way, your "spot grinding" method (**TMBR#2**) was called "slip grinding" by the old timers around here. Their main advice was not to do it upstream of any windows! .... Keep up the good work... it is appreciated.

Bill

And here's the Final Frosting on the Cake...
from **Keith Francis, Coos Bay, OR**

At Bill Webb's suggestion, I ran the above past Keith Francis, a well known chambering reamer maker in his own right, who now works with the JGS people in Coos Bay, OR. *(And guys, please note this: Keith does not accept calls from the general public; he is simply too busy to do so. The only reason I got to speak to him at all was that Bill Webb had sent me.)* Keith said that what I'd said already had pretty much covered the matter, but he then proceeded to supply the final frosting on the cake - 7 pages of handwritten material based on about 40 years of his own experience. If you want to study reamer stoning, this is Graduate School.

Dear Guy:

About the only comments I would add to what you had from Bill Webb and Cliff La Bounty are the following:

Reamer Centers
Reamer centers should be diamond lapped. Use 150 grit for roughing and 320 grit for finishing. Poor centers can be "moved over" to relocate for minimum total indicator reading (TIR). This is done in a drill press or similar machine, with a lap in the chuck, and a center set up on the table in line with the spindle centerline. A little side pressure with vertical strokes soon does the job; my old "Francis" reamers (i.e. those made before joining the JGS people) often have a rear center that has been moved over 5 thou or more to minimize heat-treat warpage.

Stoning
Stoning is simple, but requires good eyes and technique. In my opinion the fixture should have a good set of centers. (By this Keith means something not unlike a "headstock" and a "tailstock" off a set of bench centers, together with the little pointy things that go in them, to run say a reamer on; he does not mean just a couple of 60° male centers turned up from a piece of 3/8"φ CRS. GBL) A set of centers from a K.O. Lee or other small cutter grinding machine can often be had cheap from a job shop or other shop that has "junked" the machine's centers and set the machine up for permanent use as a tap grinder, drill grinder, or whatever. A suitable base is made or purchased. An old grinder table with T-slots is good, if you can find one. (You might get one from a machine tool rebuilding shop. GBL)

I use 1/4 x 1/4 x 4" stones - fine India for roughing, fine hard Arkansas for finishing. I rarely need the India, except for correcting other brands of reamers.

My "Francis" chambering reamers (1953-1977) were made with zero rake, ground primary and secondary clearances, and a stoned "primary". Those portions of the reamer that were to cut straight (zero taper) parts of the chamber (for belts, rims, freebore and straight necks) were circle ground (industry slang for *"cylindrically ground"*; easier to say, and easier to spell!) to size, and then were "backed off" to leave only a 0.002" margin.

Tapered sections of the profile were circle-ground to 0.002" over finish size. These areas were then backed off to 0.001" over the finish size. This left a sharp edge which was stoned to size. The stone only cut 4 or 5 "tenths" per side, and if stoned with a fine hard Arkansas stone, the reamer would usually produce chambers with a 7 to 11 micro-inch finish. My "production" customers usually reported a 7 to 9 micro-inch finish for about 1200 chambers (although they often got as many as 2400) before the finish dropped off to 16 micro-inch or rougher. These tools were made of AISI-07, which is an oil hardening tool steel, not HSS. They were run at 40/50 sfm with 1/2 to 1-1/2 thou chip load per flute (0.001" chip load per flute x 6 flutes = 0.006" per revolution of the workpiece).

(Keith did an article about making chambering reamers for the January 1963 *American Rifleman*. It shows, with photos and detailed captions, how a machinist/toolmaker goes about making a reamer from a blank bar of steel. Much info on steels to use, heat treatment, etc. You might want to see that article.₍GBL₎)

For the Armalite-designed 5.56 (forerunner of the M-16), I recommended 4 cutters: #1 produced the breech counterbore. #2 was a 4-fluted reamer that roughed the body and shoulder of the chamber to 0.004/0.004,5" under final chamber size. #3 was a semi-finisher (often made from dull finishing reamers) that would cut 0.003/0.003,5" under chamber size, and #4 was the finishing reamer. Oil slots under the pilot bushings and through the screw heads allowed a good grade of sulfur-based cutting oil to be pumped from muzzle to breech so the chips flowed out the rear - this minimized stops for "chip clearing" activities.

The best stone I've used is the 5/16" square Ruby #212 fine, made in Germany and sold by Gesswein, 255 Hancock Ave., Bridgeport, CT 06605; phone (203) 366-5400) for about $30 each. They now make them in 1/4" and 3/16" square, which would be even better.

I'd judge these (Ruby) stones to give an edge likely to produce a 4-6 micro-inch finish. With the right technique they cut either very fast for roughing or very slow for a super edge.

I used to use an indicator with the stoning fixture, but found I could do a good job just by eye, wearing a 6" focal length Magna-Focuser #107. One to 2 "tenths" TIR is easy, and in fact desirable, because even with flutes staggered to Machinery's H'book specs (which is the best way to do it), a reamer is less likely to chatter, especially in the stainless steels. Trying to stone a reamer to 0.000,0"TIR is a waste of time. However, I digress, so let's get back to the jig....

Guide Rod
Once you have a good base plate and set of centers at hand, the next thing to attend to is the guide rod. This can be a piece of 3/8 or 1/2"φ drill rod, and should be mounted on a vertically adjustable device (screw-adjustable if possible) mounted on the base plate. The centerline of the guide rod should be parallel to, and about 3.5" from, the centerline of the centers.

Once this is done, the guide rod can be set level to the base plate, a protractor-level (i.e. the bevel protractor head off a combination set) can be set on the stone, and with the stone on the largest part of the reamer, a clearance angle of 2° can be established (as Cliff La Bounty showed me, and as illustrated earlier;₍GBL₎) and the guide rod re-leveled. (You will have raised or lowered one end of the guide rod while setting up the 2° clearance angle, per previous sentence.)

Jig Orientation
The reamer shank should be pointed toward your solar-plexis, and the vertically adjustable guide bar should be 3.5" to the right of the centers. The stone and holder are held gently in the right hand. The stone holder is made so that it can clamp the stone against a surface about 7/16" long, so that the stone is parallel to the stone holder's attached shank of drill rod. (This will become clearer in a minute.)

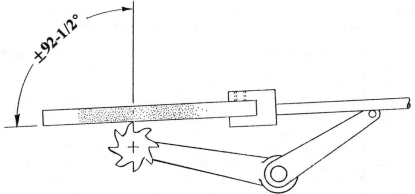

This drwg will serve 2 purposes. First, it is the "end view" for the reamer stoning jig drwg on the 2nd page of the material several pages back headed "Some Basic Reamer Information". Second, it shows the stone holder and jig orientation Keith Francis describes below.

The Stone Holder
My stone holder is about as shown in the drwg above. The stone is clamped into a block 3/8"

wide, 5/8" high, and just under 3/4" long. The slot for the stone is 5/16 x 0.450" long. The hole in the back end should be drilled parallel to the slot and centered with it. Into this hole is pressed (or Loctited) a 4" length of 3/16" drill rod. The set screw on top clamps the stone parallel to the 'bottom' of the slot, and also serves as an ideal indicator that the correct face of the stone is being applied to the reamer.

A lightweight stone holder like this gives a very sensitive "feel" to the stone.

Clearance Setting
The diameter of the reamer governs how much clearance is stoned. You can get this info by looking at a clearance chart for end mills. (For such a chart, see Machinery's H'book, or a grinding handbook from one or other of the abrasives manufacturers.GBL) Using these clearances for the grind, and using about 20% of the recommended clearance angle of the primary grind for the "stoned" clearance, you will get a good edge. (If the recommended primary clearance angle is 10°, you would want to set up to stone a 2° clearance.)

When setting up the reamer, snug it up between the centers and adjust a spring-finger (also called a tooth-rest finger) to rest on the flute-face while it is in a vertical position. This is easily done with the aid of a light straight-edge (a 3" section of a 6" steel rule works fine). The flat side of the rule is held against the flute face with a small screwdriver or similar tool, and the reamer is turned until the rule is "eyeball" vertical.

The stone is placed on the largest diameter of the reamer, the clearance is set, and the rod leveled. If the rod is level, and you have a bottleneck reamer (with or without a throating section), you will note that, since the 3.5" spacing from rod to reamer centerline remains constant, your clearance increases as the stone moves to the smaller diameters.

Stoning Technique
I stroke at about a 45° angle to the flute faces, usually 1/2" long strokes on necks, 1/4 to 1/2" strokes for throats and up to 2" strokes on bodies. The stone is held with the narrow way of the working face of the stone fairly flat on the edge being stoned, tilting slightly toward the cut. When going from the neck onto the shoulder portion of the reamer I keep the face of the stone parallel to the reamer centerline, and stone the shoulder with the leading edge of the stone, closely watching how much clearance I'm producing. The level guide bar gives an automatically changing clearance as the stone rises from the small neck to the large body. The "stoner" needs to "eyeball" the other (compound) clearance so it looks right to him, whether on the shoulder or on a big rim or belt*. (Rim and belt portions of a reamer are usually ground to give 5° clearance, regardless of diameter, and stoned at 2°-3° if required.)

> * I think what Keith means here is that the operator needs to watch what he is doing with the stone, so that the stoned area looks right in relation to the primary clearance angle ground on the reamer during manufacture. GBL

Preparing Stones for Use
Newly purchased fine hard Arkansas stones with sharp corners need to be touched up a bit themselves before use. Give each of the 4 flat faces of the stone about 4 strokes or so on a piece of 320 grit wet-or-dry paper. Then roll the stone about 45° and sand about 4 strokes on each long corner of the stone; this puts about a 2 to 3 thou radius on the edge and smooths it out - going up shoulders will soon show if you still have the "saw-tooth" 90° or a smoother edge!

Honing Lube
Kerosene or WD-40 are good to dip your stone into. I keep my stones damp and dip and rub them with my thumb as required - once per reamer for throaters, and up to maybe 30 times for 25mm cannon reamers.

Lighting the Job
Seeing what you are doing is easy if you know how. I use a 60 watt bulb in a shaded adjustable lamp set about 6" above the front of the reamer. I move it a bit while looking at the

edge I've just barely stoned: the stoned section lights up like a neon sign when the light is right, even if it's only 1/2 a thou wide.

Seeing What you're Doing
To measure widths of the stoned primary etc., I use a 7/16"φ x 4-1/2" long DuMaurier Co. "micro-mike" #2010 20x glass with a scale in it. (You can get this item from J & L, or Production Tool.GBL) The scale is 0.180" long with 0.002" graduations, and - being 20X - is rather shaky, so I cut a plastic tube to slip over it so the plastic can rest on the cutting edge and hold focus as I move along the entire edge.

What To Do
Used reamers requiring just a stone job are checked often while stoning. I complete each flute before starting the next; the wear will usually be concentric even if the reamer wasn't ground that way, so you usually wind up with better concentricity than you had when new, just by removing the wear and looking at the edge.

I prepare used reamers by removing any build-up on the flute faces with a 1/4" sq. fine hard Arkansas stone with the end tapered to a 45° or less angle to get well into the flute depth with the long axis of the stone held perpendicular to the reamer axis. I smear a dab of 600 grit lapping compound on the flat side of the stone, and - holding the stone flat to the rake - go back and forth until the face is "clean". The edge is then examined with the 20X 'scope, and is then set up and stoned, or if very dull, reground first.

Keith

And now, fellas, I think it's fair to say that, thanks to "Professors" Webb, La Bounty, and Francis, (and Assistant Instructor Lautard!) if you wanted to know how to stone your own reamers, and it ain't here, you don't need to know it.

NOTE:
Wolfe Publishing Co. has just brought out "The Illustrated Reference of Cartridge Dimensions," a book of dimensioned cartridge case drwgs (about $20). If you plan to stone your own chambering reamers, this would be handy to have, particularly if you don't have SAAMI specs on the calibers you chamber for, which you should.

SOME MORE IDEAS

Cliff La Bounty loaned me an original copy of "The Muzzleloading Caplock Rifle," by Major Ned Roberts. Unlike the Wolfe reprint of the same title, this book has in it, at page 290/291, drwgs for an extremely simple underhammer caplock rifle action that would be well suited to being built in a basement workshop, and could be a good basis for a "buggy rifle*" of the sort that Billinghurst and others made about 100 years ago. If a guy were to make such a rifle, as well as the additional items (detachable shoulder stock, loading tools, detachable scope sight, etc.) that would comprise a complete outfit, and if all of these items were of a quality suited to the carriage trade, he would have something both very nice, and very unusual. He would probably **also** have a very saleable item at almost any price he might care to name, should he be so inclined. That's one idea.

The other idea is to make a more conventional Schuetzen rifle, either on a Ruger #1 action, or a caplock (again using that same easy-to-build action noted above), and get into muzzle loaded target rifle shooting as Pope and others did it. (If this appeals to you at all, you should have a copy of my book "The J.M. Pyne Stories and other selected writings by Lucian Cary". This is available directly from us - *see inside back cover of this book. See also "The Secret of the Old Master" in TMBR#1.*)

And here's a 3rd idea: you could make just the action as above, and sell it to other guys to build up into a rifle. It's so simple as to make you laugh if you saw it.

* See Gun Digest #35, (1981), page 21 for right and left side views of a very fine example of a buggy rifle. In the previous issue of Gun Digest, i.e. #34/1980, at page 212, there is an interesting article on availability, properties, heat treatment and application of special steels for gun making.

See also p. 345 of the book "Gunsmithing Tips & Projects," for an article on making a Billinghurst-type action.

Five designs for such actions are also shown in "Gunsmith Kinks, Vol. II," p. 350-352. One of them is essentially the action shown at page 290/291 in "The Muzzleloading Caplock Rifle," as mentioned about 5 paragraphs back. (Incidentally, back at p. 344 of "Kinks II" you will find some good info on tuning up a commercial Mauser trigger not unlike the type we talked about making earlier, and also about tuning up issue triggers on military Mauser type actions. One of my guys told me that the idea in Kinks II, page 357/358 about using a ball (from a ball bearing) to modify a military trigger works extremely well.)

In "I Make a Muzzleloader", in Gun Digest #34, page 91, well known gunsmith Roy Dunlap describes how he built a black powder rifle with 4 or 5 interchangeable barrels, 2 interchangeable butt stocks, and 2 different fore-ends, all to fit a receiver he machined from a chunk of mild die steel. It can be done. As Dunlap says at the beginning of the article, he (and we) had better equipment than the old timers ever dreamed of.

Any local library can assist you, either directly or via the Interlibrary Loan service, if need be, in accessing any of the foregoing articles.

Future Possibilities with a Twist

Now I'm going to pull **a very dirty trick** on you fellas that might have inclinations towards rifle building at home... **the following is a further quote from one of Bill Webb's letters:**

"Enclosed are several photos of the 2nd rifling machine my Dad and I built, and a write-up (9 pages, and **lotsa** details!$_{GBL}$) about it. I thought you would be interested. It is a simple and relatively cheap machine to build, yet with it a basement machinist can make competitive benchrest-quality barrels.

"I've never had the desire to produce barrels for sale, and it would require more machine than this to do so in any sort of quantity. However, the little rascal will drill an accurate, smooth hole, usually with much less runout than a match grade stainless barrel from a major supplier.

"The drilling takes about 45 minutes (drilling time). Making a 6mm barrel, for example, I would run at maybe 2850/3000 rpm, with a feed of 0.7"/minute. A couple of stops are made during drilling to clear chip clogs, pour cutting oil from shoes, etc. Oil pressure is maintained at about 900 psi.

"Reaming is done in two passes at about 400 rpm, with a feed rate of about 5 thou per revolution, and with about 150/200 psi oil pressure. Reaming time is about 15 minutes per pass. Reamers are 6-flute, pull type, 1.5° cone angle, tube mounted.

"Straight holes are dependent on correct drill geometry and sharpness. Oil pressure is hard to regulate when drilling several consecutive barrels due to temperature rise in the oil system. Heat kills the viscosity, dropping pressure. A plastic bag full of ice cubes helps.

"The hole is air gaged with a Sheffield gage with 2000/1 multiplier tube, and lapped if necessary to a variation of less than 0.000,1" end to end. (And that's as good as the tolerance on the fabled "star gaged" Springfield barrels from the "good old days."$_{GBL}$)

"Rifling is done in the conventional manner with a hook cutter, rake angle varying from 3° negative to 3° positive depending on the steel, 1 to 2 "tenths" per pass until final finish, etc. There are many "ifs" and "buts" involved. Razor sharp cutters are a must! The rifling head should be heat-treated and ground.

"1-1/4"φ bars of Stressproof (1141), 41L40 or 416 stainless are available locally. 416R is difficult to obtain in small quantities, because the big barrel makers gobble it all up. All of the above materials run around 28 to 32 on the Rockwell C scale, and cut very well.

> Note: Stressproof is fine to practise on, but do not use it for making rifle barrels intended for use. Some who have tried Stressproof barrels have had them split from end to end, with sometimes very nasty results.

If you're interested, I can give you much more information on this subject, **including detailed working drwgs for an improved version of the 2nd rifling machine my Dad and I built...**"   Bill

I sent a copy of what Bill had sent me to one of my guys down in Connecticut. He showed it to a friend who is very interested in rifle barrel making, and his comment was that it was *the* most comprehensive single block of info on basement rifle barrel making that he'd seen in 50 years. **If the above interests you, let me know...**

Added to the 3rd Printing: In May 1995 I went to Kansas City and shot a video on Bill Webb's rifling machine. In the video, Bill describes the machine in sufficient detail to enable any interested machinist to duplicate it, and points out several possible improvements. He then demonstrates the machine in action - drilling, reaming and rifling - transforming a solid bar into a match grade 6mm rifle barrel. There is much detail and close-up info on deep hole drilling, rough and finish reaming, and making and using the rifling head and rifling cutters.

Added to the 5th printing: The **Rifling Machine Video** is now available, and copies can be ordered directly from me. The video is presented on 2 tapes, with a total running time of 3 hours, and comes with a 36-page Written Supplement. If you wish to purchase a copy of this video, the price is US$89.95 postpaid, to US customers. Canadian and overseas customers should check our website, at www.lautard.com, for further details, pricing etc. Or phone us at (604) 922-4909 after 8 AM West Coast Time, for further info. See also the top left hand corner of page 263, herein.

Henteleff's #9
from **Jacob Henteleff, Winnipeg, Manitoba**

It is surprising how often things arrive in my mail from you fellas that fit in with something I'm doing at the time. I had just finished putting together the section on making a good rifle cleaning rod, which begins on the next page, when I got a letter from **Jacob Henteleff** of Winnipeg, Manitoba in which he gave me the recipe for everybody's favorite gun cleaning solvent.

Apparently the commercial stuff stems from an old US Army recipe used back in the days when both Nitro and black powder were still both in use (as they are today, for different reasons). The recipe, says Jacob, is simple to make and costs far less than the name brand stuff:

100 fluid ounces of Kerosene or Varsol (preferably deodorized)*
10 ounces of mineral oil
4 ounces acetone (which is really only for black powder)
1 ounce Amyl Acetate

The last item is the "secret" ingredient which gives that wonderful smell. The common name for this stuff is "banana oil" - one whiff'll tell you why. Jacob says he got some from a well stocked veterinary supply house; "...it was expensive, but you really only need the smallest of amounts. If you have a contact at a local university, you may be able to get some there. Its main purpose in the recipe is to provide that indefinable nostrilific thrill that one associates with gun cleaning."

> * Somewhere in my writings I have commented upon the fact that a number of people have asked me, "What is Varsol?" Varsol is paint thinner. I phoned a paint store and a hardware store, and found that you can buy both paint thinner and kerosene in regular and odorless. Odorless paint thinner is just slightly more than regular kerosene, while regular paint thinner is less than half the price of regular kerosene. Odorless kero is slightly less costly than Chanel #5.

MAKING A RIFLE CLEANING ROD

There is a US Government publication, issued about 1916, called "Manufacture of the Model 1903 Springfield Service Rifle". I have a copy. Don't be jealous - you can have one too, because it was reprinted in 1984 by Wolfe Publishing Co. It contains about 330 pages of notes/drwgs on how to make every part of the legendary old '03. While no one would be likely to try to make the whole rifle today (more's the pity), the book shows numerous tools and jigs that are interesting. Also shown are full details of 2 types of GI-cleaning rods - solid and jointed - plus the thong or "pull-thru" type cleaning equipment used with the '03.

Not shown in the above book are cleaning rods of later issue, which were similar but used different (uglier) handles. Some had oval knobs or T-style handles with swivel-in-the-handle, some had the patch tip integral but drilled and tapped to accept the brush from the "pull-thru" cleaner, some had a separate patch tip with a slot instead of a button for the patch, and these too could use the pull-thru brush. About 1931 two new cleaning rods came into service, one solid, one jointed, both loaded with all the best features the American Gov't could devise. (This info comes from Lt. Col. William S. Brophy's book "The Springfield 1903 Rifles" - also from Wolfe Publishing.)

Although you can readily buy cleaning rods that are probably slightly superior today, it would be quite easy to make very good GI-type cleaning rods generally along the lines of those shown in the old book on how the M1903 was made.

One-piece rods are better for home and range use, while the jointed type is the obvious choice for trips afield. The swivel feature is a desirable refinement which should most certainly be built into either type.

If I were going to make one, I would incorporate features from both the solid and jointed forms of GI cleaning rod. Both were made from 1/4"ϕ half hard brass rod. The jointed type (Model of 1913) had a 1"ϕ round knob-style handle of aluminum, and a swivel near the tip (so the cleaning patch would follow the twist of the rifling), and interchangeable "patch" and "brush" tips. The solid rod had a "patch" tip of the same form as the patch tip for the jointed rod, but could not accept a brush tip. It had an oval ring style handle. Whether it incorporated a swivel is not clear from the drwgs in the book.

The drwgs that follow show necessary details. If someone asks about yours, you can tell them it is a rare form issued only to snipers and others particularly skilled in the use of the '03. Stamp it "1915" - that'll really drive 'em up the wall.

> NOTE: The standard thread now used on most cleaning rod accessories - brushes, jags, etc. - is 8-32. If you decide to make your own cleaning rods, this is well to know.

If you make one of these cleaning rods, be sure to lightly chamfer each end of each piece so there are no sharp corners to catch on anything.

Also, if you make a rod of this type and use it in a 270 WCF or other rifle of less than .30 cal bore, you may need to turn down the button at the end of the Patch Tip to give the right fit between the bore and the patched cleaning tip. Use good patch material. (Read "The J.M. Pyne Stories and other selected writings by Lucian Cary", and you'll know what is good patch material and what isn't. You'll also learn Harry Pope's method of removing a stuck cleaning patch from a rifle barrel.)

To make a cleaning rod for a .22 or .25 caliber rifle, use a piece of 3/16" brass rod, and adjust other dimensions as needed. A 3/8"ϕ rod would do nicely for rifles of .45 cal or more. A short rod could also be made for any handgun.

Handle

Stamp 1915
on top

1"Ø

1-1/4"

cross drill 0.086"Ø *

1/2"Ø

1/4"

drill 1/4"Ø x 5/8" deep

Not Drawn:

Brush Tip. Make from 1/4"Ø brass rod, 3/4" long, plus 10-32 threaded portion as on Patch Tip. Drill and tap other end 8-32 to take factory-made bore-cleaning brushes. Chamfer "brush" end.

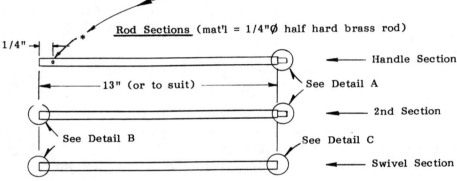

Rod Sections (mat'l = 1/4"Ø half hard brass rod)

1/4"

Handle Section

13" (or to suit)

See Detail A

See Detail B

2nd Section

See Detail C

Swivel Section

Detail A

0.150"Ø

10-32

0.52"

0.15"

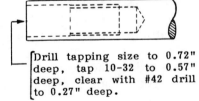

Detail B

Drill tapping size to 0.72" deep, tap 10-32 to 0.57" deep, clear with #42 drill to 0.27" deep.

Detail C

Drill #42 x 0.68" deep, tap 4-40 full depth, clear with #10 drill to 0.19" deep.

A Swivel-in-the-handle Rod...

You might want to consider making a cleaning rod with a pair of small deep-groove ball bearings in the handle. My Parker-Hale rods have the swivel in the handle, and they work exquisitely. I've never taken them apart to see what's inside.

Garrison Rod

One-piece cleaning rods were also issued, for use in barracks or similar situations. These were referred to as "Garrison Rods". If you want a Garrison Rod, make the Knob and Rod as above, but make the Rod all one piece right to the butt end of the Swivel fitting. Make this end as Detail C, to take the Swivel and the Patch Tip. A one-piece cleaning rod should be about 10" over barrel length.

Swivel

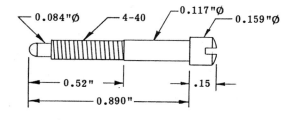

NOTE A

The Swivel is made from a piece of 1/4"ϕ brass rod 1.232" OAL. Drill #31 x 5/8" deep from one end, & then reduce OD to 0.193"ϕ for 0.187", leaving little or no radius at the shoulder. Reverse the workpiece in the chuck, & drill #21 drill x 0.85" deep. (This will leave a tapered juncture where the #21 & #31 holes meet, which would cause binding between the Swivel & the underside of the head of the Swivel Screw on the "pull" stroke of the cleaning rod in use. Therefore, make up a small boring tool or other kind of cutter to poke down the hole and machine the bottom of the #21 hole square, as drwg.) Counterdrill with a #10 drill to 0.27" deep, & then tap 10-32 to a depth of 0.57".

Next, make the Swivel Screw. Adjust its length (and possibly its head diameter) such that when it is bottomed in its 4-40 tapped hole in the end of the "Swivel" section of the Cleaning Rod (so it won't unscrew in use), you have minimal axial play of the Swivel in the c'bore in the end of the "Swivel" section of the Cleaning Rod, but the Swivel rotates nicely on the end of the Rod.

Swivel Screw (steel)

Patch Tip

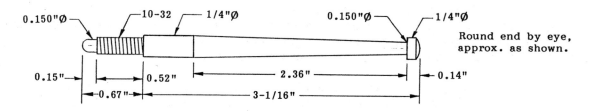

Round end by eye, approx. as shown.

Indicator Sleeve

A ring or sleeve is shown in the book "Manufacture of the Model 1903 Springfield Service Rifle". Although its purpose was not explained in the book, it would seem this ring was intended to indicate the full stroke of the rod for the rifle being cleaned. Although no locking screw or friction device is indicated in the government's drwgs, it would need to slide along the Rod and stay where put by the user. Such a ring could readily be made if wanted.

A PORTER'S BAG FULL OF TRICKS
from David L. Porter, Westland, MI

Dear Guy,

Fig. 1 is a handy pin vise I made from a Dremel replacement chuck. The chucks are available at hobby shops for $4-6 as replacements or retro-fits. If you were making such an item for yourself, even if you have a 9/32 x 40 die, screwcut the 9/32 x 40 thread on your lathe to assure straightness and a real close fit in the chuck.

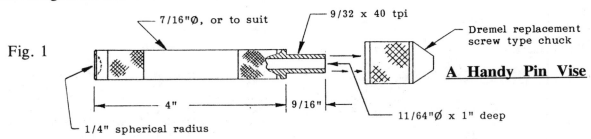

Fig. 1

A Handy Pin Vise

Fig. 2 is a small key type drill chuck on a 1/4"⌀ arbor that slides inside the 3/8" tube. After drilling a tapping size hole in a job in your mill or lathe, you can put a small tap in the chuck, pop it into the machine, and start the tap absolutely true right in the same machine. In sizes up to about #4 you can tap the entire hole this way; in larger sizes, start the tap this way, then finish with a tap wrench, off the machine. I've never broken a tap using this tool, even down to #00-90.

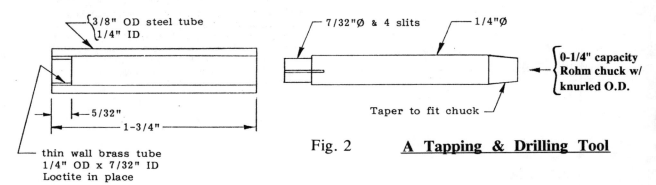

Fig. 2   A Tapping & Drilling Tool

I think taps are much more likely to break if subjected to bending as well as torsional stresses, as in hand tapping. This tool completely eliminates the bending stress. Also, you can back the tap out of the hole really fast and easy when you need to clean chips out of the flutes. (Keep several old tooth brushes around for just this purpose - they work great.)

One more use for this neat little tool is to drill very small holes. Chuck the outer sleeve in the drill press, and let oil between the sliding members transmit the torque to the drill. If the drill is larger, use heavy oil, if smaller, vice versa. I use STP (engine oil additive) as way lube on some of my machines; it's available in pint cans, and stays put forever, even on vertical surfaces. Put it in one of those squeeze-type mustard bottles.

Still on the subject of tap starting, I'm afraid I don't agree with your statement that given a chance, they'll start themselves square. I would say that given a chance, they'll start from 10-20° off square, and either break themselves or make your screws lean like the Tower of Pisa! The only time I start a tap by eye is into very thin stock. For parts too big to go in a mill or drill press, I often drill and tap holes with the aid of a block with an appropriately sized guide hole. I have a small block of aluminum about 2" square x 1" thick, with numerous holes drilled in it, corresponding to various tap drill sizes, clearance drill sizes and tap body diameters. By just clamping the block to the work, I can drill a perfectly aligned hole with a hand held drill, with no "walking" of the drill. Admittedly, the aluminum block is soft, and the holes wear oversize pretty easily. If I were making a new block, I'd use steel (slower to drill in the first place), or maybe brass (easy to drill, but still a lot more wear resistant than aluminum). The present block sprouted a couple of holes at a time as required, and never was intended to be a TOOL. Looks sorta like aircraft-grade Swiss cheese!

Other times, I use **my disappearing tapping machine**. You probably have one too, but never noticed it... About the only important difference between my Taiwanese drill press and those hand tapping machines you see in some professional shops is that the drill press has a clockspring return to suck the quill back up whenever you let go of the quill feed handle to try to rotate the spindle by hand.

The quill return feature is then a nuisance:- suppose you are trying to hold the work with one hand, feed the quill with one hand, rotate the spindle with one hand, and maybe squirt some tapping fluid on the job with one hand, scratch your nose... well, I guess you get the idea.

But if you could magically release the tension on the quill return spring, you would no longer need to keep one hand on the feed lever, which would free up at least one hand.

I modified my drill press by making a new spring case with a knurled rim (the purpose of which will become clear in a moment), and attaching a pawl to the machine's head casting via a screw, so it can swing up from the spring case. With this arrangement, I can instantly release the tension on the quill return spring and use the drill press to start a tap. Once the tap is started, I will release it from the chuck, re-tension the quill retract spring (via the knurled rim on the spring case) until the quill sucks back up, re-engage the pawl, and presto! my tapping machine turns back into a drill press!!   There are probably a lot of different sizes and configurations on various types of drill presses, but the basic idea should be almost universal in application.

Fig. 3 is what I'd call my Strokagenius Depth Gage, providing you don't intend to copyright the term "Strokagenius," Guy. It's just a piece of square keystock and a few pieces of 1/16"ϕ drill rod of various lengths.   Its tremendous utility comes from its small size; for instance, to measure the depth of a hole being bored in lathe or mill, you don't have to back the tool out 6 or 8 inches, only an inch or less. The extra hole/setscrew near one end allows you to move the rod to the end, which can be convenient when measuring a step or a hole near an obstruction.

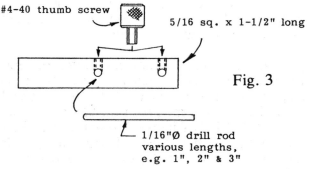

#4-40 thumb screw

5/16 sq. x 1-1/2" long

Fig. 3

1/16"Ø drill rod
various lengths,
e.g. 1", 2" & 3"

To help the setscrew positively lock the rod without mutilating its surface, make a tiny "spring" out of copper magnet wire and drop it in under the screw. This works like a charm! In the past I've laboriously made all sizes of brass and plastic pads for #0 through #10 setscrews. Wish I hadn't bothered - **this copper wire trick works way better, and is 20 times faster.**

Fig. 4, lower right, shows my preferred method for putting steel balls on machine handles, etc. Set up to do the silver soldering with the ball on a fire brick or whatever, with the tapered shank above it, as shown. Apply heat mostly to the ball, not the shank you

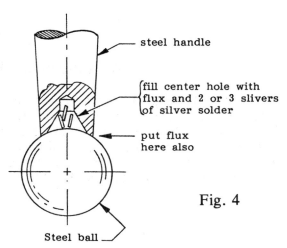

steel handle

fill center hole with
flux and 2 or 3 slivers
of silver solder

put flux
here also

Fig. 4

Steel ball

want to solder to it. I've tried your incremental cut method of generating spheres and radii, per **TMBR#1**, and while it works very well, I find it somewhat time consuming. Providing you have a ball bearing of the right size at hand, it's faster to just paste it onto a handle than to machine it all in one piece.

David

Note from GBL: I would agree, but I would almost never have a ball bearing at hand that would fit. I do happen to have a few used 1.5"φ ball bearings, and on one occasion, to make the big ball at the butt end of a couple of new motor lock levers for my B'port, I annealed two of them, chucked them in the lathe, faced them to produce a flat face of a desired size, and then drilled/bored and tapped them. They worked out just fine, but be warned - the steel they use in ball bearings is tougher'n woodpecker lips, even after annealing. I made the tapered shanks and integral small ball for each of these handles from 3/4"φ CRS, and they look good. If I'd had a couple of 3/4"φ ball bearings, plus this idea at the time, I'd have been happy to do it David's way.

On the other hand, read on and you'll see why somebody else was happy he used my incremental cut system....

Balls, Bullnoses and A Butterball Paycheck
from **Michael O'Connor, Buffalo, NY**

Dear Guy,

I am employed as a toolroom machinist by a company that manufactures co-ordinate measuring machines. I also program and set up our machining centers. Many times some idea I've spotted in HSM or one of your books has made me look like a genius at work, and left a few of the guys scratching their heads.

For example, some years ago, after reading your article "Balls and Bullnoses" in HSM, it happened that I had to generate a number of large radii on a workpiece. (This was at a different shop from where I now work.) The boss had figured on buying some commercial tooling that would have killed the profit on the job because he expected to squeeze that operation in on one of our CNC lathes between other work. As usually happens, things were running behind and the job had to get done. I took the job drawings home that night and, using the method described in your article, did the necessary calculations. The next morning I had a few hours head start; I was just finishing up when the boss walked in and saw the job finished. I spent over an hour explaining to him how I'd done it. The next week I had more money in my paycheck.

Mike

A Source of High Quality Cast Iron

Another tip **George Leverton** passed on to me was a source of very nice sand-free, blow-hole free continuous cast iron, in various cross sections, but of uniformly fine quality. He has used it for material from which to make bullet molds, and said that in spite of the small size of his business with the outfit, they fixed him up with a charge account arrangement right off the bat, and were just generally real nice people to deal with. The outfit? Midwest Metals, 2195 Lakeshore Drive, Woodstock, IL 60098-7467, Phone 1-800-526-0548

SEVERAL GOOD IDEAS
from John S. "Rusty" Rauth
Rauth Machine & Welding, Highland, MD

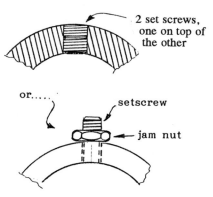

2 set screws,
one on top of
the other

or......

setscrew

jam nut

Dear Guy,

I noted that you were having trouble keeping your countershaft pulleys tight on the shaft. (**TMBR#1**, page 44.) The best way to lock setscrews without using Loctite is to use double setscrews if you have the room to do so: screw the setscrew in tight and then screw another setscrew as tight as possible down on top of the first one. It works very well. If the pulley hub is not large enough to permit this, use Service Removeable Loctite (Blue). Another method is to use a longer setscrew with a hex nut as per lower sketch.

On blueing of steel: Muriatic acid (the kind you can find in hardware and building supply stores for washing concrete and brick prior to painting) will rust steel and sharpen not-too-dull files, if all the dirt and oil is removed first. Be sure to wash off **all** the acid after use and preserve with oil to prevent further rust.

Chem-Tech Cutting Fluid mixed with humidity in the air in the summer will brown steel, usually chucks, machine spindles, machine tables, vises, etc., or any steel not protected with oil after use.

Some further thoughts on tool storage box material: There is a version of Masonite available called Benelex. This material is harder than Masonite and is made by the same manufacturer. It is available in plastic supply houses only - hardware and building supply houses will not know anything about it. It can be milled, drilled, routed, sawed, turned, tapped, glued, and sanded, but not welded. It is available from 1/4" to 2" thick; thicker pieces can be created by glueing up with epoxy or contact cement. It must be cut with carbide tools if you are using a lot of it. When milling, cut into the edge, not off the edge, as it will delaminate. When through-drilling, use a backup block clamped tight against it to keep drill breakout clean. When using machine screws to fasten anything to it, use coarse threads - fine threads will pull out. Better still, use Heli-Coil inserts to make it stronger.

When tapping plastic for machine screws, always use coarse threads, as they will take more torque before the threads pull out than will fine threads. It has something to do with the square inches of thread surface engagement available, and the width of the base of the thread.

To transfer punch deep holes using transfer screws, an item as at right is good for very deep holes, as the rod length can be as long as you need, and can be used on big diameter holes with small screws. You can also use a sleeve for any size of transfer punch you may have at hand, as shown in the lower part of the drwg.

The outside diameter of either item above can be reduced by a couple of thou by filing in the lathe.

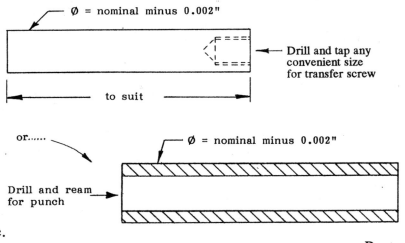

Ø = nominal minus 0.002"

Drill and tap any
convenient size
for transfer screw

to suit

or......

Ø = nominal minus 0.002"

Drill and ream
for punch

Rusty

SAWING A 0.001" SLOT
by David Coupland, Teulon, Manitoba

In 25 years as a tool & die maker, I've had many jobs that taxed my ingenuity. On one occasion a few years ago I was assigned the job of making up some specialized plastic extrusion dies. The basic machining was straightforward - a tapered bore starting with a round entrance and ending with a smooth entry into two 0.038"ϕ holes with a center-to-center distance of 0.040". This left a 2 thou web between the holes, with a designed length of 0.060". Through this web a slot of 0.001" was required. Since the tooling was to be of one piece construction, a method was needed to make this slot. While I was working on the routine aspects of the job, my mind was busy on how I was going to produce the slot. I started my weekend with no real ideas as to how to go about it.

By Monday morning I'd decided I needed a saw. I had an idea, but not much confidence. In about 10 minutes I had a slot. I could not believe how easy it had been - all that worry for nothing!

The engineering people were extremely curious as to how I'd made the slot, but I told them first that I couldn't remember, and when they persisted, that I didn't know, because I'd had my eyes closed. They asked my partner, and he said he'd gone to the other end of the shop and turned out the lights, so neither of us knew. One engineer in particular wanted to find out how I cut the slot.

I looked him right in the eye and said, "With great difficulty."

This upset him, and he asked me several more times, but I told him the same thing. After about six times he said, "I am going to ask you one last time. If you don't tell me how you did it, I'm going to kick you in the shins". If you'd like to know what he still doesn't know, read on.

I had decided that a saw blade was necessary, but the question was where to get one?

I took the 0.001" leaf from my thickness gage, clamped it between two pieces of cold rolled flat stock with about 1/32" exposed along the full length of the gage leaf. I placed another piece of flat cold rolled stock against the projecting portion of the feeler gage, and hit this with a hammer. Because the thickness gage leaf was hardened, the narrow exposed piece cracked off, giving me a 0.001" 'saw blade' about 3 inches long.

The problem now was to thin this blade so it would cut undersize. I clamped the ends of the blade down on a piece of flat cold rolled stock and stoned the side of the blade lengthwise until I got it down to about 3/4 thou. Looking at the blade with a magnifying lens, the broken edge looked quite jagged. This suited me fine, as it would be the "tooth" edge of my saw blade, which I would use in an adjustable-frame jeweller's saw.

To find the center point of the holes, I placed flat stock against two bore-diameter pins placed in the two 0.038" holes. After clamping this bar in place, I removed the pins, and with the aid of spacers, placed a second piece of flat stock at half the bore diameter plus half the saw blade thickness, and clamped it there. I then put a shim the thickness of the blade next to my second piece of flat stock, and moved the first piece of flat stock in against the shim, and again clamped it in place. This small opening between the bars would guide the saw blade..... Still with me?

I now threaded my saw blade through one of the 0.038" holes, fastened the ends in the saw frame, and was finally ready to cut the slot. I used a medium grade diamond paste (available in small jars and tubes) to do the actual cutting. The rough edge on the saw blade helped to hold the diamond paste. With a little paste dabbed along the rough edge of the saw, and light pressure applied so as not to break the delicate blade, I cut the slot in less time than it takes to tell about it. I also used a pair of tweezers in my other hand to control the blade as it entered the guide slot. After removing the guide pieces, I put a little paste on the sides of the blade, and went back through the slot carefully, widening it to a tight 0.001".

That's it - a 0.001" slot and sore shins.

TERROR FROM THE CANAL

It was a dark and stormy night... the postman arrived with a package that made my blood run cold....

The package contained several pieces of barbed wire. No, not the stuff you would find in a fence, but rather the sort you will find in your mouth if you ever need a root canal. I'd just had my first root canal a few weeks before, hence my reaction. These little roto-rooters of the dentist's trade - if you should be so fortunate as to lack familiarity with them - are little straight pieces of fine wire with very fine and very sharp barbs cut into them for about 1/2" at one end, and a little wee wrapped wire handle about 3/8" long at the other end. They are used to rake out, in pieces, the nerve in a tooth, if you have, for example, a toothache.

One of my guys had had the clever idea that these would be the ideal tool for fishing toilet paper out of small, tapped holes, and if you once handle one, you will immediately know that they would be much more effective for this purpose than the little item I made from a straightened paper clip, and described at page 19 in **TMBR#1**.

See if your dentist will sell or give you one. My dentist says they are called files, reamers, and broaches, all these terms being applied willy nilly to the same item. They cost about $3 each, up here, perhaps less in the US.

I keep mine in my Gerstner, in the drawer where I keep tap wrenches, thread pitch gages, etc... at the very back of the drawer, where I don't have to look at them unless I need them.

> Thanks to dentist Paul Eleazer, of Albany, GA for a chilling but toothsome idea.

Something else: dentists sometimes use disposable strips of thin, tough plastic about 5/32" wide x maybe 7" long, with two grades of fairly fine abrasive bonded onto one side. These are used to sand between your teeth for one reason or another. As you can well understand, these could be useful in the shop, so ask your dentist to save you some.

My dental hygenist made herself understood pretty clearly when she suggested that if I had any teeth I particularly wanted to keep, I'd better start flossing them regularly. (Nobody ever put it to me quite like that before.) She also put me onto a little wee brush that looks like a bottle cleaning brush. These are for rooting around between your teeth, and are sold in drug stores in the toothpaste section. I was certainly glad to learn about these - they look like they'll be great for very small cleaning jobs in the shop.

"THE WOODPILE"
-- An Introductory Note

In TMBR#1 I had some notes on workbench legs, as well as some on making wooden storage boxes for tools, etc. Since then I've gotten somewhat interested in certain aspects of woodworking, and have learned a little more about those topics, and about woodwork in general. I've also acquired a table saw, a bandsaw, a wood lathe, and a few other smaller items.

I don't think I'm alone, as a machinist, in feeling less than entirely comfortable with woodwork. However, as I said, I have learned a few things.

The material in the pages which follow is offered for those readers who would appreciate a little advice on one aspect or another of some woodworking jobs that are sometimes unavoidable, and/or could even be enjoyable if a fella knew how to go about them.

There will be those who read this book who could do in wood, in their sleep, things I'd show off like a new grandchild if I'd done them. This section is not for such readers. It is more likely to interest a guy like myself who might say to me, "Lautard, I have to build a new workbench - you got any ideas?" And my reply would be, "Yes, I built 2 new workbenches about 3 years ago, and I sure learned a thing or two doing it. How'd you like some help? It may be a case of the blind leading the blind, but we can do it."

So that's the perspective from which the following material is offered. To save space, I will *not* introduce each chunk by saying, "I don't think I'm alone, as a machinist, in not feeling entirely comfortable with woodwork...", but you can take it that that phrase explains why every part of "The Woodpile" section is here. I hope it is helpful to some who read this book.

MORE WORKBENCH IDEAS

We had some ideas about workbenches in TMBR#1 (pages 133/35.) I have some further ideas (some acquired at great cost to myself), which I would like to share with those readers who, like myself, don't feel entirely comfortable with woodwork.

Over the winter of '89/90 I revamped my little 12' x 12' shop quite a bit, and built a couple of new workbenches. I'm not all that hot as a woodworker, and deciding how to go about making a bench that would suit me had me going around in circles. Plywood? 2x4's? 4x4's? 2x6's? logs? Steel? Dexion? Every time I got about half way ready to take off in a particular direction I'd see another bench that seemed to hold less terrors for me to make.

And then, too, I had one workbench in place already, which gave me a minor scheduling problem. Plus the fact that the existing bench had been built many years before by my late father-in-law, and although I didn't like it - it was too low for me, for one thing - I wondered how Margaret would view its removal (= destruction). When I asked her, she said it was quite ok with her - her dad had built it (when he built the house we now live in) with what he had had in the way of materials and methods, but she wasn't going to be upset if I turned it into firewood.

When it came time to actually do it, I felt pretty bad about it within myself. I had a lot of respect for Margie's dad, and although he died some months before we were married 22 years ago, the day I pulled it apart I felt like an intruder in *his* shop. (I chose to do it on a day when Margie was out, in case she would feel unhappy seeing it being taken apart.) As I worked, I was thinking: "I'm not doin' this with no thought for what you built, Jack. I've used it for a lot of years, and I know you did a lot more things on it than I ever will - things I'd be proud to say I'd done, if I could even do them. But I need a different type of bench in here now. I'd like to think it's the sort of bench you'd like, too, if we could've built a new bench together." Now I don't want to go gettin' all maudlin, but I tell ya, guys, I took that old workbench apart and carried the pieces outside with a certain feeling of..... well, I guess "respect" would be the most appropriate word.

Now, you may be in sync with me on this, or you may not. If you're not, let me explain it from another point of view: a workbench is a very personal thing, to some degree a reflection of a man's soul. You don't buy them at the supermarket like a bag of donuts. Well, okay, there are commercially made workbenches, but guys like us don't own them.

So how do you build a good workbench? Well, one of mine was to be a sort of machine bench - a place to put the scroll saw, the die filer, the surface plate, my Strokagenius file rack, and one of my two toolboxes. In the course of my renovations to the shop, I had tacked 2x4's, flatwise, to the bare concrete north and east walls with a Hilti gun, and had screwed 3/4" G1S (good one side) plywood to the 2x4's. (Why such heavy plywood? If I want to stick something to the wall, I have some real meat to screw it to without hunting for the studs.)

To make the machine bench, I spiked a 2x4 solidly to the east wall, horizontally, and made front legs and front-to-back stringers from paired 2x4's all bolted together. These were secured to the rear 2x4 with metal brackets I made from 1/16" CRS sheet. I got the latter material sheared to size (a very fast job) at a sheet metal shop, brought the pieces home, deburred them, punched 6 holes per piece, and then bent them into 90° angle brackets as needed. Some I left flat and used to reinforce some of my other under-the-bench joints. It was going to be as solid as a brick flowerpot when I got the top on it.

Ah, the bench top. Everybody knows that laminating 2x4's together, with their wide faces stuck together, is a good way to make a bench top. I sawed off four 24" lengths of 1/2" CRS, cut a 20 tpi thread on both ends of each piece, and applied washers and nuts. I had the 2x4's all ready except for jig drilling the holes in the 2x4's, when I started thinkin', "How'm I going to make a nice flat surface on them once I get them all bolted together?" Well, plane it, of course.

"You'll never be able to hand plane it flat," said my friend Henry the cabinetmaker. Henry ain't a pessimist, he's a realist. (Or, to put that another way, I think what Henry may have meant was that *I* did not have the skill to do it. Any cabinetmaker skilled in the use of hand tools *would* be able to do it.)

I knew of an outfit in town with a thickness planer big enough to swallow my (proposed) benchtop, but they wouldn't touch it if it had so much as one nail in it. They just turned pale at the very idea of those four 1/2 x 24" tie bolts in there. "But they're centered in the depth of the 2x4's....."

"Don't make no difference, fella. No metal in it, or it don't go into our planer."

"What if I glued the 2x4's together, and then pulled out the 4 long bolts before you put it through the thickness planer?"

"That'd be ok with us, but the planer might cause the glued joints to come apart, and your benchtop could come out the other side of the planer in pieces, and you'd be no further ahead than before."

I went back to see Henry. Henry said he had a benchtop upstairs for which he had no further use, and which might suit me. It turned out to be 24" wide, maybe 4" longer than I needed, and it was in nice shape. We did a fast monetary shuffle, and I was more or less set. Shortening it to suit my purposes was no problem, and it wasn't long before I had my "machine bench" done.

Two footnotes to the foregoing:

1) Henry had made the benchtop I bought from him by gluing together a piece of 3/4" plywood and a piece of 3/4" particle board, and adding a stringer full length under the front and rear edges, for added stiffness, as drwg. The plywood he'd used had a face of thin maple veneer that shows on the top and front faces. It was as close to ideal for my purposes as one could hope for, considering the cost, and the fact that, although found rather than planned, I was readily able to adapt it to my bench foundation efforts.

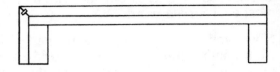

2) There is an 'out' I did not think of at the time, that would've let me proceed with my plan to laminate my benchtop from 2x4's, and it may be helpful to someone else reading these notes to know about it.

I could have glued up my 2x4's exactly as I had planned to do, c/w bolts, except for one thing: no glue at the middle joint. When the glue dried, I could have removed the tie rods and the benchtop would have self-divided into two pieces, each of a size that would go through a more modestly sized thickness planer such as would be found in some private woodworkers' shops, most any cabinet shop, and the wood shop in just about any local high school. When the 2 pieces were planed and re-joined - with or without glue - it's a pretty good bet they could have been made to show a smooth face to the world.

My knowledge of workbench building expanded somewhat through the above and other direct experiences, also by reading, observation, and to some degree by the slow, osmotic machinations of various disjointed segments of my brain. However, I doubt you could stand to hear much more about them, so I will end with details of a design I would have been glad to know about before I started on that first bench for which Henry supplied the top...

A SIMPLE, SOLID, NO-SHAKE WORKBENCH DESIGN

One day in my travels I spotted several very solid workbenches. They caught my attention particularly at the time because I had just gone through more bench building agonies of my own than I have described above. These had been built by a steel rule die maker by the name of Stephen Silver, for use in the in-house die shop of a firm making electronic membrane switches and other work. The die maker gave me some detail sketches and the ok to include info on the design (which he has taken through several "marks" in his shifts from job to job over the years) in this book.

You do not need to be a cabinetmaker, or even a rough carpenter, to build a workbench of this type. The "joints," such as they are, can be done with a minimum of the simplest hand tools, or a router, or on your milling machine. Tight fits are not required. All holes can be drilled with a hand drill; an electric drill, a drill press, or your mill can be used to make an easy job even easier.

You don't need a table saw, although the use of one would be a convenience. All of the sawing can be done with nothing more than a backsaw (or even just a hacksaw). If you are not skilled at hand sawing wood to a layout line, this can be compensated for by the simple expedient of a wooden miter box, homemade if need be. A couple of C-clamps would be helpful. All the hand work could be done on a couple of saw horses or a similar improvised work surface.

If your floor is level both ways, your job is made easier yet. If it is not, you can either ignore the irregularities, and shim each leg to bring the benchtop level on completion, or you can make legs of different lengths to suit the floor. (The latter approach will make the bench ill suited to being relocated, but if you never intend to move it, this won't matter.)

The bench frame members are mostly 2x4's. As for quality of wood to buy, the better the better, but construction grade wood, hand picked at the lumberyard, will do fine. 4x4's could be used for the legs, but paired 2x4's are also fine.

If you can put your wood through a thickness planer before you use it, buy "#2 & better construction grade fir," and let it sit in your shop for a week or 2 before you plane & use it. This way, your wood should be flatter, smoother, squarer, and of more uniform thickness, from piece to piece, which will help make everything go together more easily. Or buy kiln dried double dressed fir or similar - you'll be *almost* as far ahead, 'tho poorer.

A router jig could be made, and a router used to cut the horizontal rebates and grooves which are required in some of the leg pieces. Glue two pieces of 1/2" or similar plywood to a base of 1/4" plywood. Clamp the jig to the leg, and cut away the unwanted wood.

The "subtop" is short pieces of 2x6, which means suitable pieces will be easier to find. (A straight, unwarped 36" length of 2x6 is easier to come by than one 8' or 12' long).

The top can be laminated or solid. (I am not forgetting what I said above about the bench top I got from Henry, and about making a bench top from 2x4's. What follows has - I hope - a few other/additional kernels in it.)

3/4" medite over 3/4" plywood, glued and screwed together, is one easy way to make a very strong, solid, bench top. Any cabinetmaker can cut the pieces to size for you. A nice way to finish off such a top is to "edge" it with a strip of clear fir, glued on, pipe clamped in place, and then planed or sanded flush with the top face. Do the long front face first, then the ends, then the back if you want. Such a bench top, once varnished, can be cleaned with the palm of your hand. When you have a piece of rough work to do, toss a piece of 1/4" masonite board onto the bench, and the varnished face will stay nice for many years.

3/4" plywood topped with a sheet of 12 gauge black iron is another way to go. The edge of the steel top can be lipped by the supplier at minimal cost (down at the front, up at the rear to keep stuff from going off the back of the bench, or down both front and back, if preferred). BB had a couple of bench tops made this way. They look good, and are clean and durable.

I was talking to **Phil Lebow** one day about workbenches, steel tops for same, and so on. Phil said he likes steel tops, but pointed out that if you ever drop something heavy on one, it's almost impossible to get the dent out.

Our friend Herbert Dyer (see TMBR#1, p. 134) recommended making a bench top by nailing up 3 layers of 1" tongue-&-groove flooring, the bottom pieces running the long way of the bench, the middle layer going front to back, and the top layer again the long way. He suggested inletting a piece of 2-1/4 x 2-1/4" angle iron into the full length of the top front edge. This, drilled and tapped 1/2-13 at 6" intervals, plus a piece of 1/2 x 1-1/2" CRS, clearance drilled to match, and with a bevel milled along one edge, provides a commodious but unobtrusive built-in sheet metal bending capability*. Of course, not everyone would want this feature. Dyer was a master sheet metal worker, and made his living as such, more than 50 years ago, at a bench he made as above. (If you do want to include this feature, the actual inlaying of the piece of angle iron could be accomplished with a router - just don't put any nails/screws in the general vicinity until after you have all the wood carving done to your satisfaction!)

> * (For a cross section drwg of this idea, see **TMBR#2**, page 110.
> The item on the right, there, if rotated clockwise about 90°, would
> be as good an illustration of this as one could want.)

* SEE PAGE 264

An Easy-to-Build WOODEN WORKBENCH
Length, width and height to suit user (see text)

* = 2x4

** = 4x4 or doubled 2x4's

denotes carriage bolts
or redi rod both ways
thru leg at slightly
staggered elevations.

vertical member cut out by
about 1/2" deep on outside
only, to take horizontal 2x4.

3/4" plywood top or heavier
- see Note 1

There would be 2 of these stations
in a bench 6 to 8' long.

Note 2

END

Note 3

2x6

FRONT

TOP

182

Notes to Workbench Drwg on facing page:

Note 1: A sheet of 3/4" plywood would serve well enough as a top. A better way to make a top is to laminate a piece of 3/4" medium density fiberboard (MDF; one trade name for this mat'l is Medite) to 3/4" plywood with glue and screws. Put the screws in from the plywood side, and use screws short enough that they don't quite come right through both layers. Edge this plywood/MDF top with 3/4" solid fir or similar; glue the edging material in place, and then plane and sand it flush with the top face of the MDF after the glue is dry.

1/16" or 1/8" black iron sheet (screwed down with c'sunk screws) can be overlaid on a plywood top, with the front edge bent down, and the rear edge bent up or down, depending on whether the bench is placed against a wall or not.

Note 2: Shelf is cut to fit around legs. To install, tip it diagonally, and insert from end(s) after bench is assembled.

Note 3: Tim Smith, whose main workbench is almost identical to that shown here, says: "I added plenty of cross bracing all over the place for strength. You could park a car on mine - I can't move it at all. Materials cost me over $100, but I love it." Tim used 4x4's in some places where 2x4's are indicated in this drawing.

---//\\---

Now you may be thinking, "but I'm a metalworker, not a woodworker!" Ok, I can relate to that. Bench legs can be made by bolting angle iron to channel iron. A steel fabricating shop could supply the pieces cut to length, and it would be no trick to drill the necessary assembly holes if you have a milling machine or even a small drill press. Or the fab-shop guys could punch them for you. One could also make the frame from steel or aluminum angle, assembled with 3/8" bolts.

Now, if you want yet more ideas and info on how to tackle the job of building a workbench, may I recommend most highly a book called "The Workbench Book" by Scott Landis - this is put out by the Fine Woodworking people, and is simply tops. It is primarily about work benches built by and for woodworkers, but there is a multitude of good ideas (not only on workbench building) in it, many buried within its numerous color photos. See it at your local library along with the same author's companion work, "The Workshop Book". Like those wonderful books by that guy Lautard, these are books you cannot help but find something useful in every time you browse through them. I stayed up till 1 a.m. reading "The Workbench Book" when I first got it.

Is there a Bandsaw in your Future?

I got some info in the mail one day from Marshall R. Young, one of my guys over on Vancouver Island. (If you have a copy of **"Hey Tim..."**, you'll find therein, at page 44, reference to an article in FWW by Mr. Young, in which he tells how to make a good fence for a table saw*.) What Marsh had sent me this time was some info about bandsaws - I'd been talking to him about making or buying one.

* In late '92 I built 3 fences to Marsh's design, incorporating a few changes Marsh has come up with since his article was published in *FWW* (see page 46 in *FWW* #68, for Jan/Feb 1988). I put one on my new bandsaw (see below) and the other two went onto table saws belonging to two friends. One has been installed, and the owner and his son like it a lot. Paul Scobel got the 3rd one; he regards it as "state of the art", other than for the locking handle, which on the 3 fences I built is simply a box end wrench hung on the inboard of two nuts on the locking screw. ("Think of it as a free wrench, Paul.")

I would say that while it may be somewhat overkill for a bandsaw (there are other ways to put a usable fence on a bandsaw, including simply clamping a board across the table), Marsh's design makes a deluxe fence for a tablesaw, and any machinist could make one with no difficulty whatsoever.

The idea of making a bandsaw still appeals to me, and among the stuff Marsh sent were several photos of a big bandsaw he'd built some years previously for re-sawing. (For the non-woodworker, "re-sawing" means to saw a plank into thinner pieces, right on down to making your own veneers, which is what Marsh wanted this machine to do.)

Marsh's saw had a 15-1/2 x 21" table, and was capable of resawing material up to 12" wide (i.e. 12" from table surface to underside of the upper blade guides). It had a smaller throat capacity than you would normally see on a machine with 18.5"ϕ wheels, because he built it with a center column, for rigidity. The wheels, like most of the rest of it, were made from Medite. It stood about 6 feet tall, and was most handsomely done in all respects. Most of the metal parts were aluminum, and were machined on his 7" swing Myford ML7 lathe.

If you saw this bandsaw, you would be impressed - it is good looking, nicely and neatly made, and I have no doubt it could hump out the work in a professional cabinetmaker's shop for many years. Marsh sold it a few years after he built it, but - doggone it, it is IMPRESSIVE.

Marsh has also owned a Canadian-made General Model 490 15" bandsaw for about 10 years, and speaks very highly of it. In the package he sent me was an article reviewing several different bandsaws of this type and size - the well-known 14" Delta, the 15" General, and some of the off-shore knock-offs. Of the General, the reviewer said either that it runs like, or is built like, a 1949 Cadillac. I forget which he said, but whichever way he put it, it's true.

I wanted the bandsaw right away, and since I had paying work for it, buying one made more sense than making one, particularly at the rate I make things, so I bought a General #490. This machine normally comes with motor and stand, but the local dealer had only one in stock, with neither motor nor stand. So I thought: why not take that one, make my own stand, and rig it to share the variable speed DC motor on my Conover wood lathe? The switchover would probably take 2-3 minutes each time, and I'd have both metal and wood cutting speeds in the one machine at the twist of a knob.

This basic idea probably would have worked fine, but in the end, I decided against it, as I didn't see myself doing a lot of metal cutting on the bandsaw. But if you **did** want both wood and metal cutting blade speeds in the one machine, this is something to consider. You'd need a suitable size DC motor, plus a speed controller to suit it - the latter will allow you to dial the motor speed down to almost zero. Keep it in mind.

For wood, blade speed is typically about 3000 sfm. Faster than that gets you nothing but shorter blade life, and slower is rarely necessary. Aluminum can be cut at about the same speed, but for steel, you want to cut down to about 300 sfm or less. Brass can be cut at 300 - 500 sfm. *For how to calculate bandsaw blade speeds, see drwg next page.*

> **NOTE:** According to a letter in *Fine Woodworking* for Sept. 87, page 4, Sears sells a speed converter (mechanical, not electronic) that can be used to drop the speed of a bandsaw. Cost in '87 was about $140.

In Case You Want to Make a Bandsaw, but are Still Hesitant...

In "Hey Tim...." I give some info about a home made bandsaw I came across for sale, and some additional notes about one built by one of my guys in South Africa.

In early '92 I had a call from **Michael Martin**, one of my guys in El Dorado, Arkansas. He told me he'd built himself a bandsaw from free working drwgs offered at the end of an article in *FWW* about a big bandsaw a fella had built for himself in Ontario. Michael said it took him about 4 days to rough out the main frame, and finishing it completely took him another week. He said he'd wondered, when he started, if he was going to end up with something that was fit only to be tied to the back bumper of his car and towed to the dump, but he has found the finished machine eminently satisfactory.

After talking to Michael, I looked up the article, (*FWW* July '87, page 60; see also page 16 in "Power Saws and Planers", in the 'Best of Fine WoodWorking' series), and read it through. And if you need encouragement to tackle such a project, I'm going to give you not one but 2 very solid doses of same.

1) The prototype of the Arkansas bandsaw was built by Bill Corneil, of Thorndale, Ontario, who say says he built it with "..... a sabre saw, an electric hand drill, a borrowed belt sander, and (his own) shop built 12" disk sander". Not a "Skil" saw. Not a table saw. Not a radial arm saw. Not even a portable circular saw. A sabre saw!!! Now if that doesn't motivate you to feel you could build one, I don't know what will, but...

2) go back to the beginning of this section, and read again about the bandsaw Marsh Young built. Note also that many years before, Marsh built a 12" 2-wheel bandsaw and a 3-wheel unit based on articles in Popular Mechanics 1939 and 1942. (See also PM 1963 or earlier for a disc/drum sander.)

3) There's also a little 23 page booklet called "How to Make a Metal Cutting Bandsaw", by David Wimberley. Cost is about $4.50. It's put out by Lindsay Publications, and while it's no coffee table book, I think a fella would get his money's worth out of it, even if he didn't follow the design exactly as presented. The wheels, at only 7.5"ϕ, are a little small for my liking - the smaller they are, the sharper the blade bends in going around them, and the shorter its life is likely to be; but on the other hand, small wheels make for a compact machine.

How to Figure the Blade Speed on a Bandsaw
from **Jerry Kamp, Wooster, OH**

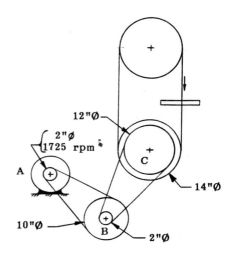

A: 1725 rpm motor & 2" pulley

B: intermediate shaft: 10 & 2" pulleys. Shaft speed =
$(2 \times 1725) \div 10$
= **345 rpm**

C: 12" pulley on 14"ϕ bandsaw wheel. Wheel speed =
$(2 \times 345) \div 12$
= **57.5 rpm**

S: cutting speed S, in lineal feet per minute (fpm), is calculated thus:

$S = $ (saw wheel rpm x wheel ϕ x π) $\div$ 12"/ft*

$= (57.5 \times 14 \times \pi) \div 12 = $ **210 fpm**

* this is inserted to convert inches to feet

MACHINE STANDS

On the surface of things, a machine stand (in my case, two of them, one for a tablesaw, the other for a bandsaw) is a simple enough proposition. But if you ain't Joe Genius Woodworker, you may not be too sure which way to proceed. Also, until you get the stand made, the machine - whether it be a saw, a milling machine, or a lathe - probably can't be used to help make the stand required to make it usable.

Below are details of a simple way to make a stand that will handle tablesaws, bandsaws, and similar machines of let's say under 300 lbs., and which should not have to be replaced ASAP with something better.

> If you want a much briefer read on the topic, get hold of the Jan/Feb. '93 issue of *FWW*. Therein, at page 74, is an excellent article on restoring an older bandsaw. In it, the author tells (in half a page!) how to make a good bandsaw stand. Like me, he uses plywood. He uses more glue, and shuns bolts, which he says will loosen due to wood movement and machine vibration. Will I regret my use of carriage bolts and no glue, as described below? We'll see.

As I have indicated above, I've recently added some woodworking equipment to my shop. One item was a Ryobi BT3000 tablesaw, which I bought without a stand, to keep the cost down.

The Ryobi is a portable "contractor's saw". It weighs about 75 lbs. Introduced in mid 1991, it has a number of attractive features, most notably a sliding table which can be located to the right or left of the blade. It seemed ideal for the sort of jobs I had in mind for it. Just about the time I was getting the Ryobi operational, my friend Paul Scobel told me his brother had recently bought one, and was very pleased with it.

Plan view of tablesaw stand w/ top removed

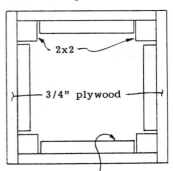

2x2

3/4" plywood

Top is attached to sides by screws into cleats glued & screwed to inside of tablesaw stand flush with top edge of side panel.

I made the Ryobi's stand in the form of an inverted "box" made from five 3/4" plywood panels variously bolted, or glued-and-screwed to an internal frame of 2x2's. It's overbuilt - 3/8" plywood would have been strong enough, given the Ryobi's modest weight. But on the other hand, the extra weight is no detriment in a tablesaw, either, so I'm not sorry I chose the thicker material.

Henry cut the plywood panels and 2x2's to my dimensions, after which I put everything together dry with a bagful of drywall screws and 5/16" carriage bolts. I then took off one screwed-on panel, smeared glue down the two exposed leg faces, & screwed the panel back in place. 3 hours later I did the other panel.

I then marked the pieces to identify their original places in the assembly, took the stand apart, sanded everything, and brushed on 2 coats of satin Varathane.

All this took much longer than I'd figured it would, but when it was done, it looked ok. If money is not a problem, buying the factory stand looks even better.

Some guys never learn
Then I picked up a bandsaw - a "General" Model 490 (made in Quebec, very nice over all, with a finish on it like a Myford lathe.) The dealer had just the one in stock, with, for some reason, no stand and no motor. I guess they figured it'd be slow to sell that way to any normal person, because they gave me quite a good deal on it.

So here I was with another stand to make. (I know - "Some guys never learn....".) However, by this time my tablesaw was operational, so I figured making another stand would be no sweat.

The bandsaw would weigh about 250# with the motor - quite a bit heavier than the Ryobi tablesaw, so I spent some time cogitating on what form of stand might be best, and eventually decided to go with pretty much what I'd made for the Ryobi, but without using glue this time.

I knew intuitively that the Ryobi's stand was plenty strong enough to support two of me, so it obviously wasn't going to collapse under the weight of the bandsaw.

I've seen a 1200 lb. milling machine on a low stand made mostly from 2x6's, and as I recall, it was bolted together. No plywood. Lotsa diagonal bracing. You don't need diagonal braces if you use plywood in a box-type stand this small. (My Ryobi stand is about a 22" cube; the b'saw stand, on its casters, is about the same height, with an 18 x 20" footprint).

This brings up the matter of **machine height**. For a small bandsaw like the 15" General, you want the saw table at about the same height as the point of your elbow. If it's higher, such that you have to lift your forearms above horizontal to place your palms flat on the table of the machine, long work sessions will be very tiring. A tablesaw wants to be a little lower - my Ryobi is about a hand width below hip height, whereas the bandsaw is about an inch or so above belly button height.

I dug out a used 1/2 HP motor I'd blown $10 on about 15 years before. It proved to be ok, wanting only to be wired to run in the right direction. It had a "C-Face" frame (hence mounts via a flange on the "drive" face), which was just fine - I would stick it right to the side of the stand using black iron pipe spacers and 3/8 redi-rod.

In the middle of all this, I had a letter from Tim saying he was going to buy a surface grinder. If he bought it without a stand, he could save some money, plus he could make the stand to suit his own height (Tim's 6'4). Did I have any ideas as to how to go about it? 2x4's? 4x4's? What? So I thought, "Tim knows more about woodwork than I do. If he's puzzlin' over the same question I am, maybe some of the other guys would also like some info on this." Which is why it's in here.

Margaret had given me a #5657 Skilsaw (portable circular saw) for Christmas. The ads said, ."...build a house or a subdivision with it. If you don't like it, bring it back for a full refund..." Knowing I was unlikely to do either, I guess Margie figured it would suit me fine, and she was right.

I bought some 3/4" plywood, borrowed BB's 50" 'Clamp 'n' Tool Guide' (see below), fired up the Skilsaw, and proceeded to whack the plywood up into pieces about 1/2" over final sizes for the panels in my bandsaw stand. By making the top in 2 pieces, I was able to get all 5 panels from half a sheet of plywood. (The 2-piece top panel has proven satisfactory. If crosswise stiffeners had been needed on the underside of the top, adding them would've been easy.)

I then cut the plywood to finished size on the little Ryobi, ditto the 2x2's for the internal frame. This was my first use of the tablesaw, and it worked like a champ. (I quickly learned one thing about tablesaws: ***Don't wave a measuring tape about over one when it's running.*** I did. The tape collapsed, hit the blade, and there went the end of the tape.)

The internal frame of the bandsaw stand was to be slightly different from the one on the Ryobi's stand, so I set up a stop on the vise on the milling machine, and drilled 5/16" holes at

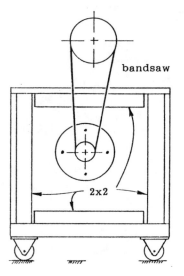
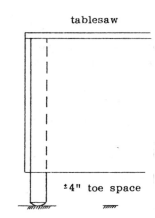

Side view of tablesaw and bandsaw stands, showing some differences between the two.

bandsaw

tablesaw

2x2

±4" toe space

predetermined points down the center-line of each piece of 2x2. These holes were then "match drilled" through the plywood. To do this, each 2x2 was clamped lightly to its piece of plywood, knocked about until it was exactly flush, clamps tightened, and a piece of scrap medite clamped against the exposed side of the plywood at each hole location, to prevent tear-out when the holes were drilled. As soon as a hole was drilled, a bolt, washer and nut was applied through both 2x2 and plywood.

I think one could cut the time for this part of the job in half by drilling holes in the plywood along a layout line, and then drilling clearing size holes - say 1/16" oversize - in the legs. If everything were done to stops and fences on the mill, it would go pretty fast. However, I'm not convinced that this is the best way to go. The match drilling approach is slower, but gives you a sturdier structure in the end, because the bolts then fit the holes so closely they act almost like dowel pins, as well as clamps.

(While I was working on my bandsaw stand, Stewart Marshall phoned, and when I told him what I was up to, he said he regards a bandsaw as indispensable in shops like ours. Certainly I see it providing capabilities I never had before.)

Voids & Splinters

There are not supposed to be any internal voids in plywood, but you will run into them some-times. Two of the 1/2" holes I drilled for my motor mounting scheme revealed voids a couple of layers in from the surface. The motor mount nuts were going to get pulled up really snug, so the voids were bad news. I filled them by stuffing the butt end of a 1/2" drill into each hole, and then forcing spackle (crack filler) into the hole with my thumb. This drove it into the voids, and when it set up, I cleaned out the drilled holes and went on with my work.

Not all makes of spade drills are ground the same way, but the ones I have cut at the periphery first, and then take out the bulk of the material in the hole. This minimizes splintering at entry, providing you take it just a little easy, but any drill can make an unsightly mess of tear-outs on exit. When I drilled the motor mounting holes in the plywood panel, I drilled part way through from the outside face, then flipped the panel over and used the pilot hole created by the center spur to finish the hole

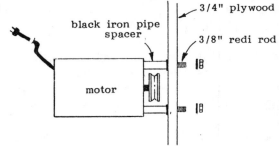

black iron pipe spacer

3/4" plywood

3/8" redi rod

motor

from the back side. No splintering occurred in the drilling of 8 such holes (16 entries).

Now if you're sharp, you're already thinking: "8 holes? There's only 4 holes on a C-face motor flange." Yes, but I oriented the motor incorrectly the first time, such that the V-belt, flaring up from the motor pulley to the bigger pulley on the saw, would've rubbed on the black iron pipe spacers. The fix was to rotate the motor 45°. The other 4 holes are now ventilation holes. Such an elegant solution.

When the stand was more or less complete, I marked every piece for re-assembly purposes, and then sanded and Varathaned them.

The bandsaw mounting holes are offset slightly to the left of the stand's centerline, to get the machine's weight more or less centered on the stand. This caused the feet of the saw to fall right on top of the heads of two of the carriage bolts securing the top to the sides of the stand. I considered moving these two bolts over an inch or two, but I found, when I installed the drive belt, that I needed a 1/4" shim under the saw feet to get the belt tension right. I made two shims from 1/4" medite, and drilled a 3/4"φ hole in each to accommodate the offending bolt head, and thus solved both problems at once.

Finally, after wiring a switch into the motor cable, installing the switch box in the side of the stand, and so on, my b'saw was operational. I didn't care to think about how many hours it had taken, but this time I'd done it 100% myself; it was a pretty sound job from end to end - if anything, better than the stand I'd made for the Ryobi. And I'd had fun doing it.

My shop space is limited, so the bandsaw had to be on wheels. The drwg at right shows how I arranged that.

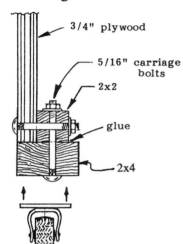

On the matter of putting machine stands on wheels for mobility, see *FWW* for Nov/Dec. '92, page 60. There are a number of other articles in this issue you might want to see, including one on tablesaw safety, one on how a reader (with the help of a machinist friend) made a disc sander with quickly interchangeable discs, and one on making a wooden locomotive model. (If this last idea appeals to you at all, and you want to make money from your shop, you might want to see this article.)

The following comments are not offered for the instruction of those who are expert or experienced woodworkers. I've learned a little from reading, watching Henry (and others), and a certain amount of doing. What follows is for the similarly inexperienced woodworker:

One key to making a job like this come out nicely lies in cutting out the plywood (or Medite) panels as I described above. If you have the use of a tablesaw, and can bust your panels down to near final size with a Skilsaw, trimming them to final size on the tablesaw is easy. Just set the saw's rip fence for a cut slightly over the final height dimension - this will most likely be common to all 4 panels. Put all 4 pieces through the saw at this setting, to produce a nice clean true edge on each. (Rough out your pieces so you have a "factory" edge on each piece to put against the fence for this initial cut.) Then reset the fence to your final height dimension, and run all 4 pieces through the saw once again, one after the other, with the new edge against the fence. The 4 pieces will now all be identical in one dimension.

Set the miter gage fence at 90°, and put each panel through the saw with one of the good edges pulled back snug against the miter fence; take off just enough to produce a 3rd good edge square to the two finished edges.

Next, set the rip fence for the correct width dimension on one pair of panels, and put your new (3rd) good edge against the fence, and trim that pair to final size. Deal with the 2nd pair the same way, and your four side panels are done. Do the top panel(s) the same way. Don't forget to allow for the thickness of the saw blade where necessary.

It goes without saying that you would never try to put two panels through the saw together.

Using a tablesaw requires full knowledge of proper safety procedures, and **absolute attention** to what you are doing. Watch an experienced cabinetmaker at work, and you will learn much.

Whatever you do, don't hurry. Don't make a cut from an awkward, unsteady or unbalanced position, and never underestimate the ability of a saw to remove a piece of wood from your grasp, and in so doing, *pull your hands into the blade*. Don't do it until you've planned out what you are going to do, and don't do it at all unless you feel confident you can do it safely.

Footnotes to the above...

If you lack, but want, a tablesaw, you can make one. At page 10 in the book "*FWW* on Making & Modifying Machines" is an article by a guy who left most of his woodworking equipment behind when he moved to Mexico. One day he spotted a homemade tablesaw in a Mexican cabinetmaker's shop. He was thus inspired to make something similar using a couple of ball bearing mandrels (from Sears), and ended up with a wood-framed machine that was not only a saw, but also a horizontal mortiser, router, shaper, drum sander and disk sander, all driven off one motor. With this he had made a very handsome dining room table and 6 chairs to go with it, a couch, a china cabinet, and 2 living room chairs. I wonder if I will ever do as much with my little Ryobi?

If you have reason to saw full sheets of plywood into smaller pieces, or if you avail yourself of the services of a friend who has a tablesaw when you do need such work done, there are a couple or three things you might like to know about.

I mentioned using a Clamp'N'Tool Guide. These are handy for many jobs involving sheet goods. They come in 24, 36, and 50" capacities. They provide an easy and convenient way to put a straight edge on sheet goods for guiding a Skilsaw or router, and are useful as clamps. You'll find them in better hardware stores, or call Mark Ashby, at Griset Industries, Inc., Santa Anna, CA; phone: 1-800-662-2892 or 1-714-662-2888.

There's another slick rig called an Exact-T-Guide; with it and a Skilsaw you can cut up whole sheets of plywood more easily, and more accurately, than on a big tablesaw. Cost is about $200. It takes up the space of a 4 x 8 sheet of plywood when in use, but that's much less than the 20 lineal feet required to run a 4x8 sheet through a tablesaw. Plus, you can rig the 4x8 table to swing up against a wall when not in use. For your nearest dealer, call the makers: BradPark Industries, Inc., 91 Niagara St., Toronto, Ontario, Canada M5V 1C3; phone: (416) 59-49455, or (416) 461-1601.

Finally, there's a tablesaw accessory called a "Jimmy Jig", which you can make from about $30/40 worth of materials, from a set of plans available from the address below. I've seen a video of this jig in use by its inventor, and, next to an Exact-T-Guide, I would say that for anyone breaking full sheets of plywood down to smaller pieces on a tablesaw, the Jimmy Jig would be an excellent addition to the saw being used. To get a set of Jimmy Jig plans, send $25 to Jim McCombie, 5273 Gertrude Street, Port Alberni, B.C., Canada V9Y 6L1; phone: (604) 723-3074.

Another way to do it
Medite (medium density fiberboard) is also excellent for projects like machine stands. It has the advantage over plywood of having no internal voids, plus a smoother surface. If you make up a stand with Medite on an internal frame of 2x2's or whatever, sink the screw heads deeply, fill over them, round over all the external corners/edges with a router, fill any gaps where the panels meet, and then paint it, the result looks like a finely finished machine casting. This (and the following paragraph) is just part of some much-appreciated advice I've had from Marsh Young, who I mentioned earlier.

Another way to finish Medite is with Flecto Varathane Natural Oil Finish. This product is very easy to get nice results with - just be sure to wipe the second coat off after 10 minutes or so. Do that, and you'll have a nice result. Don't, and you'll have a mess.

Some More Ideas on Making
Wooden Tool Storage Boxes

I have found out a few further tricks concerning the making of tool storage boxes, beyond those detailed in **TMBR**, pages 122-4, and although this section may seem to deal more with woodwork than metalwork, the info is presented with the hope that you will find some or all of it useful. If you want to do something, and you know some of the pitfalls and have some hints on how to come out ahead in doing the job, then it's likely to go at least a little smoother. (Don't overlook the info in **TMBR** - most of it is still pertinent, and little of it is repeated here.)

Why would you make a tool storage box? Well, let's say you've just bought yourself a nice new H/V rotary table. It probably came in a heavy cardboard box. That'll be ok for a while. (But watch out - cardboard can soak up moisture from the air, and cause the item within to rust.) Then you make some nice strap clamps and some T-nuts for it, and a washer and drawbolt for pulling shop-made #2MT shank fixtures into the center hole, and a brass knock-out plug to get them out again. Before long you have a pile of goodies to keep together with it. Seems about time for a proper storage box.

How to decide on the size of the box
Round up all the pieces that are to go in the box, and pile them up on the workbench in a compact arrangement. Stuff that will later be bolted to the inside of the lid of the box as at right can be either piled on top of other stuff, or ignored, or just make mental note of its effect upon the needed depth of the box. Don't be bashful about reorganizing the pile until you have an arrangement you think is optimum. When you do, measure the pile for overall height, length and width. That plus a little more can be taken as the interior dimension of the box.

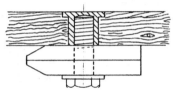

What to make the box out of?
I have made several storage boxes from medium density fiberboard (Medite and Rangerboard are two trade names for this material), and it is ok, so long as it doesn't get wet! I think good quality plywood is superior - certainly it looks better.

> **Note**: There is a material called Benelex, which is like, but even harder than, Masonite. Back at page 175 you'll find a further note about Benelex from "Rusty" Rauth, who recommends it as being worth looking into as a box making material.

By now you may be thinking to yourself: "We're machinists. Why not make the box out of sheet steel?" Well, in the matter of rust & damp protection, wooden boxes are better than metal boxes of the sort we could readily make. Note that I said "... metal boxes of the sort we could readily make ..." - I am well aware that the military uses lots of metal storage boxes (ammo boxes, to name just one type), but they incorporate rubber seals, air vents, desiccants, etc. that would be tough for us to match. I believe wood is the way to go.

Fir plywood is not bad, but it's not perfect: it can have, in the interior plys, knots - and what's worse, voids - that do not show up until you cut it to size, or until you try to drill a hole for a bolt to anchor some loose parts in the box. Most voids can be filled with some type of wood filler, but even if this is feasible, it is a nuisance to have to do it.

FWW Magazine had a good basic article on working with plywood in the May/June 1984 issue, from page 48 to 53. It's worth seeing.

Aside from furniture grade plywoods, which are better than your average half decent piece of fir plywood, better grades of plywood are available from plywood specialty outlets (see your Yellow Pages).

There are 2 or 3 types of really good plywood that I would regard as among the best possible materials for making boxes where long term quality is wanted. These are:

Baltic Birch: This stuff has 5 plys in the 1/2" thickness, and is made in Russia, or Finland, I think. It is very good stuff. Cost is about $3.50/sq. foot, and it comes in 5' square sheets only - or so I am told by the local seller. If you can search out a shop that makes steel rule dies (which are used to cut various sheet materials into stuff like gaskets, shoe parts, boxes, etc.) you may be able to buy some of this material in smaller than whole sheets, as it is commonly used as "die board" - the body of a steel rule die.

"Apple-Ply": Made in North America, this material has an alder core and maple face veneers (hence the name), and comes in conventional 4 x 8' sheets. Cost is maybe half that of Baltic Birch.

Brunzeals plywood: a true marine grade mahogany plywood, made in Holland, and probably on a par with Baltic Birch for cost. I've seen the stuff, but nobody around here currently sells it, so I can't get much info on it. If you know a builder of classy wooden boats, you might be able to get some from him, but not necessarily - he can probably put even small pieces to good use himself. (The same is true of Baltic Birch at a steel rule die shop, so in neither case should you expect to get it as scrap - i.e. free.) For a source, look in *Wooden Boat Magazine.*

Thickness? For a box of moderate size, 1/2" or 5/8" is good for the sides; 3/4" is probably overkill. The top can be 1/4" plywood, and the bottom might be 3/8 or 1/2".

Cutting
Decide on your box's outside dimensions. Figure out the sizes of the six parts you need. If you have a table saw, cut 'em out - who am I to give you advice? If you don't have a table saw, take your drawing and material to a woodworker friend, a custom cabinetmaker, or someone in that or a similar business, and ask him to cut them out for you. The main things you want are (1) exactly to size, and (2) exactly square.

Box Construction

Lid/Box Match up
About the most obvious way to get the lid to fit the box is to make the box as a sealed container, keeping any fasteners (screws, nails, etc) off the planned split line, and then cut the lid off. Now I don't pretend for a minute that that is news to you... But see below.

Joint Design
Glued joints are very much stronger than dry joints, whether they be nailed or screwed. Screws are better than nails, as a general observation, but not always appropriate in box construction, unless you can accommodate little gusset pieces in the corners, as at right, in which case you can readily have what is probably the very best arrangement: glued and screwed joints.

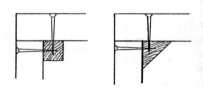

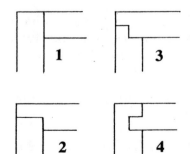

Butt joints (#1 at left) are easiest, but probably the weakest type of joint you could use. Rabbeted joints (2, 3, & 4) are not difficult, given the use of a table saw. All these joint types should be glued, and then nailed with suitable sized finishing nails.

(A more practical advantage of the corner joints named above after the butt joint is that they have greater strength, compared to a butt joint, due to the increased glued surface area in them. This is true also of most - probably all - of the other types of joints discussed below.)

Mitered and splined corners (right) are another way to go when working with plywood, and they have the advantage of hiding the edge grain, if you care about doing so. You may not; see below.

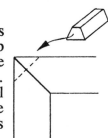

The spline is a vital feature of this type of joint, as it is otherwise difficult to nail the joint solidly, without the mating pieces sliding about, which they are even more inclined to do when they are anointed with glue. With the spline in place, they'll have slightly more difficulty getting away on you.

A somewhat fancier corner joint can be made by mitering the ends of the sides of the box, and then putting several dovetail splines across each corner. Clamp two adjacent sides together and cut a dovetail-like groove through both at once to take a little wooden key, as in the drwg at right. The L.C. Kehoe Mfg. Corp, P.O. Box 1890, Whitefish, MT 59937 offers a simple jig that will allow you to make dovetail spline joints with a router. Shop-made jigs for the table saw or router table can produce similar results, but the Kehoe jig is kinda slick, and not expensive.

If you want to get real fancy, you can dovetail the corners. Dovetails make a strong joint, and add a classy touch if you care to do it. They are usually associated with solid wood, rather than plywood projects, but they would look nice in good plywood. Hand cut dovetails are slow, and demand a certain amount of skill & practice. But if you have a dovetail jig and a router, flawless dovetails are easy, and relatively fast to do (once you get the jig set up).

To stick with the matter of hand cut dovetails for a minute or two longer, there are various dovetail marking out aids offered by woodworkers' supply houses, from about $9 on up to $40 or more. Any hsm could easily make one as good as, or better than anything I've seen offered.

Below, with permission, is a sketch of a dovetail marking tool from the British magazine *Woodworker* for October '91, page 1061. This idea was sent in by Zwi Rotem, of Israel, as an improvement over one of the commercial dovetail layout jigs. It would be easy to make one for one's own use, or for a woodworker friend.

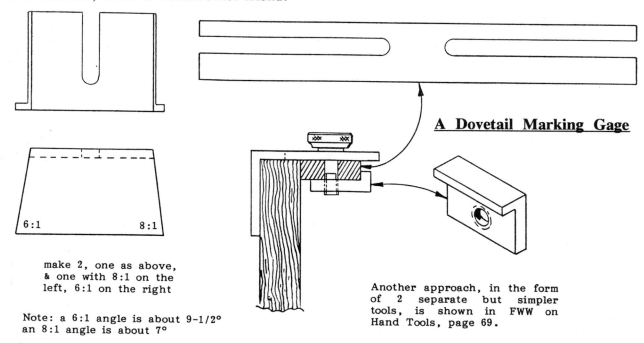

6:1   8:1

make 2, one as above, & one with 8:1 on the left, 6:1 on the right

Note: a 6:1 angle is about 9-1/2° an 8:1 angle is about 7°

A Dovetail Marking Gage

Another approach, in the form of 2 separate but simpler tools, is shown in FWW on Hand Tools, page 69.

Rather than going into more depth on hand cut dovetails here, I think it best to point the interested reader toward another *FWW* publication containing much excellent info on dovetailing, and let it go at that. Therefore, if interested, see the book "FWW on Boxes,

Carcasses and Drawers," for about 27 pages on dovetailing, plus (at page 96, right hand column) as good a summary on finishing methods for wooden boxes as the average hsm is ever likely to require. (But on that, see also *FWW* for Jul/Aug '88, bottom right corner, page 14.)

DOVETAIL JIGS: If you want to make dovetails with a router, one of the easiest dovetail jigs to use is the **Katie Jig**, which was not on the market when **TMBR#3** was first published. The Katie Jig can be quickly set to produce variably spaced dovetails. I got a Katie Jig in '99, and have so far used it just once, to see how well it works. (It works good - see below.) Starting from scratch, with one finger in the instruction manual, it took me 20/25 minutes to get set up and ready to cut the first dovetail. After doing just a couple of dovetailed corners (i.e. half of a single box project) it would go much faster - maybe 5 or 6 minutes to clamp two adjacent parts of a box in the Jig, and rout out the mating elements of a dovetail. Once you've dovetailed both ends of the 4 pieces that will be the box's sides, you're ready to start putting things together. (NOTE: Dovetailed joints need no screws or nails - just glue.) To accommodate seasonal expansion and contraction of the box's top and bottom (assuming they are made from solid wood, not plywood), it is considered good practise to set these parts into a groove in the sides of the box, *with no glue*, before you assemble the whole box. The top and bottom, if nice and square themselves, help to square up the box during assembly.

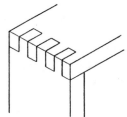

Finger joints are a variation on dovetail joints. Probably a little easier to make than dovetails if you don't have a dovetail jig, they are also stronger.

There's a jig called an Incra Jig, at about $30, which is in essence a positioning device, rather than a cutting guide. Used in conjunction with a router and router table, or a table saw, it will let you make accurate finger joints pretty readily. (The Incra Jig will also help you do dovetail joints, although perhaps somewhat less easily than the Leigh Jig.) Incidentally, the inventor of the Incra Jig is a hsm.

Recognizing me and himself as guys who fool about in our basements and make maybe two or three boxes per year, rather than professional (or dedicated) woodworkers who have reason to turn out many dovetailed joints per year, my friend Paul Scobel once said to me, "We're Incra Jig people, not Leigh Jig people." Overall, I have to say I think he's right - because of it's cost, the Leigh Jig is not something a fella would buy to make one or two boxes, but if you have a woodworker friend who you put the bite on for boxes once in a while, you might consider chipping in part of the purchase price of a Leigh Jig, which will make it easier for him to buy it, and for both of you to find the intended biting more palatable from time to time thereafter.

That's enough about dovetails and dovetail jigs.

The top can be 1/4" plywood, even if the sides are solid wood. Here again, we may want to hide the edge of the plywood for aesthetic reasons, and at the same time, we reduce the tendency of the plywood to splinter at the edges. (More on **splintering** below.) A neat trick, therefore, is to make the top slightly smaller than required for a flush fit all around, and then glue a capping piece of solid wood in place, and plane it flush with the top and sides.

If you do this, you have to trim the glued-in cap piece down flush with the plywood. Two things come to mind here. If the cap piece is very nearly flush with one surface of the box when you glue it in place, you can probably plane and/or sand that face down flush fairly easily. (Don't plane after sanding, as the sandpaper will have left grit in the wood that will do your plane blade no good.) The adjacent side of the cap piece can be brought flush with a router and a flush trimming router bit of the type designed to trim overhanging material such as our capping piece flush with the vertical face below.

Another way to do it would be to plane it down. A type of plane that would do this very nicely is an edge trimming block plane, which is designed to plane the edge of a board square to the face. Stanley used to make one called a #95, and if you find a used one, snap it up. Or, if you just like nice tools, you could be forgiven for sidetracking a few bucks every payday until you can buy a solid bronze reproduction of the #95: Tom Lie-Nielsen operates a bronze foundry & machine shop in Maine under the name "Lie-Nielsen Toolworks, Inc.", (Route 1, Warren Maine, 04864), where he makes this and several other small planes, as well as some other tools and various marine hardware, plus the nicest polished bronze chest handles you can imagine... (Did you note that? Just what you might want to finish off a really special box). Tom also makes a beautiful little scraping plane you might like - I'll tell you something about scraping wood for finishing purposes later.

Speakin' of planin' things, here's a neat tip I read in *FWW* May/June '83, page 10: if you encounter a knot, or twisted grain, when planing a piece of wood, dip a rag in turpentine and wipe it over the difficult area. Instead of tearing and fighting with you at that spot, the plane will snick through the knotty area as slick as an eel in a jelly-roll. I haven't tried it yet, and the guy was speaking in terms of using a Stanley #55, but it sounds like a good idea to keep up your sleeve.

If the box is made from plywood, the top and bottom panels can be rabbeted and glued in place. If the box is made from solid wood, rather than plywood, cut a groove for the top and bottom panels into the inner face of all 4 sides of the box before gluing up, and then set the top and bottom pieces into their respective grooves, *without* glue, as you assemble the box. Why? Because, if made from solid wood, the top & bottom panels will shrink and swell in width with seasonal changes in humidity. If glued in place, instead of being left free to move, they will likely shrink and crack, or swell and break the box.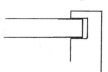

As for **glue**, the Garrett Wade (a woodworkers' supply house) catalog offers some good advice on this topic, and they offer a glue called "202GF" (GF = "gap filling") which is apparently extremely good. Whatever you use, follow the maker's directions, and then clamp up solid for as long as they say to, or more. Caution: Go easy on the clamping pressure: enough is enough; more than enough just squeezes glue out of the joint, which gains you nothing.

Excess Glue
There seems to be two schools of thought here. My friend Henry says to remove excess glue before it dries, first using a chisel and then wiping off whatever's left with a damp rag.

The other approach I've seen advocated (Ian Kirby, in *FWW*, I think it was) is to let any squeezed out glue dry completely, then pop it off with a chisel, and then clean up with a razor blade and/or sandpaper. The reason given was that if you try to wipe off excess glue with a damp cloth before it dries, you will just smear the glue into the surrounding wood, where it will mar whatever finish you may choose to apply later.

It may depend on the type of glue you use. Try it both ways with the glue you plan to use, on a piece of scrap wood, and see which works best for you.

One other thing: some white glues - I don't know what type/brand - will turn black, and stain the wood black, if they come in contact with iron/steel. If you use pipe clamps, for example, it can happen; even tiny bits of steel wool will do it. The way to avoid this problem is to interpose wax paper or similar between the glued wood and the clamps.

Cracking the Egg
Once the glued-up box is dry, saw it open. This sounds easy, but when you try it, there's a pitfall which you might not anticipate: everything goes fine at first - you set the table saw's fence to give you say a 1-1/2" deep lid, and you start pushing the box through. The first 3 sides go fine, but as you get to the very last part of the 4th side, suddenly your heretofore one-piece box starts behaving like 2 pieces, and what the 2 pieces want most of all to do is to collapse toward each other, squeezing the saw blade. Not that that bothers the saw - it couldn't care less - but it will almost inevitably flaw the cut line to at least some degree, and it may

wreck your box. Now obviously we want to avoid such flaws, but how? Well, I've seen 2 ideas for improving on this cut-the-box-open business in *FWW*, and they are these:

1st idea: Don't cut all the way through the box material. (Apparently this is pretty basic, but it didn't occur to me to do it until I read about it.) Set the saw to be say 1/32" shy of the thickness of the wood used. Run all 4 sides of the box through the saw, and then finish separating the lid from the bottom by hand, with a knife, a hacksaw, a hacksaw blade by itself, or whatever. Once the lid and bottom are separated, use a file, sandpaper, or a razor sharp plane to clean up what the table saw left.

2nd idea (instead of the above): Run all 4 sides of the box through the saw before assembly, to produce a cut on the inside of the box about 1/8" above where you want the joint line on the outside. For this cut set the saw's depth of cut to 1/2 the thickness of the box material. After the box is glued up, adjust the fence one sawblade thickness further away from the saw, and drop the blade 1/32" or less, for the same reason/effect as in the 1st idea above. Again, finish separating the lid by hand, and tidy up the cut line by sanding, files, plane, etc. Now see the result? Your box has a kind of dust seal/lip.

A 3rd way is to insert 2 or 3 shims the same thickness as the saw kerf in the already-sawn areas of the box, and then apply clamps over the shims, effectively making the box almost as solid as it was for the 1st cut.

It bears mention that the use of a table saw is not mandatory for the above operation. If you don't have a table saw, you can do it with a circular saw (commonly referred to as a "Skilsaw") or a hand saw. Most of the woodworking tool supply houses sell various types of Japanese hand saws. These cut on the pull stroke; this, plus their thin kerf, and fine, razor-sharp teeth, make them much more "controlled" than a carpenter's handsaw.

After a pull saw, I think the next best would be a back saw. Start - very carefully - at a corner, and saw to a guide line drawn right around the box, and make the cut progress along the guide line on two faces at once, possibly guided between two pieces of wood clamped to the box, one on either side of the cut line. Do this at each corner in turn, and then extend the cuts along until the job is done.

Hardware

Hinges
Please yourself. Piano hinge is good. Hinges may seem a little stiff when handled in the store, but when attached to the box, the leverage & stability that the lid/box provide overcome this apparent stiffness; I mention this in case you might find hinges you like, but notice that they seem stiff, and - if you haven't experienced this - might otherwise pass them up. Buy 'em and stick 'em on - they'll be fine (unless they're obviously junk). Also, don't overlook the possibility of making your own hinges, plain or fancy, from sheet brass, as detailed elsewhere herein.

Latches
I'm partial to my eccentric disc latch, as detailed in **TMBR**, pages 125-9. But there are some nice, effective, mass produced latches in pressed steel, e.g. Cam-Lock's #112L series latches. These can be had in zinc plated carbon steel or stainless steel for maybe $4 each. Zero Corp's #ZSP 2-204 is similar, but can only be had with a cadmium plated finish. So what? Well, we all know that cadmium fumes from silver soldering operations are bad news, but as noted in **TMBR#2**, page 116, I've also been told that one should even wash one's hands after handling cadmium plated stuff; whether it's true or not, I don't know. If I were going to use a cad

plated latch, I'd paint it first. For either brand, look under "Fasteners" in the Yellow Pages, and start asking for the name of the makers' local rep. Cam-Lock stuff comes from The Fairchild Group, in West Hasbrouck Heights, NJ. The Zero Corporation hangs its hat in Burbank, CA.

Another type of latch is as shown at right; these would be appropriate for a light duty box to hold micrometers etc., and are easily made, if you care to. They can also be bought from woodworkers' supply houses.

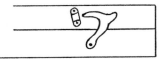

Locks hardly seem necessary on boxes to store tools in your own shop.

Corner caps
Corner caps seem to me rather high priced, considering that all they are is a metal stamping. If a fella wanted to make his own, it would take a while to tool up, but nicer-than-bought ones could be soldered up from 1/16" sheet brass on a jig. Or weld them up from CRS sheet. Re this, see elsewhere herein.

Handles
You may want to buy plain or fancy metal or other handles for your boxes. I usually make mine from a couple of 3/4 x 1-1/8 x 5" pieces of scrap oak which I get from Henry - see **TMBR**, page 124 - and attach them with shcs's and shop-made flanged nuts. Handles attached thusly will not come off under any normal circumstances (unless you want them off), yet protrusion into the box interior can be reduced to zero if you want, by counterboring the holes on the inside of the box (in advance of gluing the box up) to take the nuts' flanges.

Salty Box Handles
There are obviously many other forms of handles one could make, and many store bought ones. Sailors' chests often sported rope handles whose design demonstrated the owner's knotty virtuosity, and whose making occupied the spare time of many weeks on long sea voyages. (For an extremely well illustrated and likeable book on the topic of plain and fancy ropework, see "The Marlinspike Sailor" by Hervey Garrett Smith, if it interests you. Nowhere will you find a better illustration of how to splice an eye in a piece of rope.) And don't forget Tom Lie-Nielson's cast bronze chest handles, mentioned above.

Stewart Marshall has a number of old deadeyes hanging about in his workshop on Lopez Island, some recovered from ships wrecked off the Florida Keys in the 1700's. They're neat. In "The Marlinspike Sailor," (see prev. paragraph) the author shows how to make a pair of deadeyes, and he shows a pair with the rope rove through their 6 holes. Deadeyes come in all sizes, from as small as you might care to make them, on up to 2' or more across. (Before the advent of steam engines, big warships used BIG deadeyes.) Well, if they can be any size, why not make 4 at say 3/4"ϕ, and rig them as handles on a box?

Plywood Splinters
One shortcoming of plywood, especially that which is not of the class of Baltic Birch or Brunzeals, is that you can often get splinters at the edges. Some of these can be pretty horrible. My brother was sanding a piece of plywood by hand one time (you'd have to know Bro to really get the picture, but suffice it to say that his arm would be a blur clear up to the shoulder). At any rate, he picked up a splinter which entered the side of his thumb, and penetrated about 3" before emerging from the heel of his hand. (Sorry, but I don't have a photograph.)

As discussed above, the way to overcome this nasty characteristic is to face the cut edges (created when you saw the box open) with glued-on strips of solid wood - fir or similar. Make it oversize, glue it on, and then plane or sand it down flush with the width of the plywood. One way to bring it to uniform thickness is by running the lid and the box through the table saw again. Another way to cover the exposed edges of the plywood is with wood veneer edge banding material which you can buy and apply with an iron (set it for "linen").

When I was a little guy in the early 50's, I once watched my Dad make a box to hold the first-aid kit he carried on the "speeder." When the exposed edges of the 1/4" fir plywood were trued with a razor sharp plane (power tools? - hey, on the railway we didn't even have running water, let alone electricity!), they looked just fine, and even better when the box was varnished. I think the fir plywood we had in those days may have been better than what we see today, but if you were using some decent plywood, and cared to plane it thusly, you might like the effect. But edges faced with solid wood do look nice.

The use of solid wood - e.g. 3/4" pine or fir - for boxes of the type and size envisaged here would eliminate the need for edging, but solid wood can split or warp, which plywood won't. The box's corner joints help resist the tendency to warp, and gluing up material of the desired width from narrower pieces can also help. If you use solid wood, glued up or one piece, plan on having it planed to uniform thickness before using it. Let it come to equilibrium with the humidity in your shop before working it further.

Finishing
You can leave the box unfinished, or you can paint, stain, or varnish it. Painting and varnishing take the most time, as you need several coats, plus drying time between coats. A nicely painted box, grey with red & black painted lettering, or dark green with gold lettering, can look pretty nice. Stain - e.g. Watco oil stain - is easy, fast and looks great. Once you use it, you can't switch to another finish at some later date, but if it ever gets scuffed up, you can slap some more stain on it in short order, and it'll look good as new again.

Something else one could do, if he cared to, would be to make a box from ordinary plywood, and then surface it all over with Formica. This would give you an outer shell that is not only very durable, but that also looks great - there is a great variety of "finishes" in Formica and similar plastic laminates. Your friendly village cabinetmaker may have (scrap) pieces you can choose from, and he can tell you how to work with the stuff.

Fancy boxes
If you want to make a nice box for an expensive tool such as a master square, master vernier caliper, or anything in that class, you can glue up two slightly oversize blocks of solid wood, run them through a thickness planer to bring them to uniform thickness, and cut them to size for width & length. To produce the tool cavity(s), clamp each half to the table of your milling machine and rout out what you don't want with an end mill at high speed.

Tool Trays
At the other end of the same stick is the type of "box" that isn't a box at all, but rather simply a block with slots milled in it. Wanting protective storage for my several pairs of 1/4" x 6" Anton parallels, I bought a piece of clear, double dressed, kiln dried fir 2x4, and hand sawed it into 5 slightly overlength pieces. I stuffed a 1/2"ϕ 4-flute end mill in the spindle of the B'port, set the speed at about 2000 RPM, and milled the ends of each piece square, using the Kurt's vise stop to get them all to come out to the same 7-1/4" OAL on the 2nd cut. (This was before I bought the little Ryobi table saw mentioned elsewhere herein.)

With that done, I stuck the first piece back in the vise, and in short order milled out 4 grooves, each about 0.520" wide, slightly more than 6" long, and each sunk to the right depth for the parallels it was to hold. The shop-vac sucked up "swarf" as fast as it was produced, so I could see what I was doing. I made 5 such trays in all, some with 0.270" wide slots to take pairs of 1/8" thick parallels, and when all were done, I popped them in the oven at low heat for an hour or so to make sure they weren't going to rust my parallels, and put them into service. They do the job, and were well worth the effort required to make them.

198

The above is by no means the end of the matter. Some type of decorative work may be wanted, and can be worked into the top of the box. Lettering of one sort or another (the maker's initials, for example) can be applied: paint it on, carve it in, engrave it on a brass plate and stick that on, etc. I once saw a Gerstner tool chest to which was neatly affixed a piece of sheet steel in which was pierced out the owner's name. (If you do something like this, you want to use a handsome typeface, and get the spacing of the letters perfect before you start. Pleasing letter spacing has bedeviled printers and graphic artists for centuries. It's easy to get it wrong and not notice it - if you do, the moment it becomes irreversible, it will scream out at you like a fog-horn.)

Funny little story about boxes: *Francis Young, one of my guys down in Seattle, WA, told me he once took a micrometer-adjustable sine bar he'd made to an inventor's fair in Seattle. A lady came along and asked him about it, so he gave her the full spiel on how it worked, and all the advantages of his sine bar over normal ones, etc. When he finished, she stood there for a minute like maybe she was about to offer him a couple of million for the manufacturing rights, and then she said, "That sure is a nice box you made for it," and walked off.*

Shop-made Hinges

You can buy a wide variety of hinges at low cost in any hardware store, but there may be times when you want to make a hinge, or a pair of them, for yourself. It is not hard to do. Say you wanted two hinges 1-1/4" long, each leaf say 1" wide, and the whole job to be made from 1/16" brass sheet.

Shear two strips of sheet brass 1-1/4" wide by about 8" long. Get a couple of nails about 3/32"ϕ or maybe a little smaller, and about 2" long, and cut their heads and points off. File off any burrs at the cut ends.

Set a headless nail and one of your strips of metal in the milling machine vise, with the nail right up flush with the top of the vise jaws, and with the strip standing up at 90° to the vise. You would want about 5/8" or a little more of the brass sticking down into the vise below the nail - a little excess here does not matter. See A.

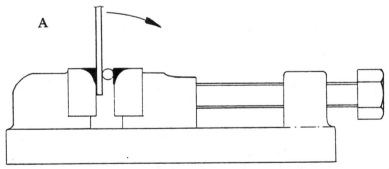

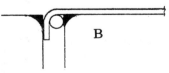

Using a hammer and a smooth strip of scrap aluminum, make the initial bend in the metal, as shown at B. Remove the strip from the vise, turn it end for end, and repeat the above procedure to begin the other half of the hinge. (Repeat these operations on the second piece of material as you go along, too, for the other hinge.)

Next, remove the strip of brass from the vise entirely and place it on the workbench with the bent ends up. Lay the nail in the bend, and run a scriber along the nail to mark the trim line. Cut off the surplus brass with a fine toothed hacksaw, and then file the sawn edge down to the scribed line. See C & D.

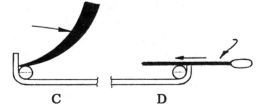

The strip of brass and the nail are then returned to the vise, positioned as at E, and the strip bent down again as at F.

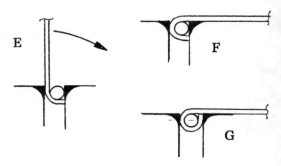

Loosen the vise, reposition the work again as at G and complete the bend. At each progressive step, do the same thing at the other end of the strip, and on the other strip, and you will now have 4 almost complete hinge halves.

Cut these hinge halves off the ends of the strips at the desired length.

To complete the hinges, file the cut edges square and clean, mark out and drill 3 screw holes in each piece, and finally mill out alternating spaces in each piece and assemble them. If the material you are working with is too thin to stand up to removal with a milling cutter, go after it with a jeweler's saw and files instead.

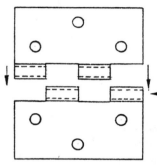

None of the above work is difficult or requires great skill - just a little attention to careful placement of the material and the nail in the vise, and a little saw and file work. Although I have shown a rectangular hinge, the tag portion of one or both halves could be sawn to whatever plain or fancy form you might like.

You can make hinges in brass, copper, steel, stainless steel, galvanized sheet metal, etc. Very small hinges are no obstacle - you might be making a model marine steam engine, and might want to put a hinged and lipped cover 15 thou thick on an oil distribution box the size of a peanut. You can do it - just work carefully.

(The foregoing is based, with permission, on an item by George Thornton that appeared in the September 1976 issue of *The Whistle,* the monthly Newsletter of the BCSME. George Thornton, a long time member of the Club, was a master sheet metal worker.)

Another application for such work would be in the fabrication of a patch box cover for the butt of a black powder rifle. There was an article on making a patch box cover in the August '79 issue of *Muzzle Blasts* Magazine. In it, the author's method of forming the hinge eyes is somewhat different from the method detailed here, and looks like it would require more effort and trouble than necessary, but I have not tried his approach. Other aspects of the article would make it worth seeing if you were planning to make a patch box.

If the hinge is long, and quite thin, another approach is to solder pieces of brass tubing alternately to the edges of two pieces of brass plate. These can be soldered up in a jig consisting of a base plate with a V-groove to seat the tubing in to the correct depth, plus a series of clamp plates notched to locate the pieces of tubing in the correct position along the hinge. This was shown by A.G. Hann in *ME* for Dec 15, 1967, page 1233.

Also, in *FWW*, there is an idea for a wooden hinge made by gluing pieces of 3/8 or 1/2"ϕ dowel, predrilled up their long axis, alternately to the lid and "box" halves of a wooden box, after routing a semicircular groove along the joint. The tricky part to making such a hinge lies in keeping the glue confined to where you want it, so that you don't end up with a hinge that is glued up solid. If I were going to do this, I would be inclined to glue pieces of dowel to the lid only, the first time through, and put some "saran wrap" stuff under the unglued mating pieces; when the first round of glue had set, I'd take the saran wrap off, and glue in the mating pieces to the bottom half of the box, this time putting the saran wrap around the pieces previously glued in place. On each trip through with the glue, you'd want to put the hinge pin in place to make sure everything was properly lined up.

MIGHTY FINE SANDPAPER
(Another way to Sharpen Fine Edged Tools)
from **George Franklin, Winter Park, FL**

Dear Guy:

... Mostly, I'm an incurable toolaholic - especially hand tools, as I don't have space for a lot of power equipment. I like to do a bit of gunsmithing, working mostly with hand tools. I am especially fond of black powder rifles and their lore.

I'm constantly amazed at the specialized things I find while nosing around in stores completely unrelated to the gunsmithing trade. For instance, I needed some touch-up paint for my wife's car one day. I had happened to spot a store catering to professional auto-body shops a few days before. So I dropped in there the next time I had a day off. I got the paint, and also found some other things, one of which I would like to share with you.

It seems that 3M in St. Paul, MN makes a wet-or-dry abrasive paper product called "Color Sanding Paper". Auto-body shops which specialize in the new super glossy urethane auto finishes use it to sand between top coats. I found the stuff in 3 grades: 1000, 1500, and 2000! Now 400 grit paper seems like pretty fine stuff; these higher numbers seem to correspond to the grit size of this stuff in the same way - and man is it smooooooth!

It's marked for either wet or dry use, and the abrasive appears to be silicon carbide. I found it in the form of 5.5 x 9" sheets (probably for a commercial sized air powered sander), 50 sheets to the pack.

When I got home, I dug out a scrap piece of 1/4" CRS plate, draw-filed one edge smooth, then pulled out an 18 x 24" piece of 3/4" plate glass. I worked on the edge with some 120, 200, 320, and 400 grit normal silicon carbide paper, both dry and wet, laid flat on the glass, and pretty soon had what most anyone would regard as a real nice finish! The 400 grit paper left a nice, frosty surface, but showed some discernible scratches if you looked real close.

I then tried the 1000 grit paper, with a few drops of water. After about 5 minutes the surface appeared clear silver. A similar amount of work on the 1500 grit stuff brought up a reflective surface that almost looked like good chrome plating - I could see myself in it. I could hardly wait to see what I could do with the 2000 grit paper.

After supper, I started on the 2000 paper. After 5 or 6 minutes, the surface went optical, with good enough resolution to count the cracks in my teeth! A few drops of water and some more effort brought the surface to a truly mirror-like quality. The plate glass made the surface flat enough to get an image good enough to shave by.

No joke, this stuff is good - as good as any of the (many) messy compounds I've used on buffing wheels and lapping plates over the years. Laid on plate glass, or on a surface plate, I think one could get surfaces almost as good as by lapping - well, maybe not quite good enough to wring to a gage block, but pretty good - and with none of the mess and trouble.

Next I tried sharpening a plane blade, using one of those cheap plastic "scissors" type jigs to hold the angle constant. The large surface area of that big chunk of plate glass makes this type of jig really easy to use - no need for doing a balancing act as when using one on a narrow stone. After a while on the 1000 and 1500 paper, I raised the angle about 1°, and used the 2000 paper to lay in the micro bevel. And Wow! What an edge!

Now Guy, I gotta tell ya, I have every kind of sharpening stone imaginable: Washita, Hard Arkansas, Japanese Waterstones, etc. This stuff beats them all.

To aid those who might want to hunt out some of this stuff, the specifications on it are as follows:

3M, 4140 Imperial Wetordry, Color Sanding Paper

Grade Ultra Fine 1000, P/N 051131-02021, 5.5 x 9", 50 sheets/pack
Grade Micro Fine 1500, P/N 051131-02023, Ditto
Grade Micro Fine 2000, P/N 051131-02044, Ditto

Note: The words "Imperial" and "Wetordry" are marked on the package as trademarks of 3M, St. Paul, MN 55144.

As I recall, the price was about $10-12 per pack. Expensive, but a pack would last one fella a long time, or could be split between several friends.

George

———————ᴧ———————

LAUTARD'S STROKAGENIUS FILE RACK, MK. 2

Since publishing "Hey Tim..." in 1990, I made a second version of the Strokagenius File Rack which is described at the beginning of that book. The new Mk. 2 File Rack is even better than the original, I think. I'll tell you about the changes here - in case you decide to make one, you might as well have the more up-to-date information:

The Strokagenius File Rack, Mk. 2, is about 15% larger than the original: the Bottom & Middle decks are now 13-3/4"φ, while the top deck is 8"φ. (The Base does not change size.) I put four 5" Pillars between the Bottom and Middle decks, and three 5-1/2" Pillars between the Top and Middle decks.

The Pillars are 1/2"φ CRS, tapped 5/16-NC each end for button head socket cap screws. The Pillar holes in the Deck are drilled 5/16"φ. I think you would be just as well off with 1/2"φ wooden dowels, which may give you a fairly strong friction fit if pushed into through-drilled 1/2"φ holes in the Decks. However, the dowels should be permanently anchored in place with glue, or one could drill into the edge of the deck and through the dowel at one go, and put a pin of some sort into each such hole.

I put my 0-1/2" keyless drill chuck, on its R8 shank, in a hole at the center of the Top deck, surrounded by my 6 R8 collets; outboard of the collets I drilled 20 uniformly spaced holes for the 1/2"φ bolts in my B'port clamping kit. I put my files in milled slots in the Middle deck, with the shorter ones on the outside, and the longer ones inboard of them, so both sizes are readily accessible.

I tapped (hammered) several pieces of 3/8"φ wooden dowel into 3/8" blind holes in the Middle deck, and on them stacked my B'port's T-nuts, flanged nuts and washers.

After all this, I still have about 3/4 of the Middle and Bottom decks to use as shelf space.

I think the Strokagenius File Rack is most aptly named, and the above changes make a significant improvement over the original unit detailed in **"Hey Tim..."**.

My friend Jake Wiebe made up a tool rack like mine as a Christmas present, with holes for files, screwdrivers, pliers, drills, etc. He set a plastic tub down into a big hole in the top deck (for stuff like nails, I suppose). The member of the (extended Wiebe) family who picked that present was just the right guy to get it, and Jake said he was just tickled to death with it.

DUST CONTROL
(this is not just about wood dust)

Added to the 6th printing: Please also see the note at the bottom of page 208

Woodworking generates a lot of fine dust. If not controlled, wood dust will travel to every room in the house. This is not appreciated by the executive department. It's also a nuisance in a shop with machine tools, because it sucks up oil that you would rather have remain where you put it. It is for this reason that woodworking and metalworking activities do not go well together in the same room.

Dust also does your lungs no good, but the stuff that's bad for you is the really fine stuff you can't see in the air. If you can smell it, it's going into your lungs. The dust of some woods is very toxic, and some people will have allergic reactions of varying degrees of severity to the dust from some species.

A vacuum cleaner will do little to trap this fine dust. Think about it: the average vacuum cleaner filter bag is barely big enough to store a chicken in. 1 square foot of surface area? 2 at most, and what size particles will it trap?

Go back into a room where, several hours previously, you were using a table saw, say, or a disk sander, and a vacuum cleaner for dust collection. Run your finger over a flat surface. You'll probably find that a noticeable film of very fine dust has settled.

In "Hey Tim..." I had a little thing based on an item in *FWW* about how to make a dust collector by putting a large polyester felt bag on the mouth of a squirrel cage blower. The idea is that you run the blower near your woodworking activity, and it pulls all the air in the room through itself every few minutes. The filter bag is supposed to catch the dust.

Several months ago I made myself exactly such an animal, from a $25 squirrel cage blower and some polyester felt I bought at a fabric store. The filter bag was almost big enough to use as a sleeping bag. It worked fine, too, or so I thought at first: I could run a few cuts on the table saw, fill the room with suspended dust, and in 2 minutes my funny little "dust collector" would have cleared the air again. But then I got down on the floor beside the filter bag and .... *I could see fine dust coming right through the filter cloth!* That I could see it was largely due to the medium green color of the filter cloth (the dust wouldn't have shown up as well against a pale grey or white material) and to the fact that I was looking towards the sun coming through the window. And if I could see fine dust coming through the filter, it was an absolute **certainty** the stuff I couldn't see was coming through too!

I contacted Kraemer Tool & Mfg. Co. Ltd., in Ontario, Canada; phone 1-800-443-6443, or see them at www.kraemertool.com. KTM makes filters and dust collection equipment. (There are American companies that do the same sort of thing, but KTM has no minimum order - they'll sell you one $1.37 clamp, if that's what you need.) I told them the size of the mouth on my squirrel cage blower, and pretty soon here comes a real serious filter bag made from a much thicker felt-like industrial filter material. Cost = about US$150. It does not completely solve the dust problem, but it sure helps, and if that is anything like what you need to take the dust out of the air, how is a vacuum cleaner's filter bag the size of a football going to do the job? It isn't.

(**Added to the 6th Printing:** I've since also bought a KTM 1 HP Model S-1 dust collector with a 6' tall bag. With this and a 4"ø flexible pipe, I can pull dust from any machine I'm running.)

In the 100th issue of *FWW* there is an excellent article on the use of small "cyclone separators" for collecting wood dust. You should see it. Paul Scobel's brother-in-law made up a small cyclone separator from galvanized sheet metal and pop rivets, per that article, and what it does is amazing. You can feed a dustpan full of wood dust into the intake and none of it appears to come out the discharge side! The bulk of the dust gets spun out and discharged out the bottom of the cyclone separator into a garbage can. Maybe the fine stuff comes through with the air, but run the discharge air through a filter bag, and you've probably caught 99.5% of the dust.

The other sources of dust that bear mention are from sandblasting and tool sharpening, or more accurately, from metal grinding operations generally. Dust from either type of activity isn't going to turn a basement machinist into a cancer victim overnight, but it isn't going to do him any good, either.

I would prefer to do my sandblasting outdoors. Any dust generated isn't going to be wood dust in the first place (it's sand = fine rock, or shattered glass beads, or whatever you're using as your abrasive medium) and you don't want *that* settling on your lathe bed, do you? If done outdoors, that problem is solved, as is, most likely, the matter of breathing the dust, since it is hopefully carried away by the breeze. And/or wear a dust mask.

I have read that cutlery factory employees who made their living hand grinding knives etc. to shape in "the old days" (say 100 years ago) rarely lived past 35. They worked more or less lying down over a large grindstone rotating in front of their faces -- long hours, 6 days a week no doubt, and a short life. Sharpening the odd lathe toolbit won't have the same effect, but remember what I said above: if you can smell it, it's going into your lungs. Why abuse your health needlessly?

Stewart Marshall told me that the dust from horn (cow horn and similar) is very toxic. Ditto for lignum vitae wood, micarta, ivory, and bone, too, I think. Added later: **"Gunsmith Kinks 3"** just arrived, and listen to this: elephant ivory may contain anthrax bacteria, which can be fatal - they are released by sanding, etc. (See page 44, Kinks 3.)

——————/——————

More On How to Scrub Down a Granite Surface Plate
(see also pages 106 and 224)

Over the years I've owned several granite surface plates, most of them used. The matter of cleaning them arises, and - being a natural worry wort - I have sought expert advice, and got quite a lot. Reviewing it all while revising things for the 4th printing of this book, in late October '97, I find much of this advice rather conflicting.

At page 46 in "Hey Tim...", I tell of buying a used granite surface plate with some metallic smears on it, possibly from someone sliding a chunk of aluminum across it. I went on to say an old T&D maker had recommended the use of Comet or similar fine abrasive household cleaner to clean it. I was told this would do no harm as a one time thing, followed by a good wash, and then drying with a towel. However, I have since been told that **the use of any such abrasive is NOT a good idea.**

One expert I contacted suggested trying paint thinner (only on the aluminum smears), and if this didn't do the job, to try proper surface plate cleaner. This source also suggested that cleaning be followed by the application of some light machine oil, *but see next paragraph re this idea.* Another source said metallic smears can sometimes be removed with a pencil eraser, but said not to use an ink eraser with abrasive bonded into it. He (or someone else) also said that dilute sulfuric acid can also be used to remove such smears, but - whatever you use - wash the plate thoroughly afterwards. In the end, I used a cloth with paint thinner plus a touch of rotten stone (an exceedingly fine abrasive), applied only on the marks, and this took them off very quickly.

However, a more recent discussion with Jim Coulson, at Starrett's Mt. Airy, NC surface plate factory yielded this advice: don't use abrasive household cleaners, erasers (any type), alcohol, solvent or acid. Soap should not be used, because it can leave a film. Paint thinner is ok, but again, not as an overall cleaner. Also, don't put oil on a surface plate. (In the past I have mentioned oil as being ok, but apparently it is not ok. Mr. Coulson said they oil surface plates during manufacture (twice, actually), and re-oiling should not be needed in service. I think there is more to this matter of oiling granite surface plates than I have yet got. Hopefully I'll be able to include more info in **TMBR#4.**)

A bottle of proper surface plate cleaner plus a padded wooden cover (cloth under plywood, or similar) is probably a granite surface plate's best protection.

My new 12 x 18" Starrett Crystal Pink has 3 pads under it. I once had a black granite plate of the same size with no rubber feet underneath. The maker (not Starrett) said that if the plate is kept on a flat surface with no great weight on it, there is no need for feet on the underside. In contrast, at an early stage during manufacture, Starrett puts 3 rubber feet - the accepted arrangement - under all their plates in this size range (and considerably beyond), and considers these pads to be an integral part of the plate - i.e. not to be removed - for preservation of the accuracy of the plate. At page 224 herein you'll find a drwg based on the U.S. Government specs regarding location of the 3 pads.

SOME IDEAS FOR LAMPS
and some other ideas to light up your financial life as well

Some readers will have seen my article in HSM (reprinted in Projects Three) on making a desk lamp in the form of a scaled up Toolmaker's Surface Gage.   And you will note the drawing of a Clamp-on Ball Handle on the back cover of TMBR#1 and the info re same at page 80-83 therein.

I had an idea for another scale-it-up project you might like, if you liked the clamp-on ball handle concept, and if you are into wood turning.

My idea was to make a table lamp in the form of a clamp-on ball handle, as at right.   The base would have the appearance of a large hex nut, surmounted by a washer.   These would be made of a dark wood such as walnut.   The rectangular part of the ball handle would sit on the washer. The ball handle would be made of maple or other light colored wood.   The clamp screw would be walnut, possibly ebonized (stained black).   It would obviously be made in two pieces, a head turning glued into a hole in one side of the handle, and a plug of the same wood turned and glued into a hole on the other side. (Alternatively, one could use a real shcs.) A dummy stub shaft would be turned from some 3rd wood and set into the larger hole in the handle base at right angles to the clamp screw.

Now this might make a lot of people retch in a bucket, but I think it'd be cute as a bug's ear.

As for how to turn a real nice smooth constant taper on a wood lathe, as would be called for in making the above lamp, here's an idea I read in *Fine Woodworking Magazine:*

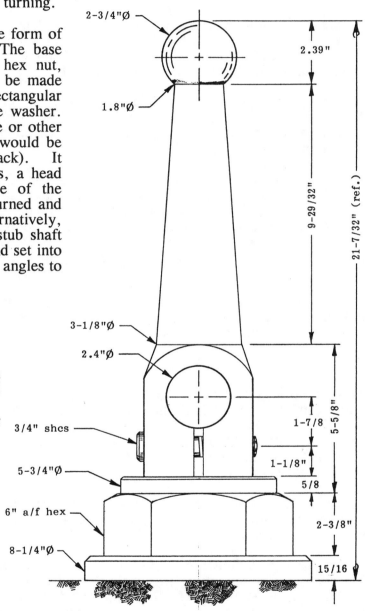

> Rough turn the taper, and neck down to a hair over finished ϕ at each end of the taper. Then lock the spindle, and use a wood plane to plane a flat along the work to exactly join the large and small ϕ's that delineate the taper. This flat wants to be *just* outside the OD of the taper for its full length by the amount you will sand off to get the final finish. Now when you restart the lathe, the flat will show up as a "ghost" of the desired taper, and you very carefully take off excess material till the flat disappears. I have a notion that this task could be done, or at least finished off, very nicely with the wood plane.   You would want to run at a moderate spindle speed, to avoid overheating the plane blade.   The plane would have to be handled carefully, and be razor sharp.

205

NOTE

For an excellent article on safe methods of making wooden lamps, see *Fine Woodworking Magazine (FWW)* for July/August 1988, page 78. At page 18 in the same issue is an idea re turning spheres which might be useful in connection with this project.

Here are two other ideas for lamps that might interest you:

(1) Suppose you were to cut, from a piece of galvanized sheet metal, a pentagon whose 5 sides were of equal length. Suppose also that the pentagon was of a size that would just fit in a 3-5/8" circle.

Now suppose you were to trace around this pentagon template onto a piece of cardboard, and trace out, and then cut out, 12 pentagons, all alike. If you then joined those 12 cardboard pentagons edge to edge with masking tape, you would end up with a 12-faceted "geometric solid" that would fit inside a sphere of about 6-1/4"ϕ.

Now suppose further that you liked the result - which is possible - and made a better one, in sheet copper, with the pieces soldered together so carefully that the joints were almost invisible. If this item were polished up nicely, and incorporated into a table lamp, you would have something most unusual.

A whole range of geometric solids can be produced from triangles, squares, pentagons, and hexagons. Some are more interesting in appearance than others. Some are beautiful (if you like that sorta thing). There is a book called "Polyhedron Models," by Magnus J. Wenninger, in which about a hundred geometric solids are pictured. See your local library. There is also a "toy" - not unlike "Meccano" and "Tinker-Toy" in some ways - called "Googolplex," from which many geometric solids can be made. This last item would give the reader who might be interested in this general idea a ready means of trying some of these solids.

(2) Stewart Marshall told me about a lamp he saw that was made from all sorts of brass pipe, pipe fittings, valves, pressure gages, etc. It was nicely done, well polished, and just begged the viewer to start twistin' handles, etc.

This whole idea of making something at some other than its normal size appeals to me. A number of other people have thought of it, too.

In Jan/Feb '85 HSM, page 44, Philip Duclos described a miniature vise he'd made, without drawings, like unto his 5" milling machine vise. Overall, it was about the size of the last segment of his thumb.

Another fella made a working model of a Bridgeport milling machine, about a foot tall. He'd used one at work for years. The model was complete right down to the vise, dividing head, etc. This project was reported in ME for February 4, 1977, pages 154-6.

Others have made functional models of various machine tools. There was an article in the Jan.'89 issue of *Modeltec*, in which Donald Strom describes and pictures his considerable efforts along these lines. His W&S turret lathe is a little over a foot long, and is fitted with a 4-jaw chuck 1-3/4"ϕ. There's a heavy duty "Cincinnati" style vertical mill about 9" tall, a horizontal ditto to go with it, with a 1-7/16" x 6-1/2" table, and a shaper with a 1-5/16" wide vise. All 4 machines are fully functional. The horizontal mill has a B&S type dividing head on it, c/w tailstock, and a twist drill blank is shown set up between the latter items to have the spiral flutes milled into it. Of course that requires the mill table to be slewed at a slight angle off 90° to the machine spindle, but that's no sweat - any "universal" type mill will do that.

Apprentices at a machine tool factory in England made a beautiful and fully operational model of a big shaper manufactured by their employer. The model was about a foot tall, and was about as handsome a piece of work as you might ever hope to see. Sure would be useful in a basement shop, for very small work.

Another guy made a model of the rotor hub of a big helicopter - a Sikorsky, I think. He worked for the maker of this particular machine which would certainly help when it came to getting drwgs. As I recall, he said he rented the finished model to the company from time to time for one purpose or another.

What's the point of all this? Well, it might get you going on an interesting line of activity, making a scaled down model of something in your shop, or in your working life. Or a scaled up item of something relatively small. If the work is of a high caliber, you will probably find it to be saleable, too - do you suppose the guy with the little B'port, or Duclos, with his little wee vise, couldn't sell them if they wanted to? There was an article in HSM for March/April '92 by a fella who makes miniature firearms - and I believe that guys who excel in that particular line of endeavor can find a ready market for their creations at handsome prices.

There are guys who do pretty well out of this type of work. I read a letter in ME - way back in the '60's, I think it was - from a fella who said he made his living as a basement model maker. He would pick a model he felt would be saleable - say a steam road roller - and make a run of say 6 of them, all at one go. I think I've mentioned elsewhere herein the magnificent museum-class model of a Civil War "Napoleon" cannon my friend Bob Eaton made. So far as I know, it is not for sale, but I have little doubt he could name any price under $10,000 and get it. Not necessarily the same day, of course, but he would get it.

Why do I think so? In the March April '89 issue of *FWW*, there's a write-up about a guy in North Battleford, Saskatchewan (i.e. not exactly New York City!) who made an equally fine model of a Wells Fargo stage coach. (There are plans for such a model in Pop. Mechanics publications of the 50's or 60's.) His was a more complex model than Bob Eaton's cannon, simply because a stage coach has more pieces in it. He'd even incorporated working wick raising mechanisms in the lamps, and had used some gold and other precious metals here and there in the model. He, like Bob, is a jeweller, which I guess explains that to some degree.

Apparently, he "reluctantly" sold the original model for $7,500, and set up to build 20 more. According to the article, these now command $50,000 each! Just how was not made clear - the article said that he gets distracted from this project by other work, so it could be that those few he has finished are now changing hands at far above what he got for them, but in any case, if they sell for that much, he would not likely sell new ones below that price. One a year would be enough to make a good living. Cut back on the detail a little, charge $40,000, and do 2 a year and you could thumb your nose at the world. 20, at $50,000 each, is a million bucks.

Hey, come back here! - you ain't finished readin' the book yet.

I have in mind another project to include in a future Bedside Reader: the making of a replica - full size, 1/2 size, or smaller - of a railway lantern of some sort, in sheet copper. I have some info that will aid this idea, and I think you will like the results. We'll see.....

My dad was a maintenance-of-way worker on the CPR's Kettle Valley Line from about '46 to '58. I was a little guy back then, and my recollections of those happy days include a standard tall glass hurricane lantern with a big reflector behind the globe. Discussions with my Dad suggest this may be an imperfect memory. We were talking about this one day recently, and Dad mentioned another type of kerosene-fueled lantern we had, called a railway signal lamp, which had interchangeable globes of different colors. These were kept in the toolhouse, clean, fueled, and ready for instant use if required. I don't suppose I'd thought of those lamps once in nearly 40 years, but his comment brought them to mind as clearly as if I were back in the toolhouse at Jellicoe - I could *see* the lamps all lined up on a shelf, with their clear, red, green, and yellow globes beside them, and I could *smell* all the old familiar smells of the toolhouse.

Today Jellicoe Section is no more. The section house, the woodshed, the bunkhouse and the toolhouse are all gone. Only the rail line and the siding remain. The switches now have red and green reflective metal disks on top, rather than a kerosene light that had to be cleaned and filled every week, and replaced in the old-style switch lamp. And there's no one there to fill the lamps, and no little boy to watch him do it. How time changes the things we remember.

Some Further Ideas for Making Money in the Shop
from **Jim Crocket, Fresno, CA**

from **Jim Crocket, Fresno CA:**

Dear Guy,

I stared my business with a little Unimat SL lathe back in 1964. Later I bought a Maximat VP-10 lathe and a BS-2 bandsaw. The Emco distributor used to send his customers to my place for demonstrations. Later, when he retired, I wrote the Emco people and said that since I'd been selling their tools for him, I might as well sell them for myself. That's how I became an Emco distributor. Without the Unimat lathe I would not have the business I have today.

About 80 percent of the Emco Maier tools I sell end up making money for their new owners.

I have one customer in a little farm town nearby, who has made his living for years with only an old Unimat SL lathe as his main tool. He buys new drive belts and a new motor from me about once a year. I asked him what kind of work he did and where he got his customers. He said he does a lot of repair jobs on small valves and sprayers for water and fertilizer. Most of his customers are local farmers.

Another Unimat owner has a sewing center; he paid for his lathe the first month he had it by turning armatures for sewing machine and vacuum cleaner motors.

Many of my customers started out as inventors, and developed a product which they or someone else manufactured for them. One chap invented a small piece of gold refining equipment. He bought a Compact 8 lathe from me so he could make parts he couldn't buy off the shelf. He advertised his product in a trade magazine and has done very well with it.

Another customer, who is both a machinist and a welder, built himself an all-steel 45-foot sailboat, which he and his family are planning to sail around the world. He bought a Compact 5 lathe from me, so he could help make money in any port by doing small repairs for boat winches and such. (You don't need to be circumnavigating the globe to seek out such work. Any hsm in a coastal location might find a demand for the sort of services he could offer.GBL)

I also have quite a few customers who have bought various size bench lathes to install in station wagons, trucks and vans to make on-the-job repairs or to do custom jobs on site. Others owned and raced cars and motorcycles.

One customer bought a Maximat Super 11 lathe to build an exact replica of an antique aircraft for a museum. The job made a profit and paid for all his tools, including the lathe.

Jim

Added to the 6th printing: According to some new info I have recently come across, even the best of the dust control methods discussed at page 203/204 herein may be inadequate. I recently got an interesting video tape on dust control. According to the tape, and from private correspondence with the 92 year old engineer who sells it, cloth/fabric filter bags on dust collectors do not take out dust particles below 5 microns. If he is right, filter bags may not provide adequate dust protection, if air from the dust collector vents back into your shop. Check out this and other offerings on this old gentleman's web site, at www.zianet.com/calexander/index.html. Note this, however: my contact at KTM (see page 203, 3rd paragraph up from bottom) tells me that their "needle felt filter bags with acrylic coating" are capable of trapping dust particles below 0.3 microns, although it also depends to some degree on how much air you put thru the dust collector. Who's right? More likely KTM. But nevertheless, the old chap's video has some interesting and valuable insights into dust collection.

From **Jim Wright, Tempe, AZ:**

Dear Guy,

You mentioned wanting to hear how others made money with shop work.

I am primarily a research technician for the Chemical, Biological and Materials Engineering Department at Arizona State University.

The machine work done in the College of Engineering is handled through a central facility of the College. At various times faculty doing outside consulting work show up there trying to get small jobs done. Our shops cannot do this, and will refer such people to outside shops. If a person is looking for occasional work for a basement shop, he should make himself known to such institutions in his area. Get on their list of "outside resources" and pass out business cards. (Business cards can be a surprisingly long-lived form of advertising.)

Mostly this sort of work will be of an off-and-on nature, and probably unreliable as an income generator, but it can produce enough to buy a fair amount of tooling and supplies.

Most of what I have done consists of building and/or modifying things used in courtrooms, or in legal consulting. I ask few questions.

A canny old machinist once told me: "Get half your money before you start the job, and the other half before you hand over the finished job." This has proven to be sound advice.

Jim

From **Charlie Carson, County Down, Ireland**

Dear Guy,

.... the idea of seeking full time employment using my model engineering experience and my workshop has paid off at last - if not handsomely in money terms, at least in a lot of satisfaction.

It all began when I was approached last spring by our local church - could I tell them who could make a couple of reproduction brass lamps? There was no one interested except in England, and then only at a very high price, so I did the job. One thing led to another via the architect on the church job, and I now have a commission for eleven lights for Queens University, plus I am also happily twisting away at wrought iron work, 3/4" square, using a hand-made portable hearth, for decorative window grilles and a fancy balcony handrail for the same job.

Some other work I have lined up is to make 30 reproduction door pulls, and - this one I'm really looking forward to - six reproduction globe lanterns for another public building getting a facelift.

There have been other small commissions in between, and if I am fortunate enough to get further work I will be very pleased and feel I have found a way of life that suits me very well.

In my spare time I have built a "Sweet Pea" (model steam loco), which I run on "The Windy Hill Railway" in my back garden.

Charlie

Dale Scherbart, of Ankeny, IA phoned me one day, and told me he makes a home shop business of making cemetery lot and pet grave markers. These are cast aluminum, typically a 4" x 2-1/2" oval, or a 4"ø disc, and they are used all over Iowa. He also casts lead weights (9.5 to 10.5 ozs. each) for antique juke box tone arms, and makes castings for toy cars and trucks, typically about 8" long, which he sells to a chap who then gets another machinist to make wheels for them.

Now here's something to make you sit up and puzzle over late at night: Dale said that archaeologists have dug up, in China, aluminum artifacts that are 3000 years old! Current aluminum smelting methods involve the use of massive amounts of electricity, so how did the Chinese do it 3000 years ago?

Lindsay Publications sells some sort of history of metallurgy by some Russian guy which tells about the above.

Turnin' a Buck on a Light

Phil Lebow is one of my guys who lives in L.A., a village in southern California with approximately the same population as Canada.

Phil phixes things phor people phor a living. He is a qualified machinist and welder. Among the things Phil phixes are old printing presses and paper shears, printers' cameras and so on. He also fixes movie camera equipment, and makes cameras (he once told me he was making (from scratch) a pair of cameras to shoot 9" x 18" negatives; these were to be used for taking promotional "stills" for movie promo posters. He added that when these were completed and paid for, he would be able to go out and buy a brand new Hardinge toolroom lathe. (Phil was at the time in the market for a new lathe, and we had had some discussions about what type/make/size of lathe to get.)

Anyway, Phil turns his hand to whatever he can to make a buck. Not everyone could do exactly the same as Phil does - he happens to live in an area with a very large and concentrated population, and certain industries (e.g. movie making) also happen to be concentrated around him to some degree. This helps.

One thing Phil makes is a weatherproof work light, which he told me has become a sort of a "staple item" for him - i.e. a standard item that he makes and sells as a money maker - and he sells on average 2 a week, at $75 each.

That's $150 per week x say 45 weeks/year = 150 x 45 = $6,750 per year. Now I didn't do that little piece of arithmetic to estimate Phil's income. (He's a machinist, so obviously he spends most of his time in sunglasses on a lawn chair with a cool drink near at hand.) What I'm thinking/suggesting/intimating here is that if a fella put his mind to it, he could probably make and sell enough of these lamps to make some serious money - if not enough to retire on, then certainly enough to significantly improve his eating habits, or his shop tooling, or his kids' education, or where he spends his vacation, or whatever. Now of course ya gotta sell the completed lamps.......

Phil sells his at garage sales, and otherwise - one time (at night) two guys were putting tar paper on a house being built across the street from his place. For light, they had one "trouble light" between them. Phil took a pair of his worklights over, and lit up their lives. They finished the job that night, and then came over to talk to Phil about could they buy these lamps? But of course.

Now what's good about them?

They're completely weatherproof. A 70 mph wind won't knock them over. The 25', 14 gauge, 3-conductor cord stores on the base, which can be a car wheel rim, or a cross made of two pieces of 1-1/2" angle iron about 24" long each. The lamp standard (post) can be extended from maybe 3' tall to say 6' tall, plus the lamp head tilts up and down. An on/off switch plus a plug-in (receptacle) for a power tool, or another light, or whatever, is housed in a waterproof outdoor connection box just below the lamp (see photo).

Here's what Phil has to say about making the telescoping stem for such a lamp:

"..... 1/2" rigid electrical conduit is a nice slide fit inside 3/4" rigid electrical conduit. Cut a 3/4" coupling in half, and drill a 1/2" hole through the side, centered. Then screw the coupling half onto the threaded end of the 3/4" conduit. Drill through the hole in the coupling and through the conduit wall. Now you have a 1/2" hole through one wall of the coupling and the conduit. Weld a 1/2-13 nut over the 1/2" hole, and then clear the threads with an old tap. Weld a short piece of 1/4" rod across the head of a short 1/2-13 bolt, for a handle, and run the bolt into the nut. Now, when you slide in your 1/2" conduit stem, you can lock it at any height with the 1/2" hand screw.

"For a base, weld the other end of the 3/4" conduit to any heavy piece of scrap iron. (Old wheel rim, angle iron cross, etc., as above. GBL)

"I use a 300 watt outdoor security light for a head. These are usually a cheap quartz-halogen fixture, with a pipe thread mounting. An outfit here had some on sale for $10.95 so I bought a bunch. Good lights are around $22-30, so it pays to wait for the sales. The first time I made lamps of this type (but with flood light heads, then) was back in the late '60's when I was working as a welder in a muffler shop. I've sold a lot of them since then!"

To attach the 3/4" conduit to a wheel rim base, weld the conduit directly to a disk of 1/8" HRS, and then weld the latter over the hub hole in the wheel rim center. Give this area a shot of zinc spray paint or similar to prevent rust, and hit the nut-collar-bolt arrangement at the top of the 3/4" pipe with some more of the same. Seal the lamp head/nipple/box openings/joints with silicon sealant. (In one photo Phil sent me, he pointed out that the connection box was salvaged (free) from a house being demolished.) Phil puts the lamp head on a short riser which permits the lamp to tilt all the way down to its own stop.

Now comes the frosting on the cake, for those who have read this far, and not gone to sleep or wandered off to seek greener pastures elsewhere herein:

> This same 3/4" plus 1/2" conduit idea **is also great for making other light-duty height-adjustable stands** for things like chip shield supports, blueprint holder stands and the like.

Another idea: Put the lamp on an arm that can tilt forward, or even just put it on an arm that sticks out parallel to the floor, and make the vertical piece adjustable as above, and you've got a lamp that would be very useful for working on stuff under the hood of a car. This opens up a whole 'nuther market segment.

Two more ideas from what Phil calls his **"Old Shop Tricks Department"**:

1. Get yourself one of these fiberglass typewriter erasers (or a drafting eraser). They can't be beat for cleaning the crud or rust (gasp!) out of the fine knurling of your good hand tools. Don't use it on blued stuff without testing first in an inconspicuous place. Also great for cleaning out stamped logos, etc., without scratching the tool. If you can't find one at a stationery store, Brownell's used to carry something similar, for more money, in a metal handle.

2. If you're putting together a shrink fit, put a short chip of 50/50 lead solder on the part you're heating. When it melts, the part is hot enough to assemble. The drop of solder is easily removed with steel wool while still molten. This also works when heating housings to get bearings out. When the solder melts, stick a wet rag into the bearing, and shake the bearing out. By using different alloy solders, you can adjust your heat level.

Another Idea....

At page 122 of the February '91 issue of *FWW* there is an item which at first glance might look like a model of the piping in some industrial plant, or an oil drilling rig from 60 years ago. On closer examination of the larger of the 2 photos, you will see a number of very tiny figures (actually HO-scale people) dotted here and there over the structure. There are 30 in all, not all visible in the photos. This will probably pique your interest sufficiently to make you read the accompanying info. The maker spent about 3 months building this thing.

He refers to it as a "kinetic sculpture" - by which I suppose is meant that it employs moving elements. The "moving elements" are in fact 1/2" steel balls. When the sculpture is "started," by pushing a button, these balls are sent to the top of the tower and released, whereupon each will follow one of several circuitous paths back to the bottom through pipes, troughs, chutes, etc. Once 8 balls have made the trip, the machine shuts itself off, and waits for the next push of the start button, for a repeat performance.

The original was exhibited at "Art In the Redwoods Show," in Gualala, CA where it took the award for the Most Popular Piece, this opinion being confirmed by the amount of nose grease on the glass walls of its case!

"Now, why would Lautard put something about that idea in here?" you may ask.

Very simple: if the idea were to appeal to you, why not make a similar rolling ball sculpture, or something equally or more entrancing, and sell or rent it to whoever might want to attract people - e.g. to look in a store window (e.g a men's clothing store) or occupy people (e.g. kids at a car dealership), etc.

I'd be willing to bet that any such item, well done, would make money for whoever built it.

212

PS: I was telling my friend Terry about the foregoing, and he said there is something similar, but much larger, at a big shopping mall in Edmonton, Alberta. Every time the ball comes down it causes something different to happen. Once in a blue moon it causes a number of things to happen. This thing was not just found at a garage sale and bought by somebody who muttered, "Now ain't that funky?!" Somebody got paid good money to design and build it.

HONDA WHEEL REPAIRS

Another possible job you might do to make money in your shop is to repair the rear wheels of Honda Goldwing (and possibly other) motorcycles.

My friend Jake Wiebe did one for his brother-in-law, whose Goldwing's rear wheel bearing went bad, to the point where it rusted and seized and spun the bearing race in its seat. A new wheel would've cost the guy Cdn $1500.

Jake slapped the wheel up on the table of his vertical mill, clamped it down on a couple of pieces of CRS for parallels, centered it under the spindle, and bored the aluminum wheel casting out to give him a seat for a ring about 3/4" deep and maybe 2-1/2"ϕ. He turned up a steel ring and pressed it into this seat using the quill of the milling machine as an arbor press. (Jake said he was a little reluctant to do it that way, but it saved him losing his centering setup, although he admitted he could've re-centered the wheel in short order if he'd had to.)

I think Jake said he put some Loctite on the steel ring insert too - for which reason he let it set overnight. The next day he bored out the steel ring and faced it off with his boring head, and pressed a new ball bearing into place. Job done. 2 hours max.

There are a couple of questions one would have to answer to determine if this would be a paying proposition or not.

First, how many Honda Goldwing rear wheels fail in this way in your area every year? Local Honda dealers may or may not tell you. The State Goldwing Owners' club (or functional equivalent thereof) may tell you, or ask its membership for you.

Second, what is the cost of a good used wheel off (say) a wrecked Goldwing? Jake did a similar job for his son's Honda trail bike, but his son was able to pick up a used wheel for $50. (A new one for a trail bike wouldn't be $1500, but you can bet they don't give them away, either.) The used parts route is the competition to the bore-out-and-rebush-better-than-new service, but if this type of failure is a common occurance on big or little Hondas, and/or other makes, the used wheel route is only a temporary fix - eventually it will fail too.

The other consideration is that of liability. If the owner falls off his bike later and sues you, even though your repair did not cause his problem, you may have a serious problem on your hands, and this aspect of the matter is therefore not to be overlooked.

A guy could start out by doing 2 or 3 ruined wheels. After that, he could tackle a wheel whose only problem was the need for a new bearing seat. He doesn't need to have any concern that he might be about to spoil a good wheel - it's scrap before he starts the job. If he makes a good fix, it won't be scrap when he's done. Nor would there be any pressure to hurry to please an impatient owner. From there on he could continue to do the job on an exchange basis, as much as possible, to maintain the zero pressure situation, if he chose to.

It wouldn't take very many wheels to pay for a Narex automatic boring and facing head, which (although not required for such work) would be about the slickest piece of tooling you could want on the mill to do a job like this. And if you had such tooling, and were doing stuff like this, you'd probably soon find yourself being asked to do other jobs where it would be useful.

Little Engines Seem to Sell

I have mentioned elsewhere herein, in connection with drwgs for an EDM machine, that the June/July '93 issue of Strictly I.C. Magazine proved to be a bumper issue. Here are the other reasons it struck me that way:

It included full working drwgs for a vertical steam engine (like a ST #10-V), complete with reversing gear and flywheel, that would fit into a box 1" x 1" x 2" tall. No castings - everything was machined from the solid. And like I said, literally just a little bigger than your thumb. Bore and stroke are 0.300" and 0.250", respectively.

This reminded me of a conversation I had one evening about 3 years ago with one of my guys. He told me he makes a small oscillating steam engine based on a past LS article, but prettied up somewhat, and sells them for about $700 each, at antique engine & machinery shows. He said he'll take an engine to a show, and usually has it sold by the end of the first day. He said he finds this product to be quite profitable, in terms of his hourly return.

I looked up the article, after I talked to him, but I don't recall the reference right now - somewhere around '78, I think, but it doesn't really matter for what I'm coming to.

Now say you have a lathe, and maybe a mill-drill*, but you'd like a better mill. And let's say you have some time to spend in the shop, but cash for new equipment is hard to come by.

> * With that, you could make an engine like my caller was speaking about.

If you made one of those little 1 x 1 x 2" engines noted above, and then maybe made a couple at one time, making parts at each end of a piece of bar stock, and flipping it end for end in the chuck, or the mill vise, to get the most outta your set-ups. You'd get a pretty good feel for the idea of making several at a time, for sale. If it looked like a go, you could sell those first ones (if they were up to the standard you wanted to offer for sale), and then make a run of 6 or a dozen. You wouldn't sell them in a week, but you probably would over a period of a few weeks or months. If a simple oscillating steam engine can bring its maker $700, I suspect you could get more for a "real" steam engine the size of your thumb, especially if it had a certain amount of polish here & there, and/or was nicely painted. The sale of a dozen such engines would pay for an extremely nice milling machine. Certainly it'd be worth thinking about.

(I know a guy here in my area who used to make miniature clocks, using ladies watch movements for the works. Every so often he would pack 40 or 50 of them in his shaving case, fly to miniaturists' shows in London and New York, and sell them for prices ranging from $500 to $1,700 each. He's gone on to other things at the moment, but it indicates that there's a market for things you might not think of trying to make or sell.)

SHIPS' WHEELS AND OBOES

I had a call one day from Stewart Marshall, from his shop down on Lopez Island, in the mouth of Puget Sound. He'd just finished making another ship's wheel, and wanted me to see it before he sold it. I had some books to mail, so I took them down to Bellingham, and then went on down to Anacortes and got over to Lopez on the ferry about mid afternoon. I found Stewart in his shop, covered from head to foot with little black specks of oil - I forget what he said he'd been doing before I got there, but he looked like he'd been havin' fun. The smell of fresh coffee on the stove, and the generally fascinating atmosphere of Stew's little shop, plus the warmth of Stewart's welcome presaged a fun visit.

We got to the wheel after a while, and a very handsome job it was, too - about 36" in diameter, flawless joints, all walnut, with plugs of a lighter color in the rim to cover screw heads used in the wheel's construction. Stewart told me it had taken him a week to make it. He planned to sell it on consignment through a marine supply store in the Seattle area. In the past he has made ships' wheels up to 80"ϕ. Sometimes a wheel will sell very quickly - one went in 2 days! Other times, they move more slowly.

Making a ship's wheel is woodwork, of course, but it is woodwork of a sort that could appeal to a machinist, involving as it does turning the spokes, fairly precise dividing of the rim, plus the making of various jigs for cutting the rim segments, and for assembly. In any case, I mention it as another example of how one of the guys is turning a buck from his shop.

Stewart also took me around to visit his friend **Sand Dalton** (another Lopez resident) who makes replicas of classic 17th century baroque oboes in boxwood, ebony, etc. Sand makes them to sell, speculatively and to order, and is also a master oboe player. He makes a nice livelihood from making and selling oboes, and giving oboe recitals. Sounds like a piece of cake, eh? But it's not all gravy. Sand is a good example of someone who has "paid his dues"; he spent years developing a reputation as an oboe maker; going to museums here and in Europe to measure old oboes in detail, then learning how to make them, making the tools he uses, etc. Not to mention the time he spent learning to play the oboe.

Oboe making also is mostly precise woodturning, but there are brass keys to make, springs, and so on - and Sand makes every part himself, right down to the reeds. Not everyone could do this, of course - you need to know what the instrument should sound like in order to fine tune it once you get it mostly done. This involves careful refinement of the bore of the instrument (using handmade reamers) to suit the characteristics of the particular piece of wood that went into its making.

> Footnote: Stewart has a considerable interest in "ornamental turning", and has pointed out to me that an ordinary engine lathe such as a hsm might have in his basement shop can be equipped with simple shop-made tooling to do much of the sort of work once done on "ornamental" lathes - lathes that were so expensive that mostly only the aristocracy could afford them back in the last century when this type of work was much in vogue. Stewart recently wrote me that he'd showed Sand an example of his ornamental turning, and Sand had proposed they collaborate on a fancy oboe - Sand would make it, and Stewart would put the ornamental work on it. Such an instrument would sell for several thousand dollars.

Phil Lebow buzzed me shortly after I got back from the above trip, and he was telling me he'd been approached by an outfit wanting him to make some underwater housings for video cameras - 50 of them, in fact. As things fell out, the job didn't materialize, but had it.... well, who would complain at grossing enough from 2 months' work to be able to take the rest of the year off?

Phil said he sometimes does darkroom renovations with a friend who is a woodworker. (Phil would make the kind of contacts that would lead to this sort of business, given his camera repair activities, etc.) Phil said when they have 3/8 or 1/2" plywood left over from a job of this sort, his partner makes up little wooden tote boxes out of the scraps, some for his own shop, and some for Phil. Some have rope handles, some have oblong holes.

Phil said his pal also made him up some heavier duty boxes one time, out of 2x8 material, with rope handles, and cleats underneath so you can get a hand truck under them easily. Phil puts steel scraps in these, and he says they don't even creak when you tilt the hand truck.

Phil also told me about something else that was news to me, anyway. He'd made up a three legged lathe stand (per my article re same in Jan/Feb '83 HSM; reprinted in "Projects *One*,"), using 4" sq. steel tubing, for a friend. It was heavier than the lathe, but that was ok. What wasn't ok was the fact that one of the legs had come out a hair short, and the stand did not sit (stand?) level. What to do? Phil said they put some "thin set" under all three foot pads, and when that set up, it was as good as you could want. I'd never heard of "thin set".

Turns out **"thin set"** is a mortar/grout/concrete kinda stuff, used for the repair of cracked concrete, and for stickin' steel wear strips onto concrete stair noses, etc. You mix it as thin as you need to, and it sets up pretty fast. It's sold in some hardware stores, and obviously also would be found in building supply places and lumber yards.

Like I said, this stuff could be mighty handy to know about.

Plans for a Nice Marine Steam Engine Model

If you look in *Live Steam Magazine (LS)* for July 1982, you will find therein the start of a serial on building a very handsome little two cylinder steam engine known as The Weaver Launch Engine. Apparently it works real good, and has a nice "bark" to its exhaust note.

Plans are available directly from the designer, Bill Weaver (P.O. Box 373, Middletown, MD 21769). This engine is no one-weekend job - Bill says you're lookin' at 100/200 hours - but when it was done, you'd really have something to crow about. The finished engine stands about 6" tall, is 6" long, and just under 2-3/4" wide. As I read the drwgs, the bore is 5/8"ϕ, and the stroke is 3/4".

If the idea of a handsome, well designed two cylinder vertical upright marine type steam engine appeals to you, **don't** overlook this one. (Photocopies of past *LS* articles are readily available from the publisher, as well as from University Microfilms (300 North Zeeb Road, Ann Arbor, MI 48106), so getting a copy of the serial on building it is no problem.

Now, let me toss a really wild idea at you: Why not build it at half size, for a real one-of-a-kind miniature? Or, if you want a small steam engine for serious use say in a light boat of some sort, build it at 2x or 2.5x original size.... There are guys who are into steam launches who are looking for stuff like that, and if you had something they liked, it oughta be saleable.

Cogent Advice

I saw a little quote in a car ad* on TV; it was so brief I didn't take in who said it, but it was quite cogent advice, I thought. It said:

"Do one thing, and do it better than anyone else."

> * nope - I saw it again: it was in an
> ad for a certain brand of popcorn.

Not a bad recipe for success. There is another, like unto it:

"Concentration is the key to success."

I don't know who said that, either, but it's true.

"Do it quickly and do it well. Perfectionists are really amateurs. Professionals get it right the first time, and they do it quickly."
Observation of contemporary English violin maker Philip Knight.

There may be some truth in that, but every pro was once a beginner. One must first learn how to do a job properly. Only then can you learn to do it quickly.

What the Kangaroo can Do, You can Do too
a footnote to page 213

When I buzzed one of the local Honda dealerships here, after putting together the page where I speak about repairing Honda motorcycle wheels, I ended up talking to a lady from Australia. She told me that a new rear wheel for a Honda trail bike could run $200 to $500, depending on which model it was for.

She then gave me another idea: It seems that here in North America, nobody thinks to bore out and rebush motorcycle master brake cylinders - the dealers just tell you to buy a new part, at maybe $200. In Australia, the normal drill is to rebuild the existing one with a kit, at about half the cost of a new part. I doubt there'd be much a kangaroo mechanic can do that we couldn't learn to do just as well, and there might be an opportunity here if a fella went around to all the motorcycle shops in his area, and offered to rebuild master cylinders for them. The shops could save them for you, and hand you a wad of them every so often. There might be money in this.

When I ran this past Jake, he said that they might only require to be re-honed, but that if need be, rebushing would not be a big job. A guy could make a jig, and do them on the mill, one after another.

Jake went on to tell me he's making new throttle shafts for a certain farm tractor carb for which replacements are no longer available. He does it for a farm equipment salvage guy, and he gets good money for a short chunk of 1/4"ϕ brass rod with a little square milled at one end, a slitting saw cut, and two tapped holes. He also rebushes the carb body with shop-made brass bushes; the carb body is aluminum, and it's usually worn pretty badly egg shaped, because there's a lotta torque on the throttle rod from the engine governor, and between that and the dusty environment farm equipment works in, there's work to be done to put them back in good order.

The idea that sprouts from my fertile mind as a result of this further discussion with Jake is this: if a hsm who lived in a farming area were to go around to all the farm machinery dealers, and offer to do the little fussy machining jobs like this that they don't want to tackle, there might be money to be made...... If you were interested, it'd be worth lookin' into.

Why is your Lathe's Tailstock "high"?

Lathe makers generally build a lathe's tailstock so it is a thou or so above the headstock spindle axis. You may wonder why. One reason I have heard of is to allow for future wear: the tailstock will wear into better alignment, rather than starting perfect and in the same period of service wearing out of alignment. The other reason, which may also be true, is to compensate for the weight of a heavy workpiece running between centers and tipping the tailstock downwards.

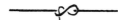

Just By the Bye

Wolfe Publishing has brought out some interesting books in recent years. **The Breech-Loading Single Shot Rifle**, by Ned Roberts and Ken Waters follows on from **The Muzzle Loading Cap Lock Rifle** (also by Ned Roberts), which I have mentioned elsewhere herein. Lotta stuff in those two books to give a guy ideas. **The Custom Government Model Pistol**, by Layne Simpson, is another, but it is not a how-to book, but rather a what-you-can-buy/who-can-do-what-for-you book.

Holding By A Thread
from **David Joly, Ashland, OR**

Dear Guy,

I just got your Treatise on Oiling Machine Tools, and a remark somewhere in the text triggered off a wrinkle you might find useful, so I'll pass it on.

At page 5 of your little booklet *How to Make a Handvise*, under Thumbscrew, you mentioned that it would be better to screwcut this thread, rather than cut it with a die (or, presumably, to adapt a standard fastener, which would be my impulse), as it would be a "truer" thread. This caused me to puzzle a moment why you cared about that, until I realized that you were going to use the end of the same part as the clamping point, and you didn't want the point of contact describing a circle as you tightened up. Further, we both know it is common for a die to run a thread which is perfectly usable for a fastener, but which is neither concentric nor parallel with the axis of the cylinder on which it was run.

If you're putting a dog point on a screw for clamping, or a cone for locating, or a cone on a needle valve (as in the sight feed oiling cup you describe in the Oiling Treatise), you had better not use the cylinder that is the crest of the threads to chuck on - it probably has a different axis than the one you're interested in.

A way around all this, I think, is similar to the three-wire measurement technique for pitch diameter: thread the threaded piece into a coil spring, and then clamp your chuck jaws on the spring. Any spring wire of a diameter smaller than the thread pitch and big enough to stick up above the crests will do. It occurs to me that Heli-Coil inserts would be tempting for this use, but I think they are very hard, and thus might damage the chuck's jaws. Just a plain old steel extension spring, or a helix wound on a smaller bolt out of music wire should do the job just fine.

I want to mention also a book I recently got from McGraw-Hill; it is called "Designing Cost-efficient Mechanisms," by Lawrence J. Kamm. It is about what I used to hear referred to as kinematic design. His bibliography cites T.N. Whitehead's book, "The Design and Use of Instruments and Accurate Mechanisms," (Dover Books, 1954) - a reprint, I think, of a 1934 original. Whitehead was better on pure Kinematic Design, Kamm is good on where, how, and why to fudge these principals. I think you'd like it a lot.

David

Since we're on the topic of books, some other books that have been suggested to me as being worth mentioning are these:

"Nuts, Bolts, and Fasteners & Plumbing Handbook," by Carroll Smith.

"Engineer to Win," from Motorbooks International. Good, easy to understand info on the technical side of metal - elasticity, strength, and so on - geared towards race cars, but still good, says Bruce Fielder, Mt. Clemens, MI.

"Fundamentals of Dimensional Metrology," by Ted Busch, Wilke Brothers Foundation; Published by Delmar Publishers Inc., Albany NY. ISBN 0-8273-2127-9. This is an excellent book that deals with just about any aspect of length and angular measurement you might ever care to know about, from the crudest positioning of a thumb somewhere along a stick to the most up-to-date means of measurement known. If this area of the machinist's business interests you, this book is worth having.

The TURNER'S CUBE
with thanks to **Ron Jeffrey, Dorset, England**

In the January 3/86 issue of MODEL ENGINEER Magazine, in the "letters" column, Ron Jeffrey wrote a letter headed "An Old Time Mechanic Remembers," wherein he described a "Turner's Square," which was made by boring undercut recesses in the six faces of a solid metal cube, to produce another cube, captive, within the first, and another within the second. (A more appropriate name for such an item would be a "Turner's Cube", and I will use that term henceforward here.)

This struck me as something that would go well in **TMBR#3**, so I set about making contact with him, via the Editor of ME, thence to the Secretary of the West Riding Small Locomotive Society, for whose Newsletter Mr. Jeffrey had penned a write-up on how to go about the job of making one.

In due course I heard back from Mr. Jeffrey, and he graciously supplied the information on which the following is based.

Making A Turner's Cube

Ron Jeffrey served his apprenticeship with a British marine engineering outfit specializing in diesel generators, electric winches and cranes, back in the mid 1930's. Much of the shop machinery was driven from overhead line shafting and leather belting. Every machinist had one or more of these little head scratchers in his pocket or tool box. Some were brass, others steel, others aluminum. Size varied according to the maker's whim. Apprentices (14 to 16 years old) were not told how to make them, but were expected to work out the necessary procedure for themselves, and were then allowed to make one (during lunch hours) on a foot-powered Milnes lathe with a 10" swing. (A very fine machine in its day, although I'd much prefer my Super 7.)

The job would have been made somewhat easier had the use of a milling machine been allowed, but this task was assigned before the boys reached that stage in their training. You may prefer to make your 'Cube partly or entirely on your milling machine. The method here described envisages the use of a lathe only.

The first job is to secure a piece of material, and make it into a cube of whatever size you please. Contemplate for a few minutes the task of making a Turner's Cube with an outside cube 1/4" on a side.... That'll put the job into a more agreeable light, when you decide to make your first one say as a 7/8" cube.

How to make a cube in the lathe will be no mystery to many readers, but we better cover a few points, for the benefit of others.

Slap the rough block in the 4-jaw chuck, and clean up one face with a facing cut, leaving a nice finish. Remove the job and reverse it in the chuck so that the just-machined face is butted up against the face of the chuck.

> TIP: when trying to chuck a piece of mat'l in such a situation, it is usually helpful to put one of the rough cube's best faces - if it has one or more that are half decent - against say the #1 jaw, and then interpose a roller between the opposite face and the #3 jaw. This roller can be a piece of say 1/4"⌀ CRS maybe 1/2" long, and having its own ends faced and lightly chamfered. Do the same with the faces contacted by the #2 and

#4 jaws. These rollers will give you more or less line, or point, contact on a rough sawn face of the cube, so that the good face on the opposite side of the mat'l can seat solidly and accurately against its jaw. I learned this little trick from Bill Fenton years ago; it works with CRS, HRMS, sawn surfaces, and castings, and is well worth remembering.

Clean up the second face, so that it will be parallel with the first one we did, and then re-chuck the block with these two newly-machined faces under the #1 and #3 jaws, rollers under the #2 & #4 jaws, and machine the 3rd face. Next, butt the 3rd machined face up against the face of the chuck, and clean up a 4th face. This gives us a block with 4 adjacent sides presumably all square to one another, and two rough ends. These are then dealt with in similar fashion, and then we can begin facing the block down to an accurate cube, equal in size on all 6 faces.

If any out-of-squareness is revealed when the block is checked with a machinist's square, get rid of it by re-machining the offending face, shimming the block in the chuck if need be.

The cube does not have to be $\pm 0.000,5"$ on size all around, but why not aim for something really close, say $\pm$ a thou or so, and have something to brag about when done? Also, use protective slips under each jaw, and shoot for an exhibition finish on all 6 faces as you bring the cube down to size.

Once you have a basic cube, the next task is to chuck the cube with each face "out" in turn, and bore a well centered 2-diameter hole in each face, to specified diameter and depth.

Dimensioning of the cube and holes is covered in Fig. 1 below.

For this job and the next, we need a chucking fixture. While we can proceed without one, it will be helpful as a means of more or less automatically centering the cube on the lathe for boring the cavity on each face.

The chucking fixture is a sort of false faceplate, which is held in the 4-jaw chuck. It is illustrated, with notes, in Fig. 2 below.

Once the flat-bottomed holes have been bored to spec in all 6 faces of the cube, plug 5 of the 6 holes with plugs of approximately the correct diameter, and having the correct length dimensions, as shown in Fig. 3 below. These plugs clamp hard against the material in the interior of the cube, and when the work progresses to a certain point, the plugs will immobilize the inner and central cubes, thus enabling the machining to be completed. If your plugs are not dimensioned just right, either the outer or inner cubes will become loose on the last undercut, which is notta lotta fun.

With the 5 plugs in place, machine an undercut at the bottom of each hole out to a specified (calculated) diameter, on the 6th face of the cube. This requires the use of a boring tool shaped approximately as shown in Fig. 4 below. Pick up the bottom of the bored hole (which partially forms the face of the innermost cube), zero the crossfeed dial, and then undercut the hole with a total outfeed of X_a on the radius, as calculated in Fig. 1.

Back the boring tool out towards the tailstock and come in again to pick up the partial face of the middle cube, and undercut it with an outfeed of X_b.

The last two steps above are repeated 5 times, each time moving one of the 5 plugs to the newly undercut hole, rechucking the Cube and undercutting the bottom(s) of the vacated hole.

On the drwg of the chucking fixture, note carefully how the 4 pieces of 3/4" square CRS (or whatever other size of mat'l is used) are arranged. Two pieces are tight against each other, and gaps are left between the others. The 2 pieces tapped for clamping screws are located 1/8" or so further out from the Cube to leave room for protective slips of sheet brass or CRS. This would tend to put the whole assembly out of balance somewhat, so I have shown these 2 pieces with more lopped off their outer corners. The clamping screws can be 10-32 set screws, located so that they bear on the centerline of the Cube. The 5th plug is at the back, trapped between the Cube and the false faceplate.

Ron said that in his day, the point of the exercise was to make the boys think, and puzzle out the method for themselves. And too, to carry out the job properly, and learn to trust the lathe to do as asked (or compensate for it's idiosyncrasies if it would not). Not wanting query letters from guys for whom the solution proved elusive, I decided to give all the necessary info herein. For us perhaps the point of the exercise is to make it, and thus have something with which to mystify the masses - or possibly to sell to them as curious paperweights.

And don't overlook the possibility of making one in plain or fancy wood, or make one with more internal iterations. Or a whole family of them, from 1-1/4" outside cube right down to say 5/8" ditto.

Fig. 1. <u>Dimensioning the Cube</u>

(1) To find the length "S" of a side of the square that can be inscribed in a circle of diameter "D", multiply D by 0.7071.

(2) The undercut "X" is equal to the difference between the radius of the circumscribing (outer) circle and the radius of the circle that can be inscribed in the square. Or, to put it down as a formula....

$$X = R_c - R_i$$

      where X = the undercut needed to create one full face
of an interior cube with sides of length "S"
and R_c = radius of the circumscribed circle for the square
          R_i = radius of the inscribed circle for the square.

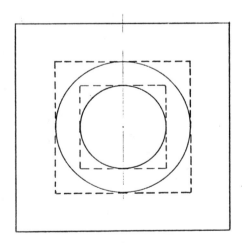

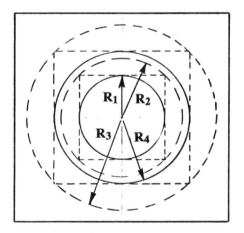

The drwg directly above shows one face of the Turner's Cube. The 2 circles indicate the 2-diameter holes bored in to specified depths to reach the faces of the 2 internal cubes. The dotted lines indicate the edges of the 2 internal cubes, which will be formed when the bottom of each hole is undercut to the extent indicated in the drwg at right above. ⟶

To create the two internal cubes, the initial holes of radius R_1 and R_2 must be undercut by an amount X (found by the formula shown in (2) above), to reach the corners of the two desired internal cubes, as indicated by the dotted circles of radius R_3 and R_4.

Notes to Figure 1, (previous page).
There is no magic ratio for the sizes of the 3 cubes in one of these Turner's Cubes. An outside cube 7/8" on a side is large enough to deal with readily, yet small enough to carry in one's pocket, and small enough to impress people.

Whatever size you pick for the outer cube, you then want the circumscribing circle for the middle cube to be of such a size so that it lies entirely inside the outer cube. That circle dictates the size of the middle cube, and you can do similarly for the inner cube. Draw it out maybe 5x full size, and it's easy to pick off your dimensions.

I figured it out via a drwg, twice, months apart, and the second time came up with a middle cube of 0.552" - i.e. pretty much the same as the 0.550" figure I'd come up with previously. So, even though I haven't made one of these Turner's Cubes, I think the numbers I've given - 7/8", 0.550", and 0.350" - will work out fine.

The diameters on the plugs (see Fig. 3 below) need not be held to tight tolerances, but the lengths do need to be very close to spec, so they actually do clamp the inner cubes against movement while the holes are bored and undercut.

The 0.163" dimension is arrived at thus: 0.875" - 0.550" = 0.325"; then, 0.325 divided by 2 = 0.1625, say 0.163".

The 0.100" dimension comes thus: 0.875" - 0.350" = 0.525. This, divided by 2, is 0.2625, or say 0.263". But we have 0.163" of length on the fat part of the plug, so 0.263 - 0.163 leaves 0.100" needed on the small-diameter part of the plug.

Fig. 2. The Chucking Fixture

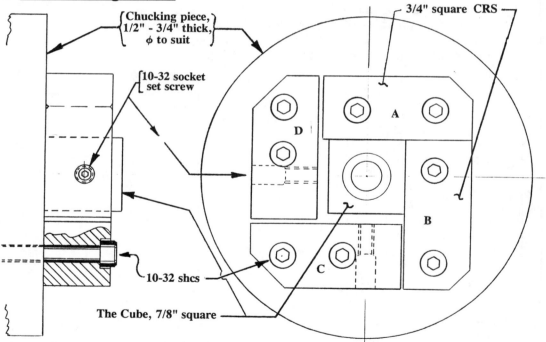

Notes to Figure 2, above.
A chucking fixture to aid the making of the Turner's Cube is shown above. The 4 locating and clamping pieces, or "fixturing pieces", on the chucking piece are made from CRS.

We need to talk about this fixture for a few minutes, so let's call the fixturing pieces "A", "B", "C" and "D", and let's call the chucking piece "CP".

222

Let's assume you're going to make a 7/8" Cube. I'd be inclined to make it from say 7/8 CRS, sawn off a little over length, and then faced back to 0.875"OAL - hence a 7/8"cube in short order.

For a 7/8" outer cube, 3/4" square CRS would be about right for the fixturing pieces A, B, C, & D, and they could be attached to the chucking piece (CP) with 10-32 shcs's.

Note carefully the way the pieces A, B, C & D are shown on the drwg. A and B are tight against each other, while gaps are left between the others. C and D are also located to give 1/8" or less clearance off the Cube. All else being equal, this would put the whole assembly out of balance, so I have shown these 2 pieces with more lopped off their outer corners. The clamping screws can be 10-32 socket set screws, located so that they bear on the centerline of the Cube. The Clamping Plugs (see Fig. 3 below) are not shown in place on Fig. 2, but they are needed as the work progresses, and would be held in place in the holes in the Cube by the 10-32 socket set screws mentioned in the previous sentence. The 5th plug is at the back, trapped between the Cube and the Chucking Piece.

The CP as drawn here is larger in diameter than necessary - anything that will clear the corners of the fixturing pieces would be fine. A piece of HRMS or similar, about 3-3/8"ϕ would be wanted, say 1/2 or 3/4" thick.

Grab this in the 4-jaw, and machine it all over on both sides. Once you've faced the second side flat and cleaned up the OD, and before de-chucking it, poke a center drill into the middle of the front face, and drill/bore/ream a 3/8"ϕ hole. This hole can be used to assist with further work required in making the fixture.

If you plan to make more than one of these Turner's Cubes, (and you may want or need to, once you start showing off the first one), then a well made fixture will help speed the job. If I were only planning to make one Cube, I'd stick the fixturing pieces on "pretty close"; I'd hold the whole business in the 4-jaw, and "indicate" a centerpunch mark or similar on each face on the Cube as I came to machine it. However, if you make a careful (accurate) job of the chucking fixture, you can thereafter hold it in either a 3- or 4-jaw chuck, center it up, and then just pop a Cube blank into it and go to work on it - no need to fuss with getting all 6 holes smack in the center of the faces of the Cube, because they'll get that way automatically off the fixture.

Assuming you want to go the latter route, de-chuck CP, and machine a 2-diameter plug (not shown). One diameter should be a good fit in the reamed central 3/8" hole in CP, and the other should be 7/8"ϕ, to match the size of the Cubes to made in the finished fixture.

Plunk the plug into CP, put "A" snug up against it, clamp down, and drill two #21 holes (tapping size for 10-32) through "A" into CP. Follow with a #8 (0.199"ϕ) or 13/64 drill to full depth of "A" plus 1/32, and c'bore the top of the hole in "A" about 0.325"ϕ x say 0.150" deep. Tap the CP 10-32 at both holes.

Blow the swarf out of the holes, put two 10-32 SHCS's in place, and tighten them down to stay.

Now put one end of "B" snug against "A", and also touching the locating plug, clamp it, and drill, tap, c'drill and c'bore as for "A". "C" and "D" go on similarly, but with a little space between them and the locating plug.

When all this is done, you have a very smart fixture that knows how to do a lot of the work for you.

Figs. 3 and 4 are shown on the next page.

Fig. 3. <u>The Clamping Plugs</u>
Dimensions of the Clamping Plugs are discussed in the "Notes to Figure 1" above.

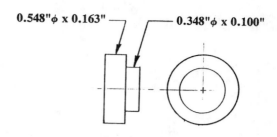

0.548"φ x 0.163"   0.348"φ x 0.100"

Fig. 4 <u>The Boring tool</u>
A boring tool along the lines of the one shown at right would do the necessary undercutting.

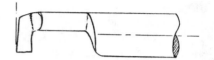

Support Pad Locations for Granite Surface Plates
(see also page 204)

If you should happen to want to put rubber pads (feet) under a granite surface plate (or a plate glass substitute) the drwgs below show how to locate them for best effect, per U.S. Government specs.

For rectangular plates:

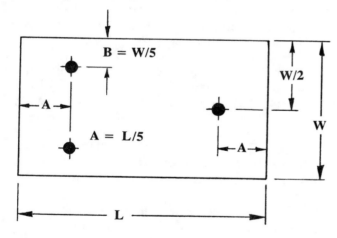

B = W/5

W/2

A = L/5

A

L

W

For a 12" x 18" surface plate,

A = 3.6"

B = 2.4"

For round plates, apply 3 equally spaced pads on a pitch circle of **plate φ x 0.7**. For example, for a 15"φ plate, the pads would be set on a pitch circle of 10.5"φ.

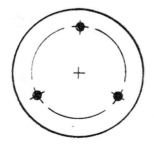

The DLS World's Best Backscratcher
from Lou Vierling, DLS Consulting Inc., East Detroit, MI

"... The goal was to produce a backscratcher with a long, stout, curved handle capable of easily delivering a reasonable force to any area of the back..."

Lou Vierling is a hsm who heads a control and instrumentation firm called DLS Consulting Inc., down in East Detroit, MI. A couple of years ago, Lou, with the help of his wife Wendy, made a limited production run of 36 of the finest backscratchers ever to come out of Detroit. Most went to family, friends and business associates as Christmas presents. Somehow, one got sent to me.

Now you must understand, this is not some silly little pink plastic hand on the end of a piece of wire. This is a *good* backscratcher. It hangs from its own little leather thong on a nail in the rafters in my shop, and the lifetime guarantee is nailed to the wall behind my toolbox. **Everybody** who's tried it likes it, because it lets you get right to the source of any itch you might feel the need to work on.

Salient features are as follows, and/or are shown in the drwg below.

The working edge of the head has 7 round-tipped fingers to do the scratching. Lou said he wanted the scratcher head to feel, in use, ".... like a nicely trimmed set of fingernails." (One does not expect to encounter such romanticism in a consulting electrical engineer.) The back taper of the head is as shown, and nicely beveled, so as not to catch on shirt collars, etc.

The handle is an 18" length of 1/2"ϕ hardwood dowel, but here's where this backscratcher parts company with the rank and file of such implements: Lou's backscratcher has a curved handle, which is the key to it's effectiveness. The handle is bent in a press after being boiled for 30 minutes in water, which treatment is preceded by a 24 hour soak in room temperature water. Lou said the slot for the scratching head should be sawn after the handle is bent, and is back out of the press.

The head and handle are glued together with 5 minute epoxy, after three 1/8"ϕ holes are drilled in the head where they will be covered by the handle. These holes are not deburred.

It takes a hsm to make a *serious* backscratcher. After this, anything else is just friction.

Head – make from 1" x 1" x 0.060" aluminum angle

The "DLS" World's Best Backscratcher

GA Drwg

Handle – make from 18" length of 1/2"ϕ hardwood dowel

1-1/2" offset after removal from the bending press

21-1/4" radius in the bending press

How NOT to Remove an Arbor from a Milling Machine Spindle
from Charles ("Bob") Moore, Walla Walla, WA

The first job I had after being discharged from the Army Air Force after WW2 was with Inland Machine Works in Walla Walla, WA. Inland was a job shop specializing in cannery machinery fabrication, harvesting equipment, etc. I was a product of the old Army Air Corp machinist school at Chanute Field, IL and I had a lot to learn about the real world.

My story is about Ed Mitchell, one of the owners of Inland, and an ex-Englishman who had served his apprenticeship in England before 1900. Ed learned how to cut threads with a hand-held chasing cutter on a lathe with no leadscrew.

One Monday morning I came to work and the first job given me required the use of our ancient horizontal milling machine. This old blister had power feed on the table only and this was provided by two 3-step, flat belt pulleys on the back end of the machine.

I needed an end mill holder in the spindle, and to arrange this, I had first to remove the arbor that was in the machine. I went through the classic procedure: loosen the draw-in bolt a turn and a half and give it several sharp raps with a large ball pein hammer. No go. That old style taper was in good and tight. I found a piece of 1-1/2" pipe in the scrap pile and faced off the ends in the lathe so that I could draw the arbor out of the spindle, or at least put a strain on it so that a rap with the hammer *would* loosen it.

About this time Ed came along and needed the mill. "What the (expletive deleted) are you doing?" he asked.

I explained that the arbor was stuck and I was trying to get it out.

"Didn't that (several expletives deleted) Air Corp teach you anything? I'll show you how to remove a milling machine arbor!"

He checked to make sure the draw-in bolt was loose and really hit it hard with the ball pein. Nothing.

He said, "Needs a bigger hammer!" He got a 6 lb. sledge and really took a swing. Well, he hit the draw-in bolt all right, but it was a glancing blow, and the sledge proceeded to take a half moon chunk out of one of the step cone pulleys that drove the table feed!

Ed dropped the sledge and muttered, "Go ahead with what you were doing," and walked off. I did, and soon had the arbor out.

I learned many things from that old man, and am proud to say that he liked me and we became the best of friends.

Chuckles by Mail

Not Quite the Right Idea

The following letter came in our mail one day from Vic Wright, of Coquitlam, B.C. (Coquitlam is a suburb of Vancouver, as is West Vancouver.)

"Dear Guy,

I recently finished thoroughly cleaning and reassembling an older, used 9" South Bend lathe. I am a voracious reader, and as I now have a larger lathe, and want to learn more about lathe operation and technique, I wanted your book "The Machinist's Bedside Reader". I gave a hastily scribbled note to my wife so she could get me a copy for Christmas.

She asked for the book at the very crowded counter of a bookstore in one of the local

226

shopping malls. The sales lady didn't hear her, so my wife passed her my note. In a very loud voice she exclaimed, "The *Masochist's* Bedside Reader!?" My wife, in an equally loud voice said, "No! The MACHINIST's Bedside Reader." Needless to say, she got some comical looks as she hurried out (without the book).

Where can I buy it locally?"

Vic Wright

I wrote back to Vic as follows: "Only a masochist would try to buy it in a bookstore when he could order it directly from the author...."

-- --

In the lower left corner of an envelope (containing an order for **TMBR#2**) was the following note:

> *"PLEASE RUSH -*
> *Leaving on boring*
> *vacation with wife"*

A SPILLPROOF CUTTING OIL BOTTLE
from **Ted Dillenkofer, Morris Plains, NJ**

Ted Dillenkofer sent me a spillproof cutting oil bottle which he made. It works, too. He bores out the top of the lid of a plastic vitamin tablet bottle to just take a smaller bottle such as prescription pills might come in. (One he sent me had a slightly larger pill bottle just stuffed straight into the open mouth of such a bottle, and the pill bottle lid was then used as the lid for the whole thing.) He cuts the bottom out of the pill bottle, and sets the pill bottle in the modified vitamin tablet bottle lid, and seals it therein with glue, epoxy, Elmer's "Squeeze 'n Caulk", contact cement, model airplane glue, or similar - as he says, if the glue you try doesn't stick, little is lost except some time. The lid, complete with inserted pill bottle, is replaced on the vitamin tablet bottle. A piece of metal tubing could also be substituted for the inside bottle.

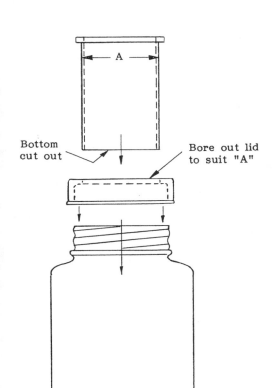

Bottom cut out

Bore out lid to suit "A"

Cutting oil is poured into the open top until the bigger bottle is maybe 1/3 or 1/2 full. Then the oil bottle is turned slightly past horizontal, and the excess oil is poured back into the storage can. Now, with an acid brush or similar dropped in through the top, your spillproof oil bottle is ready to use. If you knock it over in the lathe chip tray, or on the milling machine or drill press table, the contents will not spill.

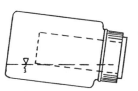

The Lamentations of Lebow
from **Phil Lebow, Los Angeles, CA**

.... I recently got 3 pieces of woodworking machinery *free* just for hauling them away from an art school where I service their darkroom and printing machines. The loot was as follows: two Rockwell/Delta scroll saws, one variable speed, one step cone, each with a 3/4 hp motor; and a DeWalt/Black & Decker 12" contractor's radial arm saw complete with a new 12" 80-tooth carbide blade.

The variable speed jig saw was missing the speed handle, and the upper blade guide bushing (a piece of turned brass) was worn out. The step-cone saw was missing the blade holders and screws, and also needed an upper bushing. The radial arm saw needed a new on/off switch, and a new angle locking piece for the radial arm. The saw motor is brand new and wired 220V single phase. The saw trunnion was missing its three set screws. That's it. All three machines are on factory-made pressed steel stands.

After I had hauled the stuff away, I asked the head of plant operations why he hadn't repaired such minor troubles.

"Minor? I couldn't find parts or anyone to put them in."

"Oh. You could have asked me," I replied.

"Too late for that now - we've already replaced them," he said. "We were going to have to pay someone to haul them to the dump, so you saved us some money!"

(nuthin' like having a happy customer, eh? GBL)

I checked with Rockwell/Delta, and they had all parts in stock, ditto DeWalt/B&D. Parts cost was about $30 all told.

I already had two radial arm saws, so I sold one of them, and kept the new 12-incher. A friend of mine who is a cabinetmaker offered me $200 for the older step-cone drive jigsaw. I agreed. I kept the variable speed jigsaw for myself. Not bad.

On further examination, I found that the step-cone drive scroll saw had a loose crank pin. I spotted this when I noticed that the lower blade clamp not only moved vertically, but also oscillated on its axis. So I tore the saw drive down, and discovered pot metal parts inside the nice cast iron housing. The crank pin was a standard shouldered die pin pressed into a Zamak crank throw. (Zamak is, I believe, a zinc/aluminum die casting alloy.GBL)

Due to a faulty lubricator check valve, the slider on the scotch yoke partly seized and yanked the crank pin loose. Then it wallowed its hole out. After I got everything apart, and cleaned off the oil, I fixed the oiling system check valve, and polished the scotch yoke until it was free. I then bored out the crank pin hole, and put in a 3/8-16 Heli-Coil with Loctite. The new crank pin is a 1/2" stripper bolt with the head cut off, Loctited and screwed into the Heli-Coil. Everything worked out OK, except I lost 30 thou clearance somewhere. The saw is now better than new. Or better than it deserves to be. The rest of the job was just paint and re-assembly.

My next door neighbor and I were comparing notes one day on the most foolish (read stupid) stunts we've ever pulled in the shop. To my dismay, I was way ahead...

Two of my best whizzers were lathe related. One time I had to re-do some bolt recesses in a turbocharger flange adapter. I had to bore through some plugs that had been welded into the mislocated bolt head recesses. The fool who made the plugs (me) had unknowingly picked up some rough finished drill rod, and when he welded them into the holes, they promptly hardened, quenched very nicely by the mass of cold mild steel around them...

Just a side note: in case it hasn't already slapped you in the face, this is a great way to add hardened wear pads to mild steel fixtures, jigs, etc. Just make the pads out of drill rod and weld 'em in with the old heli-arc! Grind to finish and that's it! Trust me, they will quench!

I heated the m.s. adaptor plate red hot, and buried it in the sweepings can from the bead blaster overnight. It was still warm the next morning. Of course, now the plugs were machinable (sort of), but still harder than the flange. And I had to make overlapping holes. The job would have to be bored instead of drilled or the holes would wander off, half in hard, half in soft.

No mills were free, so I had to do the job in a 15" lathe. I installed the 4-jaw and centered my punch mark for the first hole location. The flange was pretty far off center, but I didn't think too much of it. I was boring the new recesses with a two flute end mill (which you would call a slot drill, Guy) held in the toolpost.

I switched the motor on and things looked ok until the lathe revved up. Then it started to rock! It started out as a gentle oscillation front to back, as I bored away. Then it started moving more, and more, and more. Finally the whole machine was tipping through about a 30ø swing. At this point I stepped on the panic brake. It was real clear that the lathe hadn't been bolted to the floor! Worse, it was mounted on those rubber filled cushion leveling pads that Enco and others sell. Kind of like having a lathe on springs.

I moved the flange around so more of the mass was centered, and finished the job. But after that, that was my jinx lathe, and I later got into trouble with it again when I had to make a number of duplicate hose fittings out of 3/4" stainless steel bar stock.

The lathe was in a corner, with the tailstock end and the back of the lathe looking at two adjacent walls. To the left of the lathe headstock was a tall steel cabinet with double doors, where we kept a bunch of tooling of various kinds.

I poked a long piece of ss bar through the spindle, and supported the outboard end on roller stock rests. The lathe was equipped with a combination air powered collet closer/bar feeder, plus a turret on the carriage. When I finished a hose fitting and cut it off, I would index the turret one last time to a tool station that had a stock stop. Then I activated the collet closer and the bar would feed out against the turret stop. Then I'd re-lock the collet and start again - the human screw machine. Pretty boring - you get in the groove, and just drone along.

The chips were coming off pretty steady when I noticed some vibration (!) in the work. I glanced over at the stock supports, and to my surprise noticed that there was a big bow, like an obtuse angle, in the bar between the closest support and the headstock. It was spinning around at spindle speed. The next second, the bar flew off the supports - if I hadn't already noticed it, I certainly would have then! It made an unbelievable racket, because it was flailing the snot out of the tool cabinet next to the lathe! I got the lathe stopped a second or two later. (By this time I had an audience, too.)

There was a trench chipped in the floor, and the near support was bent over not unlike a gentleman who has been kicked where it hurts most. The right hand door of the tool cabinet was stove in with a series of long, deep dents running top to bottom. The ss bar was scrap, being folded and bent into the most fantastic angles. I guess I still looked surprised, because I took a lot of ribbing from the crew.

I turned the collet loose, and we pulled the bar out. Upon examining the bar closely, I discovered that someone had turned a neck into the bar with a parting tool right where the first bend started! When I started asking around about that, my supervisor said, "Oh, I put that there so you'd know when to quit." I said I'd gotten the message.

We took the door off the cabinet and jumped up and down on it until it was straight enough to close, and then beat the creases as flat as we could with a lead hammer and a steel plate. That made the door more or less its old width. I had to rebuild the stock support. The old one was scrap. Then I still had to finish the parts run with another piece of bar!

The next morning I came in and found the stainless pretzel mounted on the wall over my tool chests, with a small, hand-lettered sign that simply said "Sculpture, by Philip Lebow". The tool cabinet had a large Band-Aid painted over the dents in flesh colored paint. It said "Bang-Aid, Courtesy of Lebow". The shop's arch practical joker, John Ryals, had done it to me. It took me a long time to live that one down.

Next time I'll tell you about some of the things my students and colleagues pulled when I taught advanced machine work at Pierce Junior College.

Added Jan 29/92: I had a call from Phil last night, and he told me he'd done much the same thing again, this time with a length of 1/4" drill rod in his South Bend lathe. Plus he'd collected a blue thumbnail as a result of reaching over to try to steady the material just as it went all haywire! There's probably a lesson in here somewhere. GBL

ZEN AND THE ART OF MOTORCYCLE MAINTENANCE

Here and there throughout this book I've mentioned various books that you might like, or find useful to read. Here's another one: "Zen & the Art of Motorcycle Maintenance," by Robert Pirsig. Reading it may change the way you work - and think - forever.

It is not about fixing motorcycles, although that topic is sometimes used for illustrative purposes. Some of what Pirsig says, being philosophical, goes over my head - I was never trained to think about some of what he speaks about. I got more out of it on the 2nd time through, and I expect that I'll get even more next time. And yes, it's worth the time to read more than once.

The aspect of the book that guys like us can relate to* concerns one's work and one's approach to it. A couple of times he uses the example of a good mechanic troubleshooting a motorcycle. The mechanic knows why the motorcycle works the way it does. If it does not, he is thinking (deeply) about why not. If he comes to a point where he feels he is "stuck", he concentrates on why this is so, and all the ways one might get around the problem, etc., drawing upon all his knowledge and past experience. Pirsig rightly says that a lot of problems will yield to this kind of concentrated thinking.

* (Pirsig is himself somewhat of a hsm - he speaks of having a metal lathe and other tools in his basement, and how useful they can be. He must be all right.)

Again using a scene in a motorcycle shop to illustrate the opposite approach to one's work, he describes several young fellas in a messy shop, supposedly "fixing" motorcycles. Their whole demeanor bespeaks complete disinterest in their work. A radio blares out loud music (conducive to clear, careful thought? I don't think so - at least not for me). These boys are the type who, if they were digging a ditch, would let go of the pick handle the instant the quitting whistle blew, even if they'd just raised the pick over their heads for another stroke. Pirsig picked up his motorcycle and ran outta there.

He also speaks of getting satisfaction from one's work - by really putting one's own "self" into doing the best possible job, caring about - being involved in or absorbed in - what one is doing, even if that be a "boring" factory job. Peace of mind comes into it, too: if the work isn't done right, whoever's responsible for it being done right will lack peace of mind until it IS right. If you're making something - a clock, let's say - in your basement shop, and it's not right, you won't be very happy with it. But when work on the clock is coming out right, you will be happy with it, won't you? That's part of the matter of "peace of mind".

Pirsig's book is not all easy reading, but you might like it, or parts of it. And I think that if you read it all, it might in fact have an effect on your whole approach to doing any kind of work - hsm stuff, house fix-ups, whatever. The following story strikes me as an example of a guy modifying a factory job to get personal satisfaction out of it, by putting something of *himself* into it.

QUITTING TIME
from **Phil Lebow, Los Angeles, CA**

I worked for a time in a steel mill and fabricating plant in the late 70's to early 80's. Some of the guys might get a kick out of the following incident.

Back then I was something of a wiseguy (as now). I had been stuck back on the welding line after being retired from the tool and die shop. I needed more to occupy my mind than just burning rod, so I organized the welding department, and started agitating for smoke extractors, better safety equipment, and accurate pay checks. Naturally this caught the attention of my foreman, who quickly determined who the ringleader was. His approach was to isolate me from the rest of the crew, as nothing I had done was legally actionable.

The building housing the steel mill covered about three large city blocks, and the fabricating plant about the same. I was relocated (exiled) to a small, little-used bay in the fabricating plant, about 2 blocks from the next nearest human being. The only equipment was a light overhead crane and a welding machine.

The foreman walked me over and showed me a crude welding jig that held a pair of heavy foundation bolts, about 1-1/4"ϕ x 18" long. They were connected by a piece of plate 1/2" x 4" x 8".

"You make these in the jig," said the foreman with a smile.

I smiled agreeably.

The crane operator brought in a steel bin about 3 feet deep and 6 feet square. It was full of oily foundation bolts, fresh off the bar threader. Another bin full of plates followed.

"Clean off the bolts, and weld them up," the foreman told me. "And when you get through with these, there are plenty more of the same to do."

I smiled a little less agreeably.

A foundation bolt is a bar threaded at one end for a nut; the other end is bent at 90°.

There were a lot of them. I started in with rags, and a pair of plastic gloves, as I didn't want to ruin my welding gloves. I wiped down a few bolts, and set them in the jig, and welded them according to the game plan. I kept an eye on my time for each assembly, and soon figured out that I could weld up maybe 110 or 120 sets a day. Since there were about 9000 to do, I was looking at about 90 days of nothing but welding foundation bolts! Besides, my coveralls were getting soaked in extra-heavy dark thread cutting oil from reaching into the bin. I was NOT going to put up with this. What I needed was a multi-station welding jig.

The foreman came by late in the day to see how I was doing, and to gloat. He seemed satisfied with the 100 parts or so that I had done - from his point of view, the slower I went, the longer I would be away.

The next day I came to work, commandeered a forklift, and went for a tour of the scrap steel warehouse. I came back to my bay with a load of steel plate and bar and some pipe. I spent the rest of the day rather surreptitiously building a giant lazy susan out of steel plate, that pivoted on a big piece of pipe and a bunch of metal caster wheels I had liberated. It turned pretty freely, and the top was about 3 feet high when it was up on its base. The top was made out of two plates of defective steel, butt welded together on their long edges and flame cut into a disk about 8'ϕ. I had room around the edge of the top for 20 welding jigs. I had the whole thing complete by the morning of the 4th day.

The foreman hadn't been back to check on me, but one of my cohorts from the welding department stopped by to see how I was doing. He saw what I was up to and shook his head. I swore him to silence.

When the jig was finished, I torched a couple of holes in the bottoms of the bolt bins, and put out the oil fires I started in the process. The paint and cleaning shop gave me a 50 lb. bag of Oakite. I sprinkled half a bag into each binful of oily bars, and then turned a steam hose on them. The steam and Oakite degreased the bars, and the residue drained out and went into the oil sump in the floor. I was ready to boogie.

I found I could now weld up about 475 bolt assemblies a day, if I really got with the program. I completed the anchors in 24 days including building the jigs - 66 days ahead of schedule! The crane operator came and hauled away the enormous piles of bolt assemblies. I packed up my gear and walked the 2 blocks back to my old work station.

Tom (the foreman) immediately spotted me and came over. "Ready to quit?" he smiled.

"No," I told him. "The job's done. By the way, Tom, where are our new smoke extractors?" This didn't help our relationship at all.

I got chewed out for wasting scrap materials building weld jigs, but they couldn't fault my production! Best of all was the look of utter disbelief on Tom's face.

I had a few other adventures with the foreman after that, one of which resulted in the construction of the great pipe-organ-compressed-air-crane-whistle.... but that's another story.

Phil

So you see, work provides opportunities for deep satisfaction - but sometimes there is nothing so satisfying as telling the boss, "I quit!". **This further story from Phil Lebow** is a good example of that sort of situation.

After a stint of teaching machine shop at a local Junior College, I needed a decent job, as I was trying to buy a house.

I got hired on as machine shop foreman - at what I considered slave wages - at an automotive racing parts and turbocharger manufacturer. The owner proved to be both ungrateful and a liar.

I got settled in, and finally asked the boss what I could do to earn a substantial raise. He told me that if I could increase production 400% (!!) he would give me a raise to be proud of. I said I would try, and went back to work.

As I was the only toolmaker in the shop, I had to design and build new high production tooling as time permitted, while doing my regular work. However, I got the tooling built, and the crew and I even relocated the machines for a more logical parts flow. We got ahead of production, and really made the chips fly. We increased production to the point where the foundry couldn't keep up with us, for a change.

When we passed the 400% increase mark (on a Friday), I took the production sheets into the boss's office and asked for my raise. I knew our section had gone from doing $250,000/yr to over $1,000,000/yr, so I figured I'd earned it.

The boss was very complimentary, but said he couldn't give me more than 25 cents an hour raise - expenses had also gone up dramatically. 25 cents/hour amounts to about $500/year, whereas I had figured he wouldn't miss a few thousand bucks a year, and I could have used it.

I was ticked off, but decided to sit tight for a while.

The following Monday morning when I came to work, I spotted a new Ferrari V-12 coupe parked in the boss's space. I asked him whose car it was, and he said, "It's mine. I bought it on Saturday!"

I figured the SOB was driving my raise, so I told him I was giving him my notice, for that reason.

"You aren't good enough to get another job as good as this one some place else," he said.

I told him he would do well to get a replacement for me to start training, to which he replied that I could train my replacement on my last day.

Inside of 3 weeks I'd found a job in the ordnance department of Hughes Helicopter.

True to his word, on my last day at the race parts shop - at about 3 in the afternoon - the boss brought around the fella he'd hired as my replacement.

I took him out to the shop area, and pointed out the array of jigs I'd made. Some were duds, some were prototypes, some were for different machining operations on the same part on different machine tools. It was with these that we had quadrupled our output.

"What're they all for?"

"They're our jigs. You'll figure them out," I told him. Although none were numbered or identified in any way, I knew what each was for - and I was the only man there who did.

I started my new job on Monday. In a week or so, the boss phoned me at home to ask me to come back.

"No way, Jose," never fell from mortal lips with greater glee.

Phil

---------------•-----

A Note on the Shop-Made Surface Grinder discussed in TMBR#2

Dear Guy,

Your books ... are valuable sources of info for me. They are among the best books on machine work that I have (in my) fairly good library on machine work.

I am using the grinder (shown in **TMBR#2**, page 58 - 64). ...I installed some wire cloth between me and the grinder - safer. The grinder has been very successful. Thanks.

Ed LeBlanc, El Paso, TX

Coincidence Maximus

Most people can tell you of an experience of the type generally called "a coincidence," which the dictionary defines as ".... a remarkable happening together of events, apparently accidental and unconnected".

However, I'm willing to bet that few can top the following....

My friend Stewart Marshall's great grandfather came to Texas from England before the turn of the century, and soon established Marshall Machine Works, a business subsequently operated by 4 generations of the Marshall family.

When Stewart and his wife decided to move up to Lopez Island, WA in 1985, he brought with him from Texas a few of the older machines from the family business. I first met Stewart in early '91. From time to time since then, when he's been up to Vancouver on business or for fun, Stewart has visited me, and we have become good friends.

On one such visit, we got onto the topic of *Model Engineer Magazine*, which Stewart had never heard of before. I explained that I had compiled a categorized, cross-referenced Index to *ME* going back to 1920, and that not only was this so, but that I had it all in my computer, and could look things up very rapidly.

The next day Stewart went to the Vancouver Public Library (VPL), and asked them to bring up from the basement a few of the many bound volumes that comprise the Library's collection of this wonderful magazine. Several volumes soon appeared, among them the one for the first half of 1928, and this was one of the first volumes Stewart opened up.

The next time Stewart came up to visit, a few weeks later, he said he had something to show me that I was just not going to believe.

He laid 2 xeroxed pages from the January 5, 1928 issue of *ME* on the kitchen table, and said to have a good look at them. I did. What the un-named author was saying was this:

"Let us consider how to go about making a small piece of workshop equipment. I will use as my example a pillow block which was required in my own shop. ..." He then went on to detail his choice of material (gunmetal, which is a bronze/brass alloy), the laying out of the job, machining the parts (including a 5/8"ø bore, plus facing the front and back sides of the area around the hole at the same setting using a boring tool specially made for the job), numbering the screws so that each would go back into its own hole, etc. This was a fussy job done not with regard to the time put into it, but simply because "...I feel like doing it this way, and I'm going to, and if the rest of the world thinks I'm nuts, they kin go fiddle." Instead of simply drilling an oil hole at the top of the pillow block and c'sinking it, the guy had turned up a little ring, soldered that on top, and then drilled and c'sunk his oil hole through same. A working drwg was given, as well as two photos of the finished pillow block. This was NOT an item one would just go down to the store and buy - not then, and not today. This was strictly a one-off job.

When he was satisfied that I had grasped the foregoing, Stewart took from his pocket an old brass pillow block, and laid it on the table beside the xeroxed pages from that old ME article.

"I found that in my scrap box. I knew I had it the moment I laid eyes on that picture in the magazine. I went home and spent 4 days turning my shop upside down looking for it. That pillow block come all the way from Bridge City, Texas - I haven't unpacked the box I found it in since we moved up here 6 years ago."

"Yeah? What about it?"

"Don't you see? That pillow block you're holdin' is the very same one as in the photograph! The screws are numbered, the dimensions are the same, and do you see this little scratch? - well, you can see the same scratch on the one in the photo..." And so it was - we went over every single detail of Stewart's pillow block and the one in the photograph, and there was no question about it - it was the same one.

Some chap in England had made a pillow block for a special application in his own back yard garden workshop in England. Some time in 1927 he wrote it up as an example of how to mark out and machine up a small job in the amateur's workshop. An unknown number of years later this chap must have died, his workshop had been broken up (one must assume), and somehow that little pillow block had found it's way a third of the way around the globe, and into the hands of Stewart Marshall's family. Stewart had brought it all the way from the Gulf Coast of Texas to little Lopez Island in the mouth of Puget Sound. He came up to visit me, never having heard of *Model Engineer Magazine* before, and then just happened by chance to open a randomly selected issue to an article written some 65 years earlier. A few days later Stewart held in his hand the very item the other chap had made and described.

Further developments....

In the Fall of '92 (subsequent to the above events), Stewart and I went down to VPL to do some poking around in some old issues of *Engineering Magazine*, which VPL has back to about 1865. In the volume for 1877, Stewart came to a page with a picture of a steam traction engine by "Messrs. Marshall & Sons, Gainsborough, England". Stewart just about flipped!

"That's my family!! My Great Grandfather came from Gainsborough! We only knew that the Marshall family had an iron works there, but not what sort of stuff they made."

Some pages further on, we came upon an engraving of another Marshall machine.

A month or so later, I was given a collection of *Model Engineer* magazines spanning 1912 to 1992. Not every issue, but very nearly so.

Stewart was up to visit again in mid-January '93, and ended up looking through some of the earliest years of these *ME*'s. We were sitting there talking in a disjointed way, while Stewart paged through one issue after another. All of a sudden he said, "Well, I'll be.. Do you remember that pillow block I showed you?" I nodded.

"Well, look at this - I have this cutter, too." He turned the February 6, 1913 issue so I could see it.

"Oh, come on!"

"I'm serious."

The writer said that the cutter used was normally mounted on the spindle nose of a special machine built for a specific purpose, but that he had transferred it to the straight shank shown in the photo in order to use it in his lathe to flute a shop-made 3/4"ø tap with a thread coarser than Whitworth. The cutter was said to be 1-5/16"ø, ten teeth, well made, carefully hardened and ground.

We discussed the matter some, and Stewart said he owned that very same cutter - not just one somewhat like it.

Well, pretty soon he gets to the August 21, 1913 issue, and he says here's something else he has: a set of 4 split collets, an adapter 1-15/32"ø x 2-7/16" OAL, and having a female thread to go on the nose of a lathe screwed 3/4" Whitworth. The collets were about 27/32" max. OD x 1.594" long. There was a drawn steel tube drawbar, and a handwheel with milled slots 3/32" wide x 3/32" deep around the periphery, these being spaced 1/4" apart.

Same thing - Stewart had the very item in the photo. After a while he came across a couple of other things similarly illustrated, and similarly now in his possession.

Well now, about here you may be ready to chuck this whole book aside and declare that Stewart is a windbag, and Lautard is as gullible as they come, to believe any of this... well, if I hadn't seen and examined the pillow block, I would agree. But I'd seen the pillow block, and it *was* the very same one as the one in the photograph!

A Possible Explanation....
The article referred to initially above carried no author's by-line. However, the Editor (and publisher) of *ME* at the time was Percival Marshall.

One part of the *ME* operations in those days was the *ME* Workshop, in the basement of the Offices of *ME*. Items made therein were described from time to time in a section of *ME* called "Workshop Topics". A note in small italics under the main heading reads as follows:

"The principal items appearing under this heading are based upon work actually done by the students or staff of *THE MODEL ENGINEER* Workshop. The Workshop is open daily from 10 A.M. to 6 P.M., and on certain evenings for instructional and testing purposes, and readers who are interested are invited to call at 66, Farringdon Street, E.C. Particulars of subjects of instruction and fees on application."

These items - the pillow block, etc. - were therefore most likely made in (or used in) the *ME* Workshop, and although the "Workshop Topics" section carried no author's byline, Stewart pointed out to me that in those days Percival Marshall's brother Albert ran the *ME* Workshop.

I asked Stewart if his great grandfather had ever gone back to England after he emigrated to Texas. Stewart said he had, as had his grandfather, later.

My surmise would be this:

> Percival and Albert were nephews of Stewart's great grandfather; if not, they were probably similarly closely related.

> Sometime in the late 1920's or in the 30's, Grandpa Marshall went to England, and visited the family.

> Albert was by then dead (I don't know when he died, but a close study of *ME* might reveal the date).

> Albert's daughter (I surmise) said to visiting Grandpa Marshall from Texas that here were these various oddments from Albert's shop (i.e. his home workshop) or perhaps from the *ME* Workshop, and would Grandpa Marshall like to take them home? Nobody here in England was going to get any use out of them.

So (my surmise continues), off the stuff went to Texas. In due course, Grandpa Marshall begat Papa Marshall, who in turn begat Stewart, and the rest is as we see it above.

IDEA: Keep a little tray of copper pennies right near your lathe, to use when you want "soft" pads to put between the jaws of your lathe chuck and a job. Of course you ain't supposed to deface coins of the realm, so American readers should use Canadian pennies for this purpose, and vice versa north of the 49th. (Actually, that is not strictly true. Apparently, in the US at least, you are not supposed to deface ANY coins, US or otherwise. Someone brought a Canadian coin - one minted about 1939, as I recall - to one of my guys in the US who has an engraving business. The customer wanted his own name engraved on the bow of the canoe depicted on one side of the coin. Before he did the job, my guy called the Secret Service, and asked about it. He was told that to do so would be illegal. Did he do it? Only his hairdresser knows.

Some Rickshaw Repair Tips

I've been down to see a rickshaw repair technician in Sedro Woolley, WA a couple of times. I ain't makin' it up - that's what it says on **Everett Arnes'** business card. Everett's on the sunny side of 75, and will probably need another 75 years to tackle all the projects he has stashed away in boxes in his shop.

After that 16,000 mile motorcycle trip I mention elsewhere herein, Everett made a voyage to Antarctica under Admiral Byrd in 1940/41. After that, he worked for Boeing as an inspector on B17's and B29's (5000 and 2000 of them respectively) during WW2, and then for Puget Power Corporation as a substation inspector until his retirement.

Everett has shown me a number of ideas, with a view to passing them on to the rest of the guys.

For example, drawer handles - a lot of the drawer handles in Everett's shop are made from Aqua Velva bottle caps. (What girl could resist an Aqua Velva man who was also a rickshaw repair technician?)

If you want a really good heavy duty scraper for the sorts of things you might use a triangular scraper for, Everett says to get a ball bearing race (2 to 3"ϕ would be about right), anneal it, cut it through, and straighten it out from round to straight or not quite straight. Grind to taste, and stone to razor sharpness after rehardening. They use good steel in ball bearing races.

Some handy small tools - punches, etc. can be made from allen keys, which are good, tough steel. An injector rod out of a diesel engine fuel injector pump makes an excellent centerpunch, says Everett.

Everett points out that an automotive engine valve makes a good height gage - just add a scriber arm in an adjustable clamp, and machine the face of the mushroom head nice and flat. (You may need a carbide bit for that job - I'm not sure, as I've never tried it, but I do know valve *stems* can be very tough stuff. Some engine valves also contain sodium or something - I think they put it in there to aid heat dissipation. If you have one that is so made, you don't want to saw into the stem. About 30 years ago I got a scrapped valve out of a big diesel engine, and cut the head off. Luckily for me there was no sodium in the stem (at least not where I cut it - ??), but I can tell you one thing: it just plain ate two entire hand hacksaw blades for lunch, cutting it in half. I still have the head end, and I use it from time to time to push a job in flush with the front face of the jaws of my 3-jaw chuck.

A blower for getting drill and tap cuttings out of blind holes is easily bought in the form of an ear cleaning syringe. This same device is good for putting water into batteries (car batteries, etc.), and even your ears if the need arises.

Reduce a used toothbrush to a single row of bristles if you want a narrow brush for cleaning in tight quarters.

"Mouse Milk" is a product seemingly as hard to come by as its name would suggest - it's only sold in the aircraft industry. You'd find it at an aircraft supply place at your local airport. Apparently it's an excellent penetrating and lubricating oil --- "... expensive - sort of a high-class WD-40," says Everett.

Water of Ayr stone (mentioned at page 118, in TMBR#2) is available from clock repair supply houses, and will put an incredible edge on fine tools - scalpels, etc.

Everett's got some other guys down in his area he wants me to meet. One is a master telescope maker. Could be some good ideas there for **TMBR#4**....

PIERCED WORK

There is a whole field of metalwork that is largely overlooked by the average machinist, and that is what is known as "pierced work," wherein a design is produced in sheet brass or other material by sawing and filing. For the most part, the purpose is decorative. You may therefore think it lies outside the purview of the hsm, or you may say: "Aw, Nuts!! Lautard is goin' all arty on us! And he used to be such a nice guy, giving us details of tools we could make, and other good solid stuff".

Well, arty or not, I think there are some aspects of this type of work that may be worth considering. Pierced work perhaps comes closest to the province of the machinist in clockmaking: most skeleton clocks exhibit some or much pierced work, for both decorative effect and to reduce the weight of some of the moving parts.

Another application is in the making of decorative fittings for firearms - for example, a patch box for a blackpowder rifle.

A third area for the exercise of such effort would be in the making of an item such as a decorative "owner's plate", or an escutcheon (a term meaning a shield; also a fancy washer) for the lock or handles of a storage box you've made for some valuable workshop accessory. Some will of course scoff at the idea of artistic effort applied to such items: "Oh no! Just plain, utilitarian stuff for me," they grunt.

Go ahead and grunt. None of the tool storage boxes I've made up to now have anything very decorative about them either, but more and more I find myself wanting to do something along these lines, or others. I mentioned this to my friend Bob Haralson one day. Bob said, "It is simply the desire to express oneself."

It had never occurred to me that this was the case, but on reflection, I realized that he was not only correct in the specific instance we'd been discussing, but also that he'd very neatly put his finger on a little piece of my soul. (Ohmygosh --- maybe I *am* getting arty!)

I made a stylized bird to adorn the lid of my main toolbox, filing it from a piece of 1/16" thick bronze sheet material, and sticking it with double sided carpet tape to a 7-1/2"ϕ galvanized sheetmetal disc painted bright red. The bird is similar to (but not identical to) the logo of the well known U.S. firearms maker, Sturm, Ruger & Co. Inc.

I scarred the sheet while chain drilling part way around the bird. How to fix it? At the suggestion of Paul Scobel, I clamped the job upside down with the scar positioned over a shallow groove filed in a piece of masonite, and applied a nail set (punch) to the back side, over the scar. I then filed the resultant bump - and the scar - off the good side, hammered the remains of the bump out, and polished up as before with 400 grit wet/dry paper and Brasso. You can now barely see the flaw, and I could reduce it more if I cared to. I was surprised at how well it worked.

One could instead file the bird out the other way around - i.e. an "air" bird in a brass disc, the latter then being used to retain a mirror in the lid of the tool box.

This job (which I refer to as "the Watch Chicken") and the "M&J" napkin holder shown in

"Hey Tim..." show what one can do if one wants something like this. It isn't hard - it just takes time and patience; with that alone, a nice result can be achieved.

Aside from clocks, and tool box decoration, where else might one have a use for pierced work? Well, you could apply it to various projects for the home, or as gifts. I saw a very handsome lamp in a lawyer's office one day: walnut base, walnut body, and a shade. But what set it off was a simple decorative piece of sheet brass, pierced and polished and stuck on the lamp body. The brass was, I think, a character from the Japanese alphabet, although what it said I don't know. Workmanship? It looked fine from 3 or 4 feet away, where I sat, and if it was not, any hsm who wanted to could do far better.

Idea: To mount the plate, with or without "inletting" it flush with the surface of the lamp body, use epoxy, or double sided carpet tape. Or, solder 3 or 4 sharpened brass pins on the back side, then use the whole piece like a big transfer punch to mark the hole locations for all the pins at once. Drill holes in the wood to take the pins, and then push/glue the pierced and polished brass plate into place.

How about a wall hanging? Or a backing plate for a door knocker for your front door? Or a family monogram for the front door or elsewhere? Or a pierced trivet for hot serving dishes - and a nice conversation piece at the dinner table.

The photo at left shows a folding bookend I spotted in an antique store one day. Another idea: make a serving tray with handles and a seamed and soldered rim, the tray and/or rim having a nice pierced-work design, and made to just nicely fit a glass serving dish your wife has and particularly likes. The result: a much appreciated and unique family heirloom. Or one could make a lantern, complete with colored glass lenses, and electric or open flame (candle or wick/oil) light source.

Three final ideas: Why not make a car badge that says "HSM"? Why not? You'll get as much from that as from having "V8" or "455 cubic inches" or "fuel injection", or "turbo" or "AAA" signs stuck to your car, and maybe more. How? Because you might thus meet another like-minded practitioner, when otherwise neither of you'd have known of your common interest. You like the idea? Pierce out the letters "HSM" in say 3/32" brass sheet. Fasten to a piece of red painted steel or red plastic sheet, as a background, and attach to your car grille. (Don't make it too big, or the authorities will think you're a commercial vehicle.)

How about making a buckle? Pierce out whatever you like, silver solder a belt loop on the back, and a bent stud to engage the loose end of the belt, and there you have it. I had a letter from the Ruger people some while back, and was struck with the idea that their logo, which was printed on the envelope at about the right size for a belt buckle, would make a nice looking belt buckle. I mentioned this idea in a letter to Tim Smith, and he said that several guys he knew had already made such buckles.

Or you could pierce out a brooch depicting something nice - a thistle or a shamrock for your Gaelic/Irish ancestry, a treble clef sign for someone into music, or whatever you like. Saw it out, file it up to the layout lines, polish it, solder on a brooch pin, and then either leave it as is, or have it silver plated.... another family heirloom.

Pierced work involves the careful sawing and filing out of a design to a layout done on, or transferred to your material - it doesn't have to be sheet brass, but brass is easy to work, and polishes up nice. A paper work-up of your design can be stuck right to the sheet metal with Spray-Mount, a spray-on 3M adhesive which contains no water, and thus will not distort paper patterns. Buy it at an artists' supply house. After you have stuck your pattern to the material, you may find it a good idea to trace along the lines with an electric engraving tool or similar, thus making a permanent mark in the mat'l. I've had a paper pattern come adrift while I was working on the job, and mess me up to some degree or other, whence came the idea to go around my pattern with the vibrotool. These vibrotools cost maybe $15 (Brownells sells 'em), and are often used to engrave anti-theft markings on personal property.

A V-notched board is the standard sawing out aid, and is easy to make from plywood or other material; clamp it to the work bench or grab it in your bench vise. Set it to a comfortable working height or you won't last very long!

The type of saw frame to use is known as a jeweler's or piercing saw frame. They are light, and would be a good place to start. You may in due course want one with a deeper throat, if you get into doing much pierced work.

I bought a piercing saw frame for about $5, new, from a big clock making supply house in the US, and I was not impressed (understatement). I got another one locally, for maybe $20, and it seems ok. When Bill Smith (author of several books on clockmaking) visited me one day in the summer of 1991, he brought with him a gray plastic bag, from which he soon began to pull various goodies, starting with a piercing saw frame, which he said he'd picked up at a clockmakers' flea market for $3. It's a Dixon, which Bill said is the best make/type he has ever come across. Bill suggested one minor modification, as detailed at right.

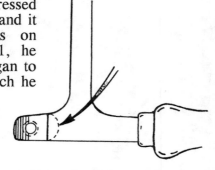

Improving a Dixon Piercing Saw Frame

Next thing out of the bag was a gross of blades for it. (I'd have been tickled to death with a dozen, but Bill don't do things by halves!)

Another type which I have seen and handled (but not used) has a light cast aluminum frame. It is quite rigid, and feels nicely balanced in the hand. The blade is held in knife edged clamps to minimize breakage. This type of saw frame is sold by some of the suppliers of posh woodworkers' tools. (It may be a little heavy for some piercing work, but you should know of its existence.)

Back a 3/8" End mill into the area behind the clamping screw hole, to create a radiused back (as dotted line) to the stepped seat for the blade clamp, rather than the straight up and down back produced by the manufacturer. Then make a new clamp piece with a radius at its heel end, to fit back into this radiused cut. The first time you put a saw blade into the un-altered blade clamp you'll see why this modification is desirable.

If you want to make a device to aid the operation of a hand powered piercing saw, see the book "Fine Woodworking Techniques, Book 7". Therein, at page 140, Ed Kempe shows an excellent idea in an article called "Improving the Fretsaw". Easily built from plywood, his device clamps to a table or bench, and gives the saw - without effort or skill development on the part of the user - a straight up and down motion which should result in better work, and lessened blade breakage.

There is another excellent idea in the book "FWW on Marquetry and Veneer"; at page 81, Ken Parker shows how he turned a piercing saw frame into a foot-operated jig saw for cutting mother-of-pearl inlays. This article first appeared in *FWW* Magazine for March 1981, pages 53-55. Any machinist could easily make a similar or better rig, and if he did, I think he'd have a very useful tool for doing pierced work. It might even be possible to put an electric motor drive to it, which would be nice.

Piercing saw blades are fine toothed, and break easily if not adequately tensioned, or if roughly used. They are installed to cut on the down stroke. A little candle wax applied to the blade from time to time as you work is a great help. Don't hurry the saw - just provide the motive power with steady, unhurried strokes, and let the saw do the work. Saw as close to your layout lines as you dare. (The old adage was: "Split the line on the waste side." It takes lotsa practice.) Afterward, a variety of needle files can be used to finish right up to the layout line.

Brownells sells both jewelers' saw blades and needle files, as well as much other stuff useful to anyone interested in pierced work. Their "Habilus" files are very nice - not cheap, but well worth the price, yea even unto having a set of both coarse and fine.

Close work of this type requires a very good view of the work, preferably under some degree of magnification. Lots of light is mandatory. Bill Smith put me onto something I'd seen, but would otherwise never have bought, and that is one of those flip up/down visors with the magnifying lenses in the hood. They are a real help in the shop many a time, and are particularly appreciated in the demanding work of trying to file or saw exactly to a layout line. I know one tool and die maker who - so far as I know - wears an Opti-Visor as a matter of habit, virtually all day, every working day, over his regular glasses. I think there may be more than one maker of these visors, but Bill recommends the **Opti-Visor** brand, which Brownells sells. Bill, in his book "Clockmaking and Modelmaking Tools & Techniques", recommends the 2.5x, 8" focal length #5 lens plate. I would say that an Opti-visor should be regarded not only as well worth having, but simply as a necessary item in the shop if close work is done. (Margie has recently taken to using mine when sewing, particularly when threading needles. The first time she tried them she exclaimed, "Now *that's* just like the old days!")

Scroll Saws
A powered scroll saw can also be used for pierced work. Robert S. Hedin showed details of how to make what looks to me like a good one in the July/August '86 issue of HSM. Hedin later showed another design in the August '88 issue of Projects in Metal. These two articles were subsequently reprinted in "PROJECTS *THREE*," and "METALWORKING Book One," respectively. Both these books are available from the people who put out HSM. Most of us could be happy with a shop-built example of either of these machines.

There are factory-made saws of this type available, at various prices from small to large. If you want to buy a really good scroll saw, take a look at the Hegner line. I lucked onto one of these saws at a very good price one time, and it is truly in the Rolls Royce category - quiet, smooth running, and extremely well built. I bought it with an eye to my interest in clock making.

About the only thing I don't like about the Hegner is the color - bright orange - just glommed all over everything including the stand. After a few days of trying to get used to the appearance of this shocking color in the otherwise conservative decor of my shop, I decided I could not, so I whipped the saw off the stand, and gave the latter a good dose of what it shoulda had in the first place: light grey. The end result was eminently satisfactory - with only the saw now orange, it.... well, actually, it kinda cheered the place up. (Eventually the Hegner got shifted to my "machine bench" (= second workbench), and the stand went out to the garage. A little splash of color on the bench don't do no harm, I always say.)

My Hegner runs faster than I would sometimes like, but there are others in the line that can be had with speed controlled motors, which allow one to slow the blade literally to a crawl. BB has one of this type, and I envy him sometimes, when mine is buzzing away like a machine gun. I believe also that most of the Hegner saws can now be had with variable speed motors... H'mmm...

When I first used my Hegner, I had a lot of blade breakage. One snapped the moment I turned the saw on. And when the blade breaks, it scares me - Chuck Yeager I ain't. Eventually I began to think I might be doing something wrong, so I called Hanns Durke, of Advanced Machinery Imports Ltd., Hegner's US distributor.

"How much tension are you putting on the blade?" Hanns asked.

"Oh, don't worry about that - they're good and tight. I spin the tensioning bolt down until the blade snugs up, and then give it 2 or 3 more full turns. The blade sings like a violin string."

"Well, that's where your trouble is. All you need is a quarter or half a turn past the point of initial tightening."

Things began to improve considerably after that.

The Hegner saws come from Germany. The top model is guaranteed for 5 years under full time commercial use, and if that means scrolling about in 2" thick oak 8 hours a day, it's still guaranteed for 5 years!

I've seen Taiwanese scroll saws that go for far less money, but on close examination, you can see why they cost less - they're not in the same league as the Hegner. If you like quality equipment, and/or want/need a jig saw that will last long enough to hand down to your grandchildren, the Hegner is likely the wiser way to go. The Delta people make equipment of the same type - in fact they offer both a low cost one and another that's priced up there near the Hegner. I don't know them from personal use, so I can't make any useful comparative comment on them, but as a general observation, Delta stuff is usually good. Skil makes a nice scroll saw, too, priced not much above the off-shore stuff. I talked to a lady who owns one, and uses it to make children's pull toys (so far about 300, she said), using lumber salvaged from construction sites. And she does this in an apartment, so the saw needs to be - and is - quiet. If I didn't have the Hegner, I would probably buy the Skil unit.

Another useful gadget in the shop of anyone interested in pierced work, as well as lots of other things, is a die filer. How about we have a look at that topic a little later on?

Getting back to the matter of what to make in the way of pierced work, a word of caution: good artistic design sense is not in all of us, me least of all. If you are going to create something for "public consumption," be sure it has that impossible-to-define quality known as "good design".

Look in books of art for inspiration. See also books of lettering and calligraphy, ditto A.A. Turbayne's book, "Monograms and Ciphers" (from Dover Publications). See also books on antique silver and brassware for ideas (see references below). If you have a Chinese friend, you might get him to draw the Chinese character for your family name, or "much swarf" or some other character/sentiment you like, and make something from same.

Other books to see for ideas along this line:

"Signs and Symbols of Japan" (Kodansha International Ltd.,, 1975), by M. Maeda and K. Matsumoto.

"Japanese Design Motifs - 4260 illustrations of Japanese crests" translated by Fumie Adachi.

"The Brass Book," (Schiffer Publishing ltd, 1978), by P., N., and H. Schiffer.

"Horse Brasses," (Country Life Books, 1981), by P.C.D. Brears.

"Victorian Alphabets, Monograms and Names for Needleworkers," (Dover, 1974), R. Weiss, Editor.

And similar stuff.....

"....This machine was a real butt-dragger"

Have you ever looked longingly at the Gravermeister (an air powered engraving machine) in Brownells catalog, but been bowled over by the price?! Then pay attention to this... (adapted from a letter to Tim Smith)

I bought a used Gravermeister! I figure it'll be good for engraving clock dials, among other things. I phoned the makers (Glendo Corporation, P.O. Box 1153, Emporia, Kansas 66801 [the same people who make the Glendo Accu-Finisher]) to ask if they might know of someone who had a used one for sale. Discussing the idea with Glendo's sales manager Rick Lee, I learned that heavier castings were used on the older machines, hence they tend to be a little quieter. Also, there's basically nothing to wear out on the Gravermeister that can't be replaced very readily by anyone with moderate mechanical abilities, and the parts (gaskets & vanes) are readily available from Glendo for about $45 at time of writing. There is a rotary valve on the machine that costs about $80, but if you find it to be worn, it's no sweat for guys like us to make a replacement in about 20 minutes (the Glendo people will send you a drwg of the part).

Rick said he would put up a note that I was looking for a used Gravermeister, and we signed off. Five minutes later the phone rings - it's Rick again - to put me onto one in Oklahoma.

I called the guy (a silversmith), who told me he'd bought it about 1985, and that he had used it 40 hours a week virtually ever since. He said it needed an overhaul, and we agreed on a price that seemed to reflect that fact.

I sent off a check for the machine, and another to Glendo for an overhaul kit and a factory owner's manual. A few weeks later the Gravermeister arrived, and I gotta tell ya, guys, if Rick hadn't already told me they can be rebuilt, I woulda cried when I saw it. It looked like it'd spent a year in the sewers of Calcutta! Part of the sheet metal enclosure at one end was missing, and what was left was bent like somebody'd stomped on it. Inside, there's a pulley whose two halves are spring loaded, so it can vary in diameter to change the stroke speed. This one's pulley would vary no more - it had a 4' piece of baling wire (or maybe what was left of the spring) wrapped into the middle of it, and the little red drive belt was about warted down to nuthin'. The handpiece was shrouded (for burial?) in electrician's tape. This machine was a real butt-dragger.

Not to worry - I had a plan. I buzzed Rick, and told him that my machine looked like it was gonna need more than a vane kit. Would they overhaul it, if I showed them how to overhaul the Gravermeister Owner's Manual, which I felt didn't do a good job for them or their customers. (I'm not as dumb as I look, ya know.) Rick liked this idea.

Pretty soon back came my Gravermeister, refurbished to good as new, except for the paint job, and a badly bent foot, which was the way it'd been when it first came to me. Glendo's service technician said it was *the worst abused machine he'd seen* in the dozen or so years he's worked there.

I poked a big screwdriver into the slotted hole in the foot, and bent it back into shape like Yahoudy Menowin bowin' his fiddle, bolted it to a little plywood base, and there it was, all ready to go. Total cost was maybe 1/3rd the price of a new machine. One of these days I think I'll paint it up like a circus wagon - dark green, red and gold - just to drive Bill Fenton nuts.

A final note: Rick said that if buying a used Gravermeister, you should ask the seller to pack it carefully, so that the oil bottle won't get busted off in transit. The machine is vulnerable in this regard, so it should be packed accordingly. If it does get busted off, it's simple to fix, (at low cost), so not to worry unduly.

More Info on Making
ETCHED BRASS NAMEPLATES
from **Ron Colonna**, McKeesport, PA

If you want to make a name plate, a small sign, or something similar, you can do it very nicely by etching your design into 30 to 60 thou thick sheet brass.

The first step is to cut and file the brass to the desired overall shape - a circle, a diamond, or whatever. Polish the plate with steel wool, and scribe borders, if any, with straight-edge, compass, french curves, etc.

For the lettering, use dry transfer letters (Letraset or similar), which comes in a great many styles and sizes. Radio Shack sells ready-made words commonly used in electronics applications.

Position each letter in turn along a penciled-in guideline, and rub on with a pencil or ballpoint pen. Scrape off mistakes with a fingernail. Don't rub too hard, or the lettering may crack, but rub over the entire letter, to make sure it all transfers without tearing. Then lay over the lettering the paper it came on, and burnish some more to ensure good adhesion.

Next, cover the borders and entire back of the job with a resist lacquer or enamel, to prevent etching of these areas. You can use fingernail polish, but Radio Shack sells special pens which give nicer results for sharp lines on borders etc. Cover the edges of the job too - don't miss any spot you don't want to etch.

Immerse the brass plate in ferric chloride etchant solution, (Radio Shack Part #276-1535) for about 1 hour. Any plastic tray slightly bigger than the job will do fine as a container; you need about 1/2" depth of fluid over the job.

After etching for an hour, remove your job and rinse it with cold water. Remove the resist with lacquer thinner, then rub over the entire plate with fine steel wool.

The result should be a good plate with raised letters & borders, and with the surrounding areas etched about 3 to 5 thou deep. Extending the etching time will cause the letters to undercut, which is generally not desirable.

Blacken the job by applying chemical blackening solution (e.g. Birchwood Casey's Brass Black) with a Q-tip. Try to get a nice even tone all over the background. Or get some sealing wax in the color of your choice, and apply it liberally over the whole plate, scrape off the excess with the edge of a credit card, polish it down off the lettering, and then heat the plate just enough to briefly melt the wax, which will then take on a nice glossy look.

Polish the letters & border bright again by rubbing the job on a dead smooth file. Don't file off the letters.

(I would be more inclined to clean the file thoroughly, and then stretch a piece of very fine emery paper over it. A file might pick up a piece of metal from the job and scratch it irretrievably. GBL)

Spray the finished plate with clear lacquer to prevent tarnishing, and then screw or glue it onto whatever you want to put it on.

(More ideas follow on the next three pages.)

ETCHING ON METAL

Based, with permission, on a brief item by the same title
by "Vulcan" in M.E. Jan 1, 1948, page 8

It is possible to mark steel items such as tools by coating the item in question with a thin film of wax, and then, after scratching the desired marking through the wax, etching the surface of the metal with acid. This process is very satisfactory when done properly, but until some experience is gained, difficulties sometimes arise.

The chief source of trouble is the wax film, which may not cut at all well under the scriber, and often shows a tendency to flake off. Another pitfall is that if the wax film is thin, one may find it difficult to see what has been written with the scriber. On the other hand, if the wax is thick, it is difficult to get the acid right down to the bottom of the markings scratched into the wax, so that the acid can attack the metal.

All these difficulties can be overcome easily by using **artists' etching ground**, and applying it in the correct manner.

"Etching ground" (also known, in general terms, as a "resist") generally consists of a hard, dark-brown, wax-like substance, readily bought at any artists' supply shop for a couple of dollars (1993 prices) for a ball somewhat smaller than an egg; a ball will last a long time.

Although it is almost certainly cheaper to buy one ball of commercially made etching ground than to make up a larger wad of it from scratch, for those who might wish to make their own, the following formula should give good results:

    White wax, 2 parts,
    Gum mastic, 1 part,
    Asphaltum, 1 part.

Melt the wax first, and dissolve the mastic in it, then add the asphaltum. When all three ingredients are completely melted & mixed, stir well and then pour the mixture into a bowl of cold water. Immediately recover the plastic mass from the water, and while it is still soft, knead it into balls about the size of a walnut. In melting the ingredients take care not to let them catch fire, as all are highly flammable. If white wax (paraffin wax) is not available, use ordinary yellow beeswax.

The next thing to do is to make yourself a dauber. This consists of a piece of silk or fine cotton cloth wrapped around a tuft of cotton wool to form a soft, smooth pad.

When you want to etch some hieroglyphics into something, heat the item to be etched until it is just a little too hot to hold in the hand, and then rub the ball of etching ground on it. Don't put it on too heavy, but make sure that the whole surface of the metal is covered.

While the metal is still hot and the ground molten, daub all over the surface with your little dauber. This will remove much of the ground, and will spread what is left evenly over the metal surface. Although the metal may appear to be inadequately covered by the ground, this is just what is wanted at this stage; every mark on the metal should be visible through the ground.

When the metal is cool, the coated area should be smoked by holding it over a smoky flame. A wax taper (a twisted piece of wax paper) will do for small jobs, but a candle flame is better. The ground will be re-melted, and will absorb soot from the flame, resulting in a hard, smooth black surface with an eggshell finish in which the writing or design to be etched will show up clearly.

The next step is to scratch the design on the blackened ground and then to etch it into the steel by covering over the design with acid.

There are various etching solutions, but for most work, dilute nitric acid will be fine. Nitric acid will attack most metals, as well as human skin and clothing, so treat it with great care. Dilute the acid as supplied by the chemist by carefully pouring it into an equal quantity of water. Apply a little to your workpiece, and let the etching go on for 10 to 20 minutes.

> NOTE: I would suggest you do all your work with the acid outdoors. Nitric acid fumes are highly corrosive to steel, and if you use the acid in your shop, you will probably later find rust on anything in the vicinity. Done outside, no harm can come to your shop goodies. GBL

Try the above procedure on a scrap piece of metal first. It is also a good idea, whenever a "real" job is to be done, to coat a piece of scrap metal also, and to etch it at the same time as the job, so that the progress of the etching on the scrap metal can be checked periodically by scraping away some of the ground. When the scrap material seems to be etched deep enough, the acid should be thoroughly washed off the real job, and then the ground removed with a solvent such as benzene.

> NOTE: Benzene is not good for you, and can be absorbed through the skin. I suspect acetone (which is more explosive than gasoline) would also remove the ground, as would heat & wiping. GBL.

LEBOW'S LUSCIOUS CAR BADGES
from **Phil Lebow, Los Angeles, CA**

I was digging in ME for something on making knurling wheels when I happened across the foregoing item by "Vulcan". A day or two later I had a phone call from Phil Lebow, and we got onto that topic. Phil told me about an interesting avenue of endeavor that he'd gotten involved in. Now I will tell you right up front what he said at the end:- that if he'd realized in the beginning how much trouble it was going to be, he wouldn't have done it. Nevertheless, I think it opens up possibilities you might not otherwise think of.

Phil wanted to make some automobile nameplates (for a Triumph), so make them he did, and he said from 3 feet away you couldn't tell them from factory originals!

He cut and formed 1/16" thick copper to shape over a wooden die, then cut the desired design through a beeswax or similar "resist", and etched the background away with nitric acid. Next he got some colored glass shards from a stained glass window maker, and made himself a kind of machinist's mortar and pestle by boring a hole in a piece of mild steel, and then turning a round nosed plug to go into the hole. In goes pieces of glass of the correct color, and thump thump thump with the plug, and the glass gets all busted up.

Phil says you want to wash the fine glass dust away, and what you want to keep is the fine grains of glass - he said "100 mesh stuff"; well, that won't mean nuthin' to most of us, so let's say about like table salt, or a little finer.

Phil mixed the desired fraction of his pulverized glass with gum arabic, (a liquid), thus producing a sort of cream, and spread that around where he wanted it to be, and then put the job on a hot plate to fire it, the object being to melt and fuse the glass. It worked, but the glass had some bumps and bubbles in it, so Phil smoothed it down with fine wet-or-dry emery paper and regular Carborundum stones. (As a general observation, emery paper will readily cut glass, and in such a case as this, the first thing it would cut down would be the bumps.) He then reheated it. This time the glass came out as smooth and shiny as you could want, all the bumps and bubbles were gone, and all that remained was to have it chromed.

Ain't that neat?

My friend Paul Scobel told me he has etched printed circuits using photo resists etc... This gave me the following idea as a means of numbering my little 3-3/4"φ Rotary Table (see photo, page 193, **TMBR#1**):

If I were to have someone in the typesetting business lay out the numbers 0, 10, 20, 30... to 350, with each number centered on a point exactly so far from the previous one, I would probably end up with a strip of black film, with clear numbers, which I could wrap around the rim of my little rotary table over the degree graduations, and through same expose a light-sensitive resist, and then etch in my numbers.

(The reason I've never numbered my little rotary table is that I want it done PERFECTLY or not at all. One could have the numbers hand engraved by a master engraver, or machine engraved with a good pantograph engraving machine. I've inquired about this, and either way, you're looking at a big bill for the job.)

Another way I've considered trying, is to scribe the numbers through a soot and wax resist as described above, but do the scribing with a mechanical device of some sort, perhaps a machine like Mac's "Rig For Small Engraving", (ME for Sept. 21/84, page 310.)*

* see next page for more info on this machine

"....and a morning's work gone up in smoke!"

Stewart Marshall told me he has etched metal signs by coating the metal workpiece with a wax and Stockholm tar resist, and then writing through the resist with a calligraphy pen. (All you need is the skill to do it nicely.) He said he once spent 3 hours penning a desired text into the resist, and then, having forgotten to bring the acid mixture to work with him, had to go back home to get it.... When he got back to the shop - this was down in Texas, right on the Gulf Coast - the heat of the sun had melted the resist, and there was his piece of metal sitting in a puddle of brown goo, and a morning's work gone up in smoke!

A Background Shading Punch

Re brass engraving: Brownells sells something called a "Delorge background punch", for shading (stippling) background areas in gun engraving. I recently made a similar punch by turning a tapered nose on a piece of 3/8" drill rod, leaving only a small flat at the tip. I then placed the tip of the punch on a fine cut file, and tapped the other end with a small hammer. This put several fine lines on the tip, but it also bulged the tip a little. I re-chucked it, and took off the bulge with a file, before hardening it. It puts a real nice textured effect on sheet brass. Other punches of various shapes besides a round flat surface could be made, and might be useful on some types of work.

PANTOGRAPHS, and other possibilities

The ability to produce, at a desired size, a good quality drwg of whatever, appeals to me, and can be very useful. Lettering, for example. Better still, engraved lettering.

This is not an ability that I have in the degree or form(s) I would like. Sure, I am aware of dry transfer lettering, which is readily bought at most artists' and/or drafting supply stores. I can use my drafting machine, and a couple of types of lettering aids I have at hand, and I can operate a reducing photocopier like Mozart. However, none of these fully satisfy me. A well-equipped pantograph engraving machine the size of a grand piano would be just about right.

In addition to the small pantograph engraving machine he built, and which is so fully detailed in **TMBR#2**, our friend Mac Mackintosh showed a very interesting item in ME for September 21, 1984, page 310, which he called "A Rig for Small Engraving".

This unit consisted of a base, below which would be set one's work; a pillar, at the top of which was a table to carry the masters; and a pivoted arm, at one end of which was the stylus for tracing around the masters, and at the other end the engraving cutter, which would scratch the desired characters into the job.

Mac's machine was intended mainly for numbering telescope dials (0-360°, etc.) and doing other 1-, 2-, or 3-characters-at-a-time engraving jobs. In Mac's version, a phonograph needle is used as a (non-rotating) "cutter" to scratch the desired markings into the workpiece. I have seen a sample of work (in aluminum) off one of these machines, and what I saw does not come up to the quality of output that would suit me for the job on which I would first want to use it, namely to number the graduations on my 3.75"ϕ rotary table. If I can't do that to perfection, I am not going to do it at all.

Aside from the less-than-perfect crispness of the engraving *per se*, the other thing that bothers me about Mac's rig is the question of what happens when the cutter gets dull - which it may do relatively suddenly. I'm probably just a worry wort, but if it did, it might make a real mess of something that you wanted to have come out just perfect.

BUT.... Suppose you were to combine the info detailed a few pages back on how to make and use "artist's etching ground" with the likely ability of one of these little machines to scratch characters not *into* the steel, but simply *through* that "artist's etching ground" (a blackened wax resist) on the steel... This would require much less pressure on the engraving cutter tip, so there would be little likelihood of it failing in the midst of cutting a character. If you then etched the characters into the steel, I suspect this approach could produce very nice work.

I have not done any experimenting along these lines, but would very much like to hear from anyone who has, or who has any other ideas as to how to do this type of work to a very high standard in a basement shop.

Working drwgs for Mac's "Rig for Small Engraving" are sold by Power Model Supply (PMS). A copy of the entire article, which included full drwgs, would be more useful, and can be obtained by writing direct to Model Engineer Magazine. (Address: see Appendix.) *If you tell them you want a copy of the article "A Rig for Small Engraving", which appeared in ME for September 21, 1984, page 310 - 313 inclusive, and send them US$5 cash to cover copying and return mail, I think you'd get what you wanted, and just maybe a sample copy of ME at the same time.)*

Bob Eaton showed me a simple but very effective little pantograph he made, and which he uses in his jewelry work. This little unit looks about like a half size version of a draftsman's pantograph. (You can buy a simple drafting pantograph for about $20 and up - see next paragraph.) Typically, the arms of these units are about 20" long. Bob has one of this latter type also, in which the arms are made from 1/8 x 3/4" aluminum, but it doesn't entirely suit him, I think because it is a little too big for the sort of work he would do with it.

You can buy a drafting pantograph from Garrett Wade, Inc., Woodcraft Supply, and probably at many local art supply stores. Typically, the 4 arms of these units are about 20" long. Ones with wooden arms seem to run about $20. The price is right, and one of these may please you enormously. One like Bob's with aluminum arms, is about $30, from Woodcraft Supply Co., PO Box 1686, Parkersburg, WV 26102, phone: 1-800-225-1153. A similar unit, with steel arms, is offered for about $21 by Craftwoods, 2101 Greenspring Drive, Timonium, MD 21093, phone 1-800-468-7070.

PMS sells drwgs for making an all-metal pantograph of about this size for enlarging and reducing drwgs, so while you can buy one for not much money, if you just wanted to make one, instead, drwgs are available.

Bob made his smaller pantograph from 1/8 x 1/2" aluminum, each arm being about 12" long. The pivots consist of brass pins, knurled nuts, washers, etc. - see drwgs. Bob said the whole thing needs to be well made to work well, but conceded that the holes in his are drilled, not reamed.

If I were going to make a pantograph like Bob's - and I may - I would cut the 4 arm strips to length, mill the ends square, and make sure there were no burrs on them. Next, I would clamp a fence and a stop to the milling machine table, and then drop each strip in turn into place against said fence/stop, and put a small center drill into each strip where a particular hole is to be, and then ditto for the next, and the next and the next. I would then move the milling machine table some certain distance, and repeat, until all the holes were spotted in with the center drill. The holes could then be drilled through and up in size to just under say 3/16", and then reamed to final size. The pivots would be pretty much like Bob's.

Here are some further notes taken directly from a letter Bob sent me:

"The brass tube bushing/spacers are a good close fit in the holes in the arms. Tightening the knurled knob down firmly onto the bushing compresses the O-ring and puts enough tension on the joints to hold them firm enough for light work.

"I checked my jewellers' supply catalogs to see if I could get any additional information on engraving machines, but they only show one pantograph type machine at $1,035. There was no description of the needles (cutters) used in it.

"When I started my apprenticeship in 1946 we had a pantograph engraver that had a motor driven cutter which "routed" out the letters. The cutters looked a lot like your home-made D-bit reamers. Later they came out with just a point - I don't know what material - which did not rotate but just burnished into the metal. The new machines are not motorized.

"Frankly, Guy, I can't see the reason for any hsm building or owning a engraving machine. Visit a local trophy shop. You'll probably be surprised with the present state of the art, in terms of equipment and what they can do with it. Let them rout it, burnish it, or etch it. Pay them and go home and put the time and effort into something more useful - like a harmonograph.

Bob

(Bob & I differ slightly on time spent making tools versus projects in the shop. We'd been talking about harmonographs last time I'd visited Bob. He'd made one, and had a bunch of drwgs off it, which he showed me. Hence the above remark. However, Bob is not without a

General Arrangement

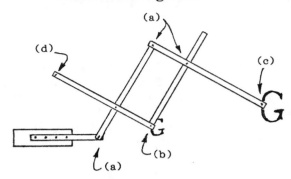

(a)   uses pivot form shown at (a) below left.

(b)   is the needle or pencil lead (shown at (b) below) that cuts or draws new work of smaller size.

(c)   is the tracing stylus (shown at (c) below) that follows your master or pattern

(d)   is fitted with item (d) below, to support the tail end of that Arm. [The bottom end of the screw in (a) can be adjusted to serve the same purpose where/if needed at pivot points marked (a).]

The short arm at lower left in the drwg above is screwed to a wooden block about 1-1/2" sq. x 5" long. A C-clamp anchors this block to a desk or bench when the pantograph is to be used.

(a) Pivot Joint Detail

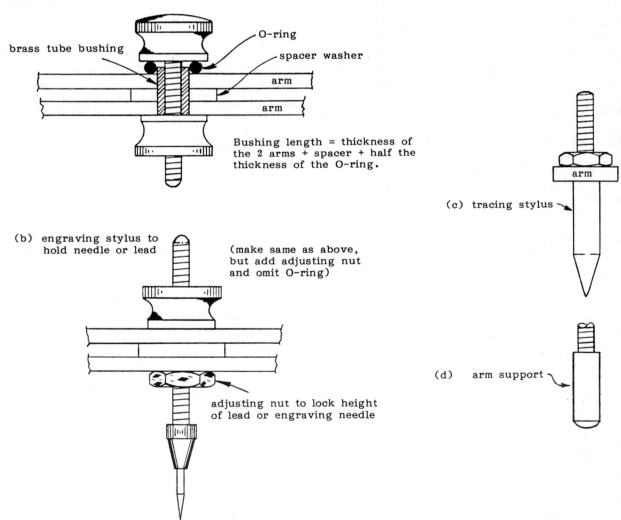

brass tube bushing

O-ring

spacer washer

arm

arm

Bushing length = thickness of the 2 arms + spacer + half the thickness of the O-ring.

(b) engraving stylus to hold needle or lead

(make same as above, but add adjusting nut and omit O-ring)

adjusting nut to lock height of lead or engraving needle

(c) tracing stylus

arm

(d)   arm support

250

sense of humor - his letter then went on to detail how he'd recently spent 3 days making a pair of cabinetmaker's clamps from some pieces of firewood, when he could have bought them for $35 and gotten on with his main project, which was to carve a pair of life size blue herons in walnut, with brass legs. He'd already made heron #1, and it was beautiful.)

Now, you may be interested to hear what Bob does with his little pantograph.... He uses it to cut designs into wax, then from the wax master, makes molds to produce investment castings for jewelry - pins, earrings, etc. One of his lines is a whole range of aircraft pins - Piper Cubs, Mustangs, etc. His simple little shop-made pantograph lets him carve extremely fine details into his wax masters from say a drawing in a book, a photo, or whatever.

Let's say you want to make a builder's plate for your latest steam engine model. Pick a style of lettering from a book of alphabets. I like Roman lettering (e.g. everything in this book) and I have a book called Classic Roman Alphabets. Set up your pantograph, position each letter master in turn in the right place, and go to work. You may choose to use your pantograph to make a pencil drawing, which you will then carefully ink, and have photographed at some desired size, or you may choose to scratch the letters directly into a piece of wax. In any case, you're on the road to a handsome, professional looking builder's plate, using this and other info in this book.

Of course, working up a master (commonly called "the art work") for such a sign is no 10 minute job - it's going to take time. But you can do it, and I suspect you can do a pretty nice job.

There are other ways to get to the same point - a small printing shop may be willing to photo-typeset your masters while doing some other larger job, and give the pieces to you for paste-up into final form, after which you can hand the whole thing back to them for photographing etc. Such shops will likely have computer equipment that can lay out lettering around a circle or oval. If you give them an idea of what you want, they can cook it up on their equipment in a fraction of the time it would take you or I to do it, and it will look better - the matter of proper spacing of letters is what makes or breaks most lettering jobs, and the computer does this work automatically. A printing shop's hourly charge is likely to be $30/hr or more, but if you make it clear you'll be happy to have it done at their convenience, they may be able to slip it in with other work and the cost may not be bad at all.

(Or, let them know you're a machinist. They may ask you to do some small repair for them, and may happily do your job for you for nothing.)

Several guys have sent me photos of examples of pantograph engraving machines built to the drawings in **TMBR#2**, and report nice results from them. Another chap sent me drwgs of a computer-controlled engraving machine he built largely from Radio Shack parts. The actual engraving spindle is a die grinder, appropriately mounted and fitted with an engraving cutter.

As I said above, if you have built anything that other basement machinists can build for themselves and thereby gain the kind of capabilities I've spoken of here, and would care to share it with the brethren through TMBR#4, I'd certainly like to hear from you about it.

Still More re Lapping an Edge Finder (see page 94)

Added to the 2nd printing: The procedure for lapping an edge finder given at page 94 is conjectural on my part, in that I have not done it. I did, however, discuss it with Herman Schmidt, and he agreed that my approach would probably work. He suggested the use of a Norton "4000 grit XXX emery paper," the correct designation for which turned out to be "Norton *4000 XXX Metalite Lapping Film, Q-135, #3 micron*", which is a special order item, 3/4 weeks delivery, $35/sheet, minimum order 3 sheets. I think my suggestion to use 2000 grit wet/dry paper is more practical, and should provide an improvement, even if not so great as might the 4000 grit lapping film. Re 2000 grit paper, see page 201 herein.

Low Cost Master Type
for your Pantograph Engraving Machine
from **James Schmidt, Menomonee Falls, WI**

Dear Guy,

There was one thing that bugged me about the Pantograph Engraving Machine in **TMBR#2**. The $130-plus cost of master type sets fairly froze my imaginary Scots blood, but I think I've found a way to thaw out.

I recently saw a set of master type for $20. The catch was that these masters were about 2" tall! They are the templates used by woodworkers to letter signs with a router. The width of the lines in the letters is about 5/16" or 3/8". I don't have exact dimensions because I only had a moment to look over a set of these templates at my local Sears store. I would assume they are also available through most big woodworkers' supply houses. I suspect one could also use interlocking brass stencils of the type used for labelling shipping crates.

I think a hsm could create his own master type from these templates. If one were to make a guide pin (P/N #14 on your Pantograph drwgs, in **TMBR#2**) with a bottom diameter that closely fitted the grooves in the large templates, the machine itself should be able to engrave all the master type one could want. A heavier engraving cutter might be in order. You'd probably have to attach some of the templates to a sub-base, and remove the bars that hold the center of some letters in place. Double sided sticky tape and some sheet metal plates should take care of that. Mounting the templates and master type blanks in fixtures which are pre-aligned one to the other would be a good idea. If you left a little lateral play in one of the fixtures, you could even play with "kerning" letters (= varying letter spacing for best appearance).

I think I would make my master type blanks 3/4" square, using 1/16" or 3/32" brass sheet. I'll bet you could even dig up some cursive script templates. An evening of engraving would give you all the master type you might need, and in any size you want.

I want to build the Pantograph Engraver. I've promised myself I will as soon as I finish the special flycutter for the arbor press gear, and build the miniature pistol, and measure the speed of a sewing machine motor I scrounged for the pantograph engraver, and design the QC gearbox for my lathe, and finish my shaper, and ...well, you get the idea. Until then, the idea is good, but unproven.

When are you going to bring out "The further adventures of John Kelly and his Machinist's Mate First Class"?

Jim

I've thought about it, but we haven't had enough feedback as to whether or not you guys like/want that kind of stuff. Some people have told me they liked **"Strike While the Iron is Hot"** better than Nevil Shute's "Trustee from the Toolroom". Others have said the opposite. One guy said, "Well, you're no Ernest Hemingway, but....". I have another fiction piece like unto "The Bullseye Mixture" that I had planned to put in this book, but we decided to use the 25 pages it would've occupied for other stuff. ("The Bullseye Mixture" was 34 pages) We'd like to hear from everybody... What do **you** want? GBL

SOME MORE WELDING ADVICE

As noted in TMBR#1, my cousin Garth Richter is a superlative welder, and knows my every welding sorrow. I was talking to Garth one day about MIG and TIG welders. These are touted as easy to use, anybody-can-weld rigs, and I was asking him which is which, are either any good, which one to get, etc.

What had sparked the question off was another dig at my conscience, you might say, that I ought to have (i.e. acquire) welding skills. I'd been reading some back issues of *FWW*. In one, a guy wrote a letter saying something along the lines that ".... all these shop-made woodworking machines are great, but why make a machine frame or machine stand out of wood? The cost of a little electric "buzz box" welder, and the effort required to learn to use it, are quickly repaid in the things you can weld up - machine stands, etc. - from scrap materials and junked machine parts."

Maybe I was missing something. Maybe a little MIG outfit would be a good thing to get... Apparently it is easy to weld even thin sheet metal, according to more than one seller's pitch for these machines.

Garth's response boiled down to this: it's true enough that it is easy to weld with a MIG welder, but how good is the weld? In fact, probably not very good, if the user is inexperienced. If you want to learn to make good welds, take a 50-hour course in oxy-acetylene welding, and first learn that basic method of welding. Then, when you know how to control your puddle, go on to electric welding, MIG, etc.

(I had already done this: afterwards I was not much better at it than before, aside from having gained a modest skill in fusion welding. More on that below.)

Now you may smile about that business about getting control of your puddle - to the non-welder it may sound kinda funny, but it is deadly serious.

Suppose I have two pieces of 1/8 x 1" CRS laying here, and my oxy-acetylene torch all lit up and ready to go, and you come along, and you let on like you know something about welding. I don't, so I say, "Great! How about sticking these two pieces together for me?"

You agree, so I hand you the torch and a piece of welding rod. Now maybe the flame on the torch is not adjusted right, but if you don't know what you're doing, you won't notice. And so you go to work slathering the flame at the pieces in question, and poking the rod into the flame to make yourself some real hot-melt glue. If you don't know how to weld, I'm soon going to know it. (Any welder would know even sooner.)

If you think it's just no trick at all to melt some of both pieces of metal simultaneously, and keep the melted area just so big and dip the welding rod into the puddle of molten metal to add more metal to help fill the gap, and pull the rod away, and bring the flame in again to increase the heat, and then pull it off and bring the rod in again, and do this in such a way as to produce a nice smooth fillet that progresses along the intended joint, well, either you ARE a welder, or you've never tried it.

If the flame isn't just right, the first thing that will happen is you will melt some metal and it will spit back at you (not necessarily burn you, just spit back at you with a "snap" like a toy cap gun - just about suddenly enough and loud enough to make you wet your pants. As I've said before, Chuck Yeager I ain't. That's ok. You guys who know how to do this can go ahead and laugh - I don't mind admitting I'm a complete duffer at welding.

But to get back to this matter of controlling your puddle... From direct personal experience I know that the only time I had any success at it was when I exercised intense mental concentration, and usually my best efforts lasted not more than a minute or so. I expect that with practise I would get better at it.

Quite a number of years ago I asked Garth to weld the crossbar into its hole in the stem of my 4-jaw chuck key, said crossbar having fallen out once too many times for my liking. I watched him do the job, and although I'm no welder (did I mention that already?) I know when I am watching someone do a job with skill - usually they make it look easy. Garth's manipulation of the fine, pale blue flame (in the first place, to get it to burn properly with the right mix of oxygen and acetylene gas at the right pressures), his absolute mastery of the molten puddle, his co-ordination of flame and rod - alternating them at just the right rate - all this was done with consummate control and grace. Have you ever watched a champion figure skater skate backwards? *That's* what I'm talking about - the way they do it versus the way I would do it. Katerina Vit I ain't either.

Okay, now maybe I've got you grinning! Or maybe you think I own shares in cousin Garth's welding shop, and I'm trying to attract more business for him.

Obviously some people are good at one thing, some at another. In my own case, which doesn't really matter to you at all, I suspect that if I were to go back and take another shot at that oxy-acetylene course, I'd do better. Why? It takes time to learn new skills. If you're studying something and you have a chance to step back from it for a few days or weeks, and then come back to it, you'll likely find that a lot of things have fallen into place and now seem easier to do. Or at least so I have found for myself.

Everett Arnes is one of my guys down in Sedro Woolley, WA. He invited me down to visit him, so one day I saddled up the Harley and paid him a visit. (Everett liked the Harley, having made a 16,000 mile trip around the US and clear down to Mexico City on one himself before WW2.) Everett is full of jokes, ideas and projects, and had many things to show me. At one point he asked me if I had an oxy-acetylene torch. I said I had, but hadn't used it much. (A typical piece of Canadian understatement that Everett missed completely!)

"Well, sell it, and get one of these. They're the greatest thing since sliced bread," Everett said, handing me an oxy-acetylene welding torch about the size of a frankfurter - say about 6" long x 3/4"ϕ. Its size was obviously much better suited to the sort of stuff I was likely to do than my existing torch, so some while later I called the makers, and talked to a guy by the name of Joel Fleming. They don't normally sell direct, but they made an exception in my case, and pretty soon along comes a little box with their #11-011-C torch, 5 tips, and hoses in it.

It's cute. It's also capable (if I were) of the finest imaginable welding: it can fuse two 0.002"ϕ copper wires together inside a cigarette filter cavity without scorching the cigarette paper, and can weld/solder/braze right on up to 1/8" or 3/16" steel. The 3 smallest tips have synthetic sapphire orifices in them, down to 0.001"ϕ. There are 7 tips in all (as well as other specialty tips) - I may yet get the 2 largest ones, but the ones I have will handle anything up to about 3/32" thick.

"The Little Torch" is made by Smith Equipment, 2601 Lockheed Ave., Watertown, SD 57201-5636; phone 1-800-328-3363. If interested, call Joel Fleming, and tell him I sent you. He'll send you a catalog, and tell you the location of their nearest dealer for you.

Not long after I got this little sucker, Jake Wiebe came to visit, so I showed it to him. Jake took a real shine to it - he figured it would be perfect for some of the fine silver soldering he does. He also invited me to bring my regulators and my Little Torch out to his place some Saturday, and he'd get me going on how to use them. So far there hasn't been time, but pretty soon we'll get to it.

Now such a small torch is obviously no good for building something like a motorcycle trailer, but for small welding jobs and for heat treating and silversoldering and similar "bench top" activities, such a torch as this would be just the cat's meow.

I'll tell you about some of Everett's other ideas elsewhere herein.

I've often wondered how does one weld up something and get all the pieces lined up and sitting still in proper relationship to all the other pieces. I mean, if it's not welded together, all you've got is a heap of pieces, right? I asked Garth about this one day and sure enough there is a little trick to it: what to do is just "tack" a couple of joints together and then push the pieces into the alignment you want - with the aid of a carpenter's framing square or whatever, and when everything is copacetic, make your weld. Clamps can also be used to hold pieces where you want them. Ain't that clever?

"Ah, that's nothing. You must be a half-wit not to have figured that out for yourself, Lautard!"

Okay, I wouldn't disagree with that - it does seem pretty obvious once it's pointed out, but I'm coming to something else...

I've mentioned Steve Cannon, elsewhere herein. Steve runs an outfit that imports magnetic stuff from Europe. One day he sent me a package, and in it were a couple of magnetic welder's helpers that I think you might like to know about if you don't already.

These things each consist of 2 pieces of blued steel, about 3/16" thick and a little bigger than the whole palm of your hand. Sandwiched between these plates is some yellow plastic material.... And I hope to tell ya, guys, whatever is inside that yella stuff is very strongly magnetic! Because of their shape, you can use them for positioning parts for welding at 90°, 45°, 135°, and 22-1/2°. Steve's note emphasized that they should be removed from the job immediately after tacking the parts, whether you are welding electrically or with oxy-acetylene.

> Note: The use of these magnetic positioners will not eliminate the constant bugbear of most welding, namely distortion - you're still going to have to stay on top of making sure the job comes out the way you want it - but these critters ought to be helpful.

A pair of them runs something under $20, and I'd say if you have inclinations to do much welding at all, they'd be handier than a monkey's tail. I very soon came up with a use for them which I'll tell you about below. For where to buy them, contact Steve at Baermann Magnetics, Inc., P.O. Box 4413, Spartanburg, South Carolina 29305; phone (803) 582-2814.

MIG welding is good for general welding of steel, stainless steel, and aluminum. Both thick and thin work can be done, given the use of filler wire of the appropriate size for the work at hand.

MIG means "Metal Inert Gas" - basically you are doing electric arc welding in a cloud of inert gas piped from a tank (just like an oxygen or acetylene gas tank) to the weld area. The inert gas excludes the normal atmosphere and prevents porosity in the weld. The wire welding rod is fed automatically to the arc at a pre-selected rate; all the user has to do is set the current to a level appropriate to the job at hand, set the wire feed rate, and go to work. When he welds, the machine feeds in wire filler rod. But if you haven't developed a feel for welding, via all those hours with an oxy-acetylene torch, you won't likely be able to do as good a job with this type of automatic equipment as you might otherwise.

TIG ("Tungsten Inert Gas") welding is good for much the same classes of work as MIG welding. One welding supply house salesman told me TIG is the manual or non-automatic version of MIG welding, but this is not actually correct: in the MIG process, the arc is struck with the wire filler rod, which then feeds into the weld, whereas in TIG welding, the electrode is a tungsten rod that wears away very slowly. In TIG welding, the filler material comes from the introduction, into the arc, of the appropriate rod, which is held in the welder's other hand. So you might say that TIG welding is the electric version of oxyacetylene welding, but I suspect that is not entirely correct either.

The temperature generated in the arc is approximately 35,000°F. The arc is analogous to the flame of the oxy-acetylene torch (an oxy-acetylene welding flame is about 6,300°F), but the intense heat has the advantage of getting the area of the weld up to a welding heat virtually instantaneously, so the weld is done very quickly, without pouring in heat from an oxy-acetylene torch for several minutes, with attendant spread of heat from the weld area, resulting in expansion and distortion of the work.

Welding is a useful skill. The man who can weld can do many interesting and useful things. Learn the basics, and learn them right, from a competent source of instruction, preferably a qualified welding instructor. With enough practice you should be able to develop a good level of skill. A small oxy-acetylene torch is about right for a lot of small jobs. MIG and TIG welding makes electric welding relatively easy, but again, get suitable instruction.

Steel Corner Caps: In the section on making wooden storage boxes, I mentioned the idea of making corner caps by soldering 3 pieces of sheet brass together. Here's something I want to try: I want to get some 1/16" CRS sheet sheared into appropriately sized triangles at a sheet metal fabricating shop, and then fusion weld them (see **TMBR#1**, page 187/8) into corner caps. I think one could stick a couple of these triangles on a magnetic welding positioner, thus setting them at 90° to each other. A couple of taps with a light hammer to get them properly aligned, and you could then tack them with the Little Torch, pull them off, reposition them and a 3rd triangle, tack it in, and then carefully proceed to weld down each seam towards the corner, in short increments, cooling frequently as you go, just as I described for making a steel box in **TMBR#1**.

– ——————— – ——————— –

A Fine Toothed Hacksaw Blade

If you've been snappin' up hacksaw frames when you come across good used ones cheap, as recommended at page 100 in **TMBR#2**, you might want to note the following, if you have occasion to saw thin walled material.

Lou Verling, creator of the World's Best Backscratcher (page 224), says he used a Nicholson 32 tpi bimetal hacksaw blade to cut off the 1" x 1" x 0.060" aluminum angle for his backscratchers. He says these blades cost $30 each, but are worth every penny. He got his from Production Tool Supply, which happens to have a store near his home. (For an hsm, such a situation would be on a par with living down the street from the Hershey Chocolate factory.)

And since we're talking about where to get interesting goodies, here's another outfit you should know about: Federal File Co., Inc., P.O. Box 161026, Memphis, TN 38186. I can't comment on the quality of their products, because I've never used them, but they seem to stock just about everything you might ever want in the way of **files** - regular stuff, plus die sinkers' riflers, needle files, parallel machine files, etc. Their catalog is quite informative, too, and can be had by calling 1-800-238-3146.

MORE IDEAS ON BENCHTOP JIGS FOR
SHARPENING VERY SMALL DRILLS

On one of my visits to Bob Eaton's place in Blaine, WA, Bob showed me something he'd come up with for sharpening drills from say #60 on down to #80. He'd mentioned this to me on the phone, and I was expecting to see a half size version of the Drill Sharpening Jig shown in TMBR#1, page 26. Not so.

 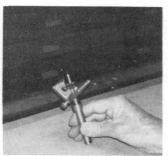

Bob had combined a small drill chuck with a piece of about 3/8 or 1/2"φ aluminum rod for a handle, and a small piece of 3/4 x 3/4 x 1/8" aluminum angle. An Eclipse #160 pin chuck, sold by Traverse Tool under Part #57-070-160; or by MSC under Part #06569040, would be about right; Bob had used a slightly larger chuck that he had on hand.

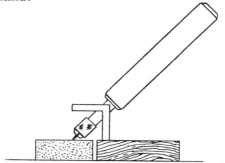

To use it, he chucks a drill in the pin chuck, and orients it correctly with the aid of a magnifying glass. He then sets one leg of the aluminum angle on a block of wood about the same thickness as his bench oil stone, and, holding the handle in his right hand, tips the handle towards the stone until the drill tip comes down onto the upper face of the stone.

He then draws the stone (which is sitting on a sheet of paper) towards himself with his left hand. He can thus control the contact pressure of the drill on the stone to a nicety.

After a couple of passes of the stone under the drill, he flips the whole handle/angle/pin chuck/drill 180°, so the other leg of the aluminum angle is down on the wooden block, and the other lip of the drill is down on the stone, and repeats the process for the second flute of the drill.

Two things to note:

1) Orientation of the drill in the pin chuck is fairly critical - if it is wrong, you will stone the drill wrongly.

2) The only thing that rubs on the stone is the point of the drill. This is a nice feature of Bob's rig, in contrast to "my" drill sharpening jig, the body of which also rubs on the abrasive stone. In TMBR#2, page 147, you will find some improvements to the DSJ suggested by Ted Lusch; his arrangement also eliminates contact between the DSJ body and the stone, but does so via little wheels and the interposition of a strip of thin feeler gage stock between the jig body and the stone.

So.... if you wish to have the capability to sharpen very small drills at home, you can lay out something like $1500 for a purpose built machine (see the SPI catalog) or you can make yourself something like what Bob Eaton made, in under an hour, and with a little care in its use, I believe you would get results satisfactory for most purposes.

Note too that a well surfaced stone will aid in producing the best results. Re-surfacing of abrasive stones is dealt with in TMBR#2, page 119. Also, obviously, one will want to use a fairly fine grit stone.

Don Carr, of Don Mills, Ontario suggested some further improvements to the original (**TMBR#1**) version of the Drill Sharpening Jig. I'll quote from his letter, and add my own comments *in italics* following each of his points.

Dear Guy:

I've just finished building the drill honing jig (**TMBR#**1) but not quite the way described, as I think (?) some of the design/drawings need revision.

1) The spring as shown on the GA drawing for the body will do nothing. The spring has to push against the spindle. This means the lower counterbore 1/2" deep should be 3/8" and the 3/8" top counterbore should be 1/2". (See marked up drawing.)

The spring is correct as shown in my drwg, and will work as intended. Its purpose is to pull the "Top"/Spindle portion down into the Body of the Jig, thus putting a firm spring grip on a drill inserted in the cross hole in the Spindle, so that the drill stays where you put it while you tighten up the Nut (Part #5). If the spring were re-located as you have shown it, it would force the jig OPEN, not closed.

2) What does the dimension 0.128" mean, and what use is it? For a 1/8" drill (0.125"), the gap between the clamp nut and body of the jib works out to 0.107", per attached calculations (not given here. GBL) for a 0.031" groove depth. This is measurable. The 0.128" dimension is not.

I used the dimension 0.128", because I was thinking in terms of using a 0.128" drill (rather than a 0.125" drill) per text at top of page 31, under FINAL ASSEMBLY, while Loctiting the "Top" to the Spindle. As you say, it is not a measurable dimension. It is a what-to-use dimension.

3) Why does the 1/8" register pin need to be located so precisely (0.138")? The clamp nut and body will line up automatically, because this hole is drilled before the clamp nut is parted from the body. I would just indicate either 1/8 or 9/64" from the corner.

The dimension 0.138" suggests a degree of precision which, as you say, is not required. 0.140" would be as good, but if I had said 9/64", there is an implied invitation to sloppier work. If you use 0.138 or 0.140 or 9/64 it'll work fine. However, you have to translate 9/64 into decimals to crank it into your milling machine feedscrew, so 9/64 becomes 0.1406" = 0.141, which is no better than 0.138"!

4) The Spindle (Part 3) page 28, Instruction 4.: There will be a tendency for many people to Loctite the spindle to the clamp nut at this point before the cross hole has been aligned with the V-groove in the body. As this is critical to the operation of the jig, I would omit it here and add it to the final assembly section.

Ok - good point.

5) A properly cut V-groove is essential to the correct location and sharpening of the drill. I cut the V-groove using a single point flycutter in the boring head of my vertical mill.

Excellent idea!

6) Hand filing the flat along one corner of the Jig body did not appeal to me. I milled it, after clamping it in the center T-slot of my milling machine table with one diagonal (as viewed from the end) horizontal.

High precision here is not essential, as the flat is just for clearance purposes. You can mill it, but filing might be quicker, if you count the setup time. A faster way to get the Jig Body oriented the way you did it would be to set a V-block in the milling machine vise, lay the

Body in it, and grip the Body by its ends. The V-block merely holds it the way you want it until the vise tightens up.

To another of Don's questions, my reply was:

Part No. 4 should NOT be a press fit in the hole in the Body of the Jig. Also, you have crossed out my bevels on Part #3. However, I understand why (and forgive you) because of your mistaken desire to have the spring push against that shoulder.

Suggestion: don't overlook the improved version of the Jig shown in **TMBR#2.**

TWO NICE DIE FILERS

Thanks to a phone call from Tim one day several years ago, I bought a used "Milwaukee" die filer that was advertised in a national "tool trader" paper. When it arrived, it showed no evidence of wear, nor even of much use. I decided to clean it up and repaint it before putting it into service.

I glass bead blasted all the loose pieces of bare metal - bolt heads, etc. They came out all clean and frosty-lookin'. I got the engraved brass degree scales off by tapping the "drive nails" out from the back side, and then polished them up better than new.

I painted the castings grey, with some parts bright red, for contrast. I cleaned the motor, painted it black, and gave it a new cable. I replaced the maker's plate, which goes over the round (and now red) cover plate on the main housing, with a decorative polished brass plate about 1/16" thick. I bored 8 rows of holes of increasing size radiating from the center of my plate; the final effect was quite nice, or at least I thought so. (This item is not shown in the photo). I made the two new gaskets it needed, and blew $3 on a new oil seal for the shaft.

When I got everything done and put the machine back together, I filled the main drive housing with oil, and fired it up. It ran as smooth as silk! I'm enormously pleased with it. I ordered a couple of sets of parallel machine files for it from MSC, and hope to make considerable use of it in the near future. (Die filers can also be used with Starrett screw slotting blades, hack saw blades, etc.)

Subsequently, Tim also bought a used die filer. His was pretty dirty when he got it, but he refurbished it pretty much as I did mine, except that he painted his all grey. He sent me "before" and "after" photos of it.

When I showed Bill Fenton the photos of Tim's die filer, he said, "Oh, my, he's done a fine job." Then I showed him a photo of mine. "Humph! Looks like a circus wagon!" says Bill.

THE MACHINIST'S BEDSIDE READER

contains a wealth of know-how and info (some of it obscure, but all useful) that would take you countless hours to dig up on your own, PLUS..

Working drawings and detailed instructions for making 15 useful and practical machinists tools and lathe accessories.

Dozens of hints, tips, and tricks to help you get things done faster, easier and better in your shop.

A collection of about 2 dozen machine shop anecdotes.

2 highly readable machinists' short stories:

- The fascinating (true) account - and photos! - of a beautiful little lathe secretly built and used in a Japanese POW camp.
- "The Secret of the Old Master" by Lucian Cary. A young toolmaker attempts to learn the secret behind the legendary reputation of a master rifle barrel maker.

Partial Table of Contents

The Scrap Patrol
Standards of Workmanship
A Source of Welding Rod for Fine Welds
Filing in the Lathe and Elsewhere
Two Not-so-Common Ways of Measuring Holes
Some Notes on Reaming and Tapping
Drilling to an Exact Depth
How to Sharpen a Centerpunch (properly!)
How to Get More Work Done in the Shop
A Sharpening Jig for Drills from 1/8" to #60
A Small Depth Gauge and a Small Tap Wrench
Swivelling Base Fixture for a 2" Wilton Vise
How to Make Your Own Reamers
How to Make A Floating Arm Knurling Tool
A Tool for Straight Knurling
Fruit Acids and Fine Tools don't Mix
Holding Thin Work in the Chuck
A Graduated Handwheel for the Lathe Leadscrew
  (and a "Rubberdraulic" Slip-Ring Lock)
Ball Turning without Specialized Attachments, and
an Improved Type of Ball Handle
Oddleg Artistry
A Finger Plate
Designing and Fitting Split Cotters
A Small Scribing Block
Cutter Blocks and Shop Made Cutters
Some Info on Silver Soldering

A Toolroom Grade Sling Swivel Base for
  Tubular Magazine Rifles
Gun Making (and some tales from one who did)
Making Bullet Molds
The Ultimate Box Latch
Working Drawings for a Machinist's Toolchest
How to Make Imitation Ivory
Why not Build a Harmonograph?
A Set of Heavy Brass Napkin Rings
A Hand Beading Tool
Some Notes on Spring Making
3 Useful Accessory Faceplates for your lathe
Metal Polish - Bought and Home Made
A Corrosion Preventive Cutting Compound
A Fair Return for One's Work
A Chinese Tool Steel that Outperforms HSS
Good Advice on Getting Ahead
"Stealing the Trade"
Helping the War Effort
How Not to Get a Welding Ticket
Delphon and the Adding Machine
The Sleepy Apprentice Boy
A 19th Century Machinist's Apprenticeship
How to Impress Your Mother-in-Law
One Way to Ruin a Lathe
Removing a Jammed-on Chuck - the method of last resort
Dimensional Matrix for 6 sizes of Toolmakers Clamps

Comments from machinists who have read The Machinist's Bedside Reader

▪"Absolutely superb. Thoroughly enjoyable, with much knowledge." (D.S., Vallejo, CA)

▪"..an excellent book I would recommend to anyone even remotely interested in working with metal." (J.S., Custer, WI)

▪"..enjoyable and informative ... I love the manner in which the material is presented." (C.A.P., Jonesville, VA)

▪"Great. I teach machine shop, and the kids read it without being asked." (D.B., Ann Arbor, MI)

106 clear and detailed illustrations, 210 pages
8½ x 11", soft covers

THE MACHINIST'S SECOND BEDSIDE READER
and THE BULLSEYE MIXTURE
More great plans and shop info in this
all new sequel to TMBR#1

omplete working drawings and instructions
r:

A small **Pantograph Engraving Machine** you
can build. (The commercial equivalent would
cost you over a thousand dollars!)

A **Toolmaker's Block** - a rectangular relative
to Lautard's Octopus.

A **Bench Vise Accessory** for holding flat work
by its edges.

The Poor Man's Jig Borer - a WWII-era
device which allows high accuracy hole
location on your drill press.

An "**Overhead Drive**" for milling and drilling
spindles on your lathe, plus details of a
couple of milling & drilling spindles.

"**Zero Wear**" improvements to the Drill
Sharpening Jig shown at page 26 of
TMBR#1.

A small steam or air operated whistle (after
an original "**Lunkenheimer**" whistle).

A **kerosene-burning blowtorch** which can be
set low enough to silver solder 1/16"Ø
copper pipe, or turned up enough to melt
1/4"Ø solid copper rod in one minute. (Try
that with the average propane torch!)

LUS complete details on:

A super-simple **indexing device** which can be made
and/or elaborated to suit your particular purposes and
equipment.

An easy way of originating high accuracy **master
division plates.**

Two methods of originating a **master reference square**
(or trueing up an existing square).

A simple **shop made sine bar** and a **sine fixture** for
your milling machine.

How to make a **true square** (commercial ones sell for
about $800), plus info on how to make your own
"space blocks" and "angle blocks".

**An old gagemaker's methods and basement workshop
equipment** for spot grinding and lapping. You can
duplicate his simple arrangements for surface grinding
and lapping for flat, square, and finish tolerances far
beyond the limits of conventional machining
techniques.

IN ADDITION TO THE ABOVE, there's also:

- Five ideas for projects (3 complete with working
  drawings) that could be used as gifts. (Want to have a
  very well equipped shop? Can't afford it? Three of
  these projects have the potential to be serious money
  makers.)

- Full info on **how to sharpen straight razors and other
  fine edged tools** to perfection.

PLUS more drawings/info for:

- an offset tailstock center for taper turning
- a fixture for cutting multiple-start threads
- an elegant between-centers boring bar
- two designs for shop-made hand hacksaw frames
- filing buttons and toolmaker's buttons
- a toolpost fixture for rounding the ends of small parts
- a simple lathe tracing attachment
- adjustable-for-balance wheel flanges for your bench
  grinder, **and**..... info on where to get:
  - working drawings for an EDM machine
  - a kit of working drawings and materials for a
    bench gear cutting machine, and
  - a super article on how to make your own small,
    fine quality aluminum castings.

And much more, including several anecdotes and **a
multitude of hints and tips.**

PLUS ... "The Bullseye Mixture"

An enjoyable story which incorporates complete details of
old time carbon pack color casehardening methods.
(Have you ever read the comment, "These parts should be
casehardened", in an article and felt like grabbing the
author by the neck and saying, "Well, why don't you give
us some idea of HOW to do it, then?!!) When you've read
"The Bullseye Mixture" you will understand not only
WHY you would caseharden certain parts, but you will
also know, in detail, HOW to do it yourself, at home.)

Illustrated, 213 pages, 8½ x 11", soft covers

**All our publications and numerous other fun goodies for machinists are described
in our catalog, which can be obtained by sending $1 to the address below. Be sure
to indicate if you are interested in our clock-related plans and/or the Hemingway
plans, and/or the TINKER Tool & Cutter Grinding Jig.**

GUY LAUTARD, 2570 Rosebery Avenue, West Vancouver, B.C., CANADA V7V 2Z9

OTHER PUBLICATIONS and PROJECTS from GUY LAUTARD

A VIDEO from Guy Lautard:
BUILD YOUR OWN RIFLING MACHINE

This 3 hour video shows full technical details on the construction of a rifle barrel making machine the basement machinist can build, and then use to produce match-quality cut rifled barrels. Watch also - often close up - as a barrel is drilled, reamed, and rifled from the solid bar. Thorough explanations are given at every stage - deep hole drill geometry, reamer details, making your own rifling heads, etc.

If you want to make your own rifle barrels, on a non-commercial basis, this is the way to do it. The machine shown in this video is in use today making match-winning 6mm benchrest barrels. Its owner/builder is a mechanical engineer who is also a gunsmith of some 40 years experience. This video will save you **endless** frustration and wasted hours. Just the part on sharpening your own deep hole drills will quickly save you the entire cost of the video. Comes with 36-page written supplement loaded with tips, drawings, info and suppliers' addresses.

Concludes with a 15-minute "short" showing several very fine old-style single shot rifles made entirely from scratch by a retired tool and die maker! This part will have you on the edge of your seat for sure!

THE J.M. PYNE STORIES AND OTHER SELECTED WRITINGS BY LUCIAN CARY

If you've ever read any of Lucian Cary's J.M. Pyne stories*, you'll want this book! Destined to become a classic and a collector's item, this handsome volume is a MUST HAVE book for machinists, gunsmiths and shooters. These wholesome and heartwarming fiction stories - ripe with the fragrance of gunpowder and cutting oil - combine humor, suspense, insights into human nature, and a certain amount of shop wisdom. They sprang from the pen of master story teller Lucain Cary more than 50 years ago, in part as a direct result of his long friendship with legendary rifle barrel maker H.M. Pope. Also included are several other of Cary's equally entertaining, instructive and uplifting stories and articles. You will read and re-read them all, with fresh enjoyment each time.

■ Almost a Gun Crank ■ H.M. Pope - Last of the Great Gunsmiths ■ The Rifle Crank ■ The Big Game Hunting of Rufus Peattie ■ Madman of Gaylord's Corner ■ The Old Man who Fixes the Guns ■ Forty Rod Gun ■ Johnny Gets His Gun ■ Center Shot to Win ■ J.M. Shoots Twice ■ The Secret of the Old Master* ■ No Choice ■ Revenge in Moderation ■ Harmless Old Man ■ Let the Gun Talk ■ The Guy Who had Everything ■ How's That? ■ I Shall Not Be Afraid.

* See TMBR#1, page 104. "No Choice" and "Revenge in Moderation" are Parts 2 and 3 of a trilogy, of which "The Secret.." was but Part 1. *6" x 9", 337 pages, quality printed, soft bound.*

TABLES & INSTRUCTIONS FOR BALL & RADIUS GENERATION

Part 1 of this handy little book shows you how to make ball handles, ball end mill blanks, etc. in your lathe, without special tooling - just a parting tool - by the 'incremental cut' method shown in somewhat less detail in TMBR#3. **Part 2** contains full instructions for generating male or female radii with a ball end mill, in a vertical mill. Say you want to machine a 1/4" radius on the corner of a piece of material. If you've got a 1/4, 5/16, 3/8, 1/2"ϕ or other size ball end mill you can do it easily, quickly and accurately. You won't believe your eyes, the first time you try it! **Part 3** consists of 70 tables (one per page) for the generation of both ball and radii from 1/64 to 1" radius.

On the inside back cover there's a memory aid you can flip to if you need to, then flip back to the table that suits your job, and go right to work.

Also includes notes and a photo showing how to rig an indicator on your lathe, which gives you direct number readout of carriage position when carving out a ball. The incremental cut system can be used to carry out some part-circle machining jobs, as well, and is also useful for making handles for feedscrew dials, levers, etc. This handy little book therefore includes some info, a drawing, and numbers for a very nicely proportioned feedscrew dial handle - if you make one, you'll fall in love with it the first time you use it!

Durable, soft covers, coil bound to lay flat in use; 104 pages, 4" x 6" to fit conveniently in your toolbox.

A BRIEF TREATISE on OILING MACHINE TOOLS

Here's the 'mother lode' on a topic of major importance to every machinist - 25 info-packed pages on ■ How to make a *really* good oil gun, in less than an hour, that will put oil where you want it, at 10,000 psi if need be! (it's a dream come true for Myford lathe owners!) ■ How to arrange for centralized oiling of machine tools using a one-shot oil pump, or the shop-made oil gun described above, how to remove spring ball oiling points, how to make and mount a centralized oiling manifold, what to use for oil lines, etc.. ■ Info on oiling machine tools - what types of oil are appropriate where, and why. ■ Complete working drawings for a handsome variable feed drip oil cup to fit lathes, mills, etc. Permits you to see that oil is definitely being fed, and shuts off tight so you don't waste oil when the machine is off. You'll like it. ■ Complete drawings for pea-sized drip feed oil cups for model work. ■ And more, including a recipe for a good leadscrew lube, and how to go about grinding in a lathe without harming it.

A UNIVERSAL SLEEVE CLAMP

An interesting exercise in machining, boring and screwcutting, the Universal Sleeve Clamp provides a clever means of hinging and clamping two rods together. It is the ideal accessory for your magnetic dial indicator base and/or surface gage, and can also be used as the hinge joint for a large capacity compass, or inside or outside caliper, much like a firm joint caliper. Scale up the design just a little and use it as the basis for a universal lamp for your lathe, mill or workbench. Made slightly larger again, you could use it to make a handsome and unusual living room lamp. (The drawings are arranged in such a way as to make designing new ones at any size an easy matter.) Also includes working drawings for a height adjustable tool holder for small boring tools and internal screwcutting jobs, plus a sheet showing the sort of boring tools I make. These incorporate a minor but significant feature not everybody thinks of when making tools of this type.

3.75"ϕ ROTARY TABLE

Even if you own a larger, geared rotary table, you will find this little ungeared Rotary Table (shown in **TMBR#1** at page 193) a most useful accessory for small jobs in your vertical milling machine: ■ mill a radiused end on a part, or round the ends of a set of loco conrods ■ space out holes accurately around a pitch circle ■ mill a circular slot ■ mill two or more slots at an angle to each other... and so on. If you make it at the sizes given in the drawings, the base will be 4" square, and the table will be 3.75"ϕ. Of course, you may want to scale it up, (or down, for the ultimate paperweight). Whatever size you make it, it's an interesting and instructive exercise in machining, and will be a prized addition to your shop tooling when done.

A LEVER FEED TAILSTOCK DRILLING ATTACHMENT

This item allows the drilling of small holes from the tailstock with ease, as it gives you a very sensitive feed to the drill, rather than the relatively insensitive feel you have when feeding a small drill forward by means of the tailstock barrel handwheel. This is not a major project, but it is one I think you'll like. It plugs into the tailstock barrel's taper socket; when you're not using it, you can whip it off the lathe and hang it on a nail on the wall.

A TILTING ANGLE PLATE (and more) FOR YOUR MILLING MACHINE

A tilting angle plate is a very useful addition to your milling machine. Most tilting angle plates - bought or shop-made - involve the use of castings. This one is made from aluminum plate, which eliminates the need for castings, and allows you to make it at the right size to suit your equipment. (An accessory of this sort would not last long in a commercial shop if made from aluminum, but it'll do fine in the hands of the guy who makes it for his own use.)

Also includes drawings for 5 other simple workholding aids for your milling machine: ▪ a pair of locating/stop buttons ▪ 2 types of stops for the front edge of the table ▪ a "catch-the-corner" work stop ▪ a set of small planer-style toe clamps, and ... ▪ a big hold-down for use on a drill press or milling machine - this one's from an engineering professor who does "impossible" motorcycle repairs for his friends.

LAUTARD'S OCTOPUS

Part bench block, part V-block, part vernier protractor, part direct indexing fixture, part tilting angle plate, part rotary table, part layout jig, this multipurpose workholding fixture has more ways of holding a job than an octopus. Not for heavy work, but just right for small (and some not-so-small) and tricky jobs that come up every so often in every shop. Fully detailed drawings, machining instructions and how-to-use info are provided. Several 'extras' are included, like how to make a lathe sub-faceplate that thinks it's a magnetic chuck: if you have a piece of 3/8" or 1/2" aluminum plate handy, it'll cost you about $2 for the "power supply", and it'll hold non-ferrous as well as ferrous materials with a "ferro-cious" grip - this info alone is worth the price of the drawings! We can also supply an octagonal casting, if you can't scrounge up a suitable piece of steel.

A MICROMETER BORING HEAD FOR YOUR MILLING MACHINE

Based on a classic design by the well known George Thomas, this boring head's generously proportioned dovetail slide and dowelled gib, coupled with a large diameter graduated dial, make for precise and silky smooth adjustments; if well made, its repeatability is phenomenal. Takes 3/8"ϕ shank cutters, and, via adaptor sleeves, smaller ones as well.

You can use this boring head with simple shop-made cutters to machine washer seats, cut counterbores for socket head cap screws down to 0-80, bore holes for which you don't have a drill of the right size, and bore/counterbore holes in jobs too big to swing in your lathe. If you make it with a straight shank, you can mount it in a collet in your vertical mill, *and* in the 3-jaw chuck in your lathe. The drawings give all dimensions for making the Boring Head in 2 sizes, to suit both larger and smaller mills. This is a quality project that will serve you for a lifetime. The instructions are very detailed, to enable anyone to make this valuable and versatile milling machine accessory, *providing he is willing to take the necessary time.* You can be justifiably proud of yourself when it's done, and you'll learn a few things in the course of making it, too.

The TINKER TOOL & CUTTER GRINDING JIG

The TINKER is a simple, practical, and compact Tool and Cutter Grinding Jig that teams up with your existing bench grinder to sharpen end mills, side and face milling cutters, slitting saws, lathe toolbits, twist drills, reamers - in fact, just about any cutter you'd find in a small machine shop, a gunsmith's shop* or home workshop.

You can build the TINKER in your own shop from our detailed working drawings and instructions.

The basic TINKER consists of only 27 parts in all, most of them very simple to make. The drawings consist of over 60 pages, with many helpful notes throughout to make the project easier. They include notes for building, as well as complete user instructions. We can supply castings, or you may prefer to make your own welded substitutes for the castings, and that is easy enough to do.

NOTE: We want you to know what the TINKER will and will not do before you buy the plans and/or castings. Therefore, please send $1 for our 5 page illustrated brochure which provides full details about the TINKER Tool and Cutter Grinding Jig. Or ask for it free of charge if you are ordering other items from us at the same time.

* Not suitable for sharpening chambering reamers. Chambering reamers should be returned to the maker for sharpening... **but see TMBR#3 for a goldmine of info on how to hand stone your own chambering reamers!!**

The COLE DRILL

A versatile, hand-cranked drill that will do most things that your drill press can do, and quite a few things your drill press won't do, including such feats as drilling right through a 1" square HSS lathe toolbit! Broken engine block studs, wheel studs, farm, logging and mining machinery, and vehicle repairs, grain elevator maintenance work, boat building -- if you are involved with any of these types of activities or problems, you may find the Cole Drill very useful. For more details and ordering information, see TMBR#3, or contact us.

VERNIER PROTRACTOR

We now stock and sell very nice quality 2-minute vernier protractors made in mainland China. As one customer said, "Every other protractor I own was designed wrong in the first place." He likes this one a lot better.

CLOCK PLANS

We also sell plans for 16 clockmakers tools, 7 clock movements, and 2 sundials, also from C.J. Thorne, of England. These are detailed working drawings with helpful notes, but no detailed how-to instructions. Anyone with the appropriate basic knowledge of clockmaking - or the above mentioned book on **Clockmaking for Model Engineers** - will be able to work from them. For more details, ask for our "clock catalog".

And see inside back cover for our newest book,

CLOCK MAKING FOR THE HOME SHOP MACHINIST

For our complete catalog, showing all of the foregoing and more, phone, or send $1 to

Guy Lautard,
2570 Rosebery Avenue
West Vancouver, B.C.
CANADA V7V 2Z9
(604) 922-4909

OUR NEWEST VIDEOS...

EXAMINING A LATHE AND MILLING MACHINE
- A MACHINE TOOL REBUILDER SHOWS YOU HOW

This 3 hour video will show you how to examine a lathe or milling machine you may be thinking of buying - and **it could save you a lot of money!** Dennis Danich, a machinist and machine tool rebuilder with some 30 years experience, shows you what to look for in a preliminary examination of a lathe and milling machine, and the few tools you should take with you to carry out several simple but crucial tests. He further shows how to determine the accuracy of the machine's alignments, feedscrews, etc.

You'll also learn how to adjust and lubricate both types of machines, so this video will be useful to you whether you are considering purchase of a new or used machine, or just want to tune up your own machines.

Also on this video:
■   An introduction to the art of scraping and frosting of machine tool surfaces; an explanation of how and why it is done, and a demonstration of several techniques, including power scraping.
■   A demonstration of how to carry out grinding operations in the lathe without harming it.
■   How to restore a bench oilstone to virtually as-new condition on a cast iron lapping plate. (Sharpening scrapers wears grooves in the surface of the combination stones typically used for this task. Eventually the stone will become virtually useless, unless the surface is dressed flat again. We show you how to do it.)
■   An inexpensive surface grinder well suited to the needs of the home shop machinist, and an important warning about buying a used surface grinder. (A worn out surface grinder is no bargain at any price, even if it carries a famous maker's nameplate.)
■   Various shop kinks, and several handy shop accessories you can make.
■   An 1890's 6-foot-stroke Gray planer in action, machining a 3' straightedge casting.

Besides the above, we'll also show you briefly the beautiful little Chinese-made 2-minute vernier protractor we sell (see my website), show you how it works, and turn it this way and that so you can have a really good look at it from all sides.

AND.... at the end of the video, there's a 12-minute "short" about a privately owned P40 Warhawk, bought in 1946 for $50!! We get a detailed look inside the cockpit, and all around the aircraft. The owner started it up for us - wait'll you see the smoke and flames roar from the exhaust stacks!!

HOW TO MAKE SIMPLE, SOLID, UTILITARIAN WORKBENCHES

This 2 hour video shows how to make 3 types of workbenches, including the type shown on page 182 in this book, and one that can be set up in about 30 seconds, will support about 400 lbs, and stores flat against the wall when not in use.

Also shown on this video is a router surfacing jig you can make, which solves the problem of flattening a heavy, laminated workbench top, without access to a large planer, as discussed at page 179/180 herein. This very versatile jig is not difficult to make. It stows in a space about 5" x 5" x 9' when not in use. When you do need it, it will let you put a known taper on a popsicle stick, or surface an 8 foot long benchtop laminated up from 2x4's. A smaller version of this surfacing jig, about 3' long, is also shown. It ought not to escape you that a jig of this sort would be quite saleable, if you wanted to make several and sell them to woodworkers you may know. See my website for further details, pricing, etc.